PENGUIN BOOKS

WANDERLUST

Rebecca Solnit is the author of more than twenty books, including *Orwell's Roses*, *Recollections of My Nonexistence*, *A Field Guide to Getting Lost*, *The Faraway Nearby*, *A Paradise Built in Hell*, *River of Shadows*, and *Wanderlust*. She is also the author of *Men Explain Things to Me* and many essays on feminism, activism and social change, hope, and the climate crisis. A product of the California public education system from kindergarten to graduate school, she is a regular contributor to *The Guardian* and other publications.

Praise for *Wanderlust: A History of Walking*

"Idiosyncratic and inspiring, Solnit channels the urge to stroll from such famous peripatetics as Aristotle, Rousseau, Thoreau, and Baudelaire. Along the way she wonders whether suburban angst is due to the depletion of walking space, and why mountaineering has become a luxury sport. . . . The history of walking is not just about evolution, she argues, but also the unfolding of philosophy, the fomenting of revolution, and the quest for God. . . . *Wanderlust* is an anecdotal *On the Road*, a rambling woman's paean to the mind/body connection, to the path where no SUV travels."
—*Voice Literary Supplement* (One of the 25 Best Books of 2000)

"*Wanderlust*'s real pleasures resemble the pleasure of walking. It doesn't systematically press on toward a goal, but savors detail and varied perspectives, stopping to consider the nature of mountaineering, the life of the London streetwalker, the conflict between public right of way and private property in nineteenth-century England and twentieth-century Las Vegas."
—*The New York Times*

"Solnit's is a sinuous course propelled by abandon yet guided by a firm intelligence. . . . Not content to send postcards from the past, she jumps gladly into the fires of the mass procession and the revolutionary outbreak, deftly joining hands with the Mothers of the Plaza de Mayo, the Velvet Revolution, Critical Mass, and the ever-militant French. It makes one see the events of 1789 and 1989 in a completely different light. But Solnit is no Pollyanna of the footpath. She has a fine sense of paradox that keeps her from proselytizing."
—*San Francisco Chronicle*

"Solnit is at her very best when her passion for history and landscape meets her progressive politics. Her mini-chapter on the late nineteenth- and early twentieth-century right of way battles in England's Peak District is rich with brilliant observation and detail. In the end, the guiding spirit of *Wanderlust*, the lonely traveler always in view on Solnit's horizon, is not Wordsworth or Rousseau but Walter Benjamin, whose rambles through the streets of Paris had the sense of wonder, the air of open-minded exploration and imminent discovery of Solnit's own journey. In describing Benjamin's writing she seems to be half-consciously describing her own: 'more or less scholarly in subject, but full of beautiful aphorisms and leaps of imagination, a scholarship of evocation rather than definition.'"
—*Salon*

"[Solnit is] a rigorous polymath capable of stunning flashes of original thought. While the connections she makes often seem reckless, when they work they are more thrilling for the risk. The subject of *Wanderlust* may be mundane, but what Solnit draws from it is fascinating."
—*Los Angeles Weekly*

"Solnit mulls over the annals of walking, intermingling her own stories of trailblazing with profiles of noteworthy walkers like Kierkegaard and Wordsworth. She moves easily from discussing theories of early human skeletal structure's evolution to detailing how the Sierra Club formed. Her words remind us of walking's simple joy and return us to a time when an aimless contemplative stroll was a daily activity, not a guilty pleasure."
—*The New York Times Book Review*

"Solnit may be a pedestrian, but her writing, for the most part, is not. An anti-nuclear activist, she has saved some of her most passionate writing for the essays that deal with walking as a form of social expression—from demonstrations through walkathons, something she calls 'a mutant form of pilgrimage.'"
—*The Baltimore Sun*

"An erudite history of walking studded with arresting insights that will make you want to throw the book down and hit the open road."
—*Elle*

"Solnit's thoughtful, thought-provoking, and delightful exploration of the seemingly mundane topic of walking offers an abundance of new ways to think. . . . An entertaining and utterly compelling read, filled with facts and observations, written with elegance and eloquence."
—*San Francisco Bay Guardian*

"Solnit covers all sorts of ground in her inspiring look at walking. She studies putting one foot in front of the other as it applies to protests, call girls, fundraisers, Greek philosophers, poets, urban discoveries, rural pleasures and the evolutionary steps we took as bipeds."
—*The Seattle Times*

"An impressive and inventive thinker, Solnit clearly relishes the nearly encyclopedic task she has set for herself, wandering amid 'religion, philosophy, landscape, urban policy, anatomy, allegory, and heartbreak' . . . that there is an increasing disjunction between the body and the modern environment it inhabits is the tragedy that Solnit chronicles. Yet she does not succumb entirely to heartbreak. She ends the book taking her own subversive stroll down the Las Vegas strip. . . . In this unlikely neon oasis, she finds a reinvented understanding of the formal garden, crossbred with the boulevard, that spells hope for pedestrians."
—*The Ruminator Review*

Wanderlust

Rebecca Solnit

PENGUIN BOOKS

PENGUIN BOOKS
An imprint of Penguin Random House LLC
penguinrandomhouse.com

First published in the United States of America by Viking, an imprint of
Penguin Random House LLC, 2000
Published in Penguin Books 2001

ILLUSTRATION CREDITS
Page v: Detail of postcard, ca. 1900, showing passersby in front of Notre Dame de Paris.
1: Linda Conner, Maze, Chartres Cathedral, France, 1983. Courtesy of the artist.
85: Cedric Wright, Sierra Club Mountaineers on Mount Resplendent in the Canadian Rockies,
1928. Courtesy William Colby Library, Sierra Club.
183: March of the Mothers in the Plaza de Mayo, April 1978. Courtesy of the Asociación
Madres de Plaza de Mayo, Buenos Aires.
269: Marina Abramović, from The Lovers, 1988, black and white photograph. Courtesy
Sean Kelly Gallery, New York.

ISBN 9780140286014 (paperback)

THE LIBRARY OF CONGRESS HAS CATALOGED THE HARDCOVER EDITION AS FOLLOWS:
Solnit, Rebecca.
Wanderlust : a history of walking / Rebecca Solnit.
p.   cm.
Includes bibliographical references (p.) and index.
ISBN 9780670882090 (hardcover)
ISBN 9781101199558 (ebook)
1. Walking History. 2. Hiking— History. 3. Voyages and travels. I. Title
GV199.5.S65   2000
796.51 09—dc21   99–41153

Printed in the United States of America
41st Printing

Contents

Acknowledgments

I owe the origins of this book to friends who pointed out to me that I was writing about walking in the course of writing about other things and should do so more expansively—notably Bruce Ferguson, who commissioned me to write about walking for the catalog accompanying his 1996 show *Walking and Thinking and Walking* at the Louisiana Museum in Denmark; editor William Murphy, who read the results and told me I should think about a book on walking; and Lucy Lippard, who, while we were on a trespassing stroll near her home, clinched the idea for the book for me by exclaiming, "I wish I had time to write it but I don't, so you should" (though I have written a very different book than Lucy would). One of the great pleasures of writing about this subject was that, instead of a few great experts, walking has a multitude of amateurs—everyone walks, a surprising number of people think about walking, and its history is spread across many scholars' fields—so that nearly everyone I know contributed an anecdote, a reference, or a perspective to my researches. The history of walking is everyone's history, but my version of it particularly benefited from the following friends, who have my heartfelt gratitude: Mike Davis and Michael Sprinker, who supplied fine ideas and much encouragement early on; my brother David, for luring me long ago into street marches and the pilgrimage-protests at the Nevada Test Site; my bicycle-activist brother Stephen; John and Tim O'Toole, Maya Gallus, Linda Connor, Jane Handel, Meridel Rubenstein, Jerry West, Greg Powell, Malin Wilson-Powell, David Hayes, Harmony Hammond, May Stevens, Edie Katz, Tom Joyce, and Thomas Evans; Jessica, Gavin, and Daisy in Dunkeld; Eck Finlay in Edinburgh and his father in Little Sparta; Valerie and Michael Cohen in June Lake; Scott

Slovic in Reno; Carolyn from Reclaim the Streets in Brixton; Iain Boal; my agent Bonnie Nadell; my editor Paul Slovak at Viking Penguin, who took to the idea of a general history of walking immediately and made this one possible; and particularly Pat Dennis, who listened to me chapter by chapter, related much on mountaineering history and Asian mysticism to me, and walked alongside me for the duration of this book.

Wanderlust

Part I

THE PACE OF
$\mathcal{T}$HOUGHTS

Chapter 1

TRACING A HEADLAND:

An Introduction

Where does it start? Muscles tense. One leg a pillar, holding the body upright between the earth and sky. The other a pendulum, swinging from behind. Heel touches down. The whole weight of the body rolls forward onto the ball of the foot. The big toe pushes off, and the delicately balanced weight of the body shifts again. The legs reverse position. It starts with a step and then another step and then another that add up like taps on a drum to a rhythm, the rhythm of walking. The most obvious and the most obscure thing in the world, this walking that wanders so readily into religion, philosophy, landscape, urban policy, anatomy, allegory, and heartbreak.

The history of walking is an unwritten, secret history whose fragments can be found in a thousand unemphatic passages in books, as well as in songs, streets, and almost everyone's adventures. The bodily history of walking is that of bipedal evolution and human anatomy. Most of the time walking is merely practical, the unconsidered locomotive means between two sites. To make walking into an investigation, a ritual, a meditation, is a special subset of walking, physiologically like and philosophically unlike the way the mail carrier brings the mail and the office worker reaches the train. Which is to say that the subject of walking is, in some sense, about how we invest universal acts with particular meanings. Like eating or breathing, it can be invested with wildly different cultural meanings, from the erotic to the spiritual, from the revolutionary to the artistic. Here this history begins to become part of the history of the imagination and the culture, of what kind of pleasure, freedom, and meaning are pursued at different times by different kinds of walks and walkers. That imagination has both shaped and been shaped by the spaces it passes through on two feet. Walking has created paths, roads, trade routes; generated local and cross-continental senses of place; shaped cities, parks; generated maps, guidebooks, gear, and, further afield, a vast library of walking stories and poems, of pilgrimages, mountaineering expeditions, meanders, and summer picnics. The landscapes, urban and rural, gestate the stories, and the stories bring us back to the sites of this history.

This history of walking is an amateur history, just as walking is an amateur act. To use a walking metaphor, it trespasses through everybody else's field—through anatomy, anthropology, architecture, gardening, geography, political and cultural history, literature, sexuality, religious studies—and doesn't stop in any of them on its long route. For if a field of expertise can be imagined as a real field—a nice rectangular confine carefully tilled and yielding a specific crop—then the subject of walking resembles walking itself in its lack of confines. And though *the* history of walking is, as part of all these fields and everyone's experience, virtually infinite, *this* history of walking I am writing can only be partial, an idiosyncratic path traced through them by one walker, with much doubling back and looking around. In what follows, I have tried to trace the paths that brought most of us in my country, the United States, into the present moment, a history compounded largely of European sources, inflected and subverted by the vastly different scale of American space, the centuries of adaptation and mutation here, and by the other traditions that have recently met up with those paths, notably Asian traditions. The history of walking is everyone's history, and any written version can only hope to indicate some of the more well-trodden paths in the author's vicinity—which is to say, the paths I trace are not the only paths.

I sat down one spring day to write about walking and stood up again, because a desk is no place to think on the large scale. In a headland

---
be walks, how he walks, if he has ever walked, if he could walk better, what he achieves in walking . . .

just north of the Golden Gate Bridge studded with abandoned military fortifications, I went out walking up a valley and along a ridgeline, then down to the Pacific. Spring had come after an unusually wet winter, and the hills had turned that riotous, exuberant green I forget and rediscover every year. Through the new growth poked grass from the year before, bleached from summer gold to an ashen gray by the rain, part of the subtler palette of the rest of the year. Henry David Thoreau, who walked more vigorously than me on the other side of the continent, wrote of the local, "An absolutely new prospect is a great happiness, and I can still get this any afternoon. Two or three hours' walking will carry me to as strange a country as I expect ever to see. A single farmhouse which I had not seen before is sometimes as good as the dominions of the King of Dahomey. There is in fact a sort of harmony discoverable between the capabilities of the landscape within a circle of ten miles' radius, or the limits of an afternoon walk, and the threescore years and ten of human life. It will never become quite familiar to you."

These linked paths and roads form a circuit of about six miles that I began hiking ten years ago to walk off my angst during a difficult year. I kept coming back to this route for respite from my work and for my work too, because thinking is generally thought of as doing nothing in a production-oriented culture, and doing nothing is hard to do. It's best done by disguising it as doing something, and the something closest to doing nothing is walking. Walking itself is the intentional act closest to the unwilled rhythms of the body, to breathing and the beating of the heart. It strikes a delicate balance between working and idling, being and doing. It is a bodily labor that produces nothing but thoughts, experiences, arrivals. After all those years of walking to work out other things, it made sense to come back to work close to home, in Thoreau's sense, and to think about walking.

Walking, ideally, is a state in which the mind, the body, and the world are aligned, as though they were three characters finally in conversation together, three notes suddenly making a chord. Walking allows us to be in our bodies and in the world without being made busy by them. It leaves us free to think without being wholly lost in our thoughts. I wasn't sure whether I was too soon or too late for the purple lupine that can be so spectacular in these headlands, but

---

questions that are tied to all the philosophical, psychological, and political systems which preoccupy the

---

milkmaids were growing on the shady side of the road on the way to the trail, and they recalled the hillsides of my childhood that first bloomed every year with an extravagance of these white flowers. Black butterflies fluttered around me, tossed along by wind and wings, and they called up another era of my past. Moving on foot seems to make it easier to move in time; the mind wanders from plans to recollections to observations.

The rhythm of walking generates a kind of rhythm of thinking, and the passage through a landscape echoes or stimulates the passage through a series of thoughts. This creates an odd consonance between internal and external passage, one that suggests that the mind is also a landscape of sorts and that walking is one way to traverse it. A new thought often seems like a feature of the landscape that was there all along, as though thinking were traveling rather than making. And so one aspect of the history of walking is the history of thinking made concrete—for the motions of the mind cannot be traced, but those of the feet can. Walking can also be imagined as a visual activity, every walk a tour leisurely enough both to see and to think over the sights, to assimilate the new into the known. Perhaps this is where walking's peculiar utility for thinkers comes from. The surprises, liberations, and clarifications of travel can sometimes be garnered by going around the block as well as going around the world, and walking travels both near and far. Or perhaps walking should be called movement, not travel, for one can walk in circles or travel around the world immobilized in a seat, and a certain kind of wanderlust can only be assuaged by the acts of the body itself in motion, not the motion of the car, boat, or plane. It is the movement as well as the sights going by that seems to make things happen in the mind, and this is what makes walking ambiguous and endlessly fertile: it is both means and end, travel and destination.

The old red dirt road built by the army had begun its winding, uphill course through the valley. Occasionally I focused on the act of walking, but mostly it was unconscious, the feet proceeding with their own knowledge of balance, of sidestepping rocks and crevices, of pacing, leaving me free to look at the roll of hills far away and the

---

world.—HONORÉ DE BALZAC, *THEORIE DE LA DEMARCHÉ*   *An Eskimo custom offers an angry person*

abundance of flowers close up: brodia; the pink papery blossoms whose name I never learned; an abundance of cloverlike sourgrass in yellow bloom; and then halfway along the last bend, a paperwhite narcissus. After twenty minutes' trudge uphill, I stopped to smell it. There used to be a dairy in this valley, and the foundations of a farm-house and a few straggling old fruit trees still survive somewhere down below, on the other side of the wet, willow-crowded valley bottom. It was a working landscape far longer than a recreational one: first came the Miwok Indians, then the agriculturists, themselves rooted out after a century by the military base, which closed in the 1970s, when coasts became irrelevant to an increasingly abstract and aerial kind of war. Since the 1970s, this place has been turned over to the National Park Service and to people like me who are heirs to the cultural tradition of walking in the landscape for pleasure. The massive concrete gun emplacements, bunkers, and tunnels will never disappear as the dairy buildings have, but it must have been the dairy families that left behind the live legacy of garden flowers that crop up among the native plants.

Walking is meandering, and I meandered from my cluster of nar-cissus in the curve of the red road first in thought and then by foot. The army road reached the crest and crossed the trail that would take me across the brow of the hill, cutting into the wind and downhill before its gradual ascent to the western side of the crest. On the ridgetop up above this footpath, facing into the next valley north, was an old radar station surrounded by an octagon of fencing. The odd collection of objects and cement bunkers on an asphalt pad were part of a Nike missile guidance system, a system for directing nuclear mis-siles from their base in the valley below to other continents, though none were ever launched from here in war. Think of the ruin as a souvenir from the canceled end of the world.

It was nuclear weapons that first led me to walking history, in a trajectory as surprising as any trail or train of thought. I became in the 1980s an antinuclear activist and participated in the spring demon-strations at the Nevada Test Site, a Department of Energy site the size of Rhode Island in southern Nevada where the United States has been detonating nuclear bombs—more than a thousand to date—since 1951. Sometimes nuclear weapons seemed like nothing more

---

release by walking the emotion out of his or her system in a straight line across the landscape, the point at

than intangible budget figures, waste disposal figures, potential casualty figures, to be resisted by campaigning, publishing, and lobbying. The bureaucratic abstractness of both the arms race and the resistance to it could make it hard to understand that the real subject was and is the devastation of real bodies and real places. At the test site, it was different. The weapons of mass destruction were being exploded in a beautifully stark landscape we camped near for the week or two of each demonstration (exploded underground after 1963, though they often leaked radiation into the atmosphere anyway and always shook the earth). We—that *we* made up of the scruffy American counterculture, but also of survivors of Hiroshima and Nagasaki, Buddhist monks and Franciscan priests and nuns, veterans turned pacifist, renegade physicists, Kazakh and German and Polynesian activists living in the shadow of the bomb, and the Western Shoshone, whose land it was—had broken through the abstractions. Beyond them were the actualities of places, of sights, of actions, of sensations—of handcuffs, thorns, dust, heat, thirst, radiation risk, the testimony of radiation victims—but also of spectacular desert light, the freedom of open space, and the stirring sight of the thousands who shared our belief that nuclear bombs were the wrong instrument with which to write the history of the world. We bore a kind of bodily witness to our convictions, to the fierce beauty of the desert, and to the apocalypses being prepared nearby. The form our demonstrations took was walking: what was on the public-land side of the fence a ceremonious procession became, on the off-limits side, an act of trespass resulting in arrest. We were engaging, on an unprecedentedly grand scale, in civil disobedience or civil resistance, an American tradition first articulated by Thoreau.

Thoreau himself was both a poet of nature and a critic of society. His famous act of civil disobedience was passive—a refusal to pay taxes to support war and slavery and an acceptance of the night in jail that ensued—and it did not overlap directly with his involvement in exploring and interpreting the local landscape, though he did lead a huckleberrying party the day he got out of jail. In our actions at the test site the poetry of nature and criticism of society were united in this camping, walking, and trespassing, as though we had figured out how a berrying party could be a revolutionary cadre. It was a revelation

---

which the anger is conquered is marked with a stick, bearing witness to the strength or length of the rage.

---

to me, the way this act of walking through a desert and across a cattle guard into the forbidden zone could articulate political meaning. And in the course of traveling to this landscape, I began to discover other western landscapes beyond my coastal region and to explore those landscapes and the histories that had brought me to them—the history not only of the development of the West but of the Romantic taste for walking and landscape, the democratic tradition of resistance and revolution, the more ancient history of pilgrimage and walking to achieve spiritual goals. I found my voice as a writer in describing all the layers of history that shaped my experiences at the test site. And I began to think and to write about walking in the course of writing about places and their histories.

Of course walking, as any reader of Thoreau's essay "Walking" knows, inevitably leads into other subjects. Walking is a subject that is always straying. Into, for example, the shooting stars below the missile guidance station in the northern headlands of the Golden Gate. They are my favorite wildflower, these small magenta cones with their sharp black points that seem aerodynamically shaped for a flight that never comes, as though they had evolved forgetful of the fact that flowers have stems and stems have roots. The chaparral on both sides of the trail, watered by the condensation of the ocean fog through the dry months and shaded by the slope's northern exposure, was lush. While the missile guidance station on the crest always makes me think of the desert and of war, these banks below always remind me of English hedgerows, those field borders with their abundance of plants, birds, and that idyllic English kind of countryside. There were ferns here, wild strawberries, and, tucked under a coyote bush, a cluster of white iris in bloom.

Although I came to think about walking, I couldn't stop thinking about everything else, about the letters I should have been writing, about the conversations I'd been having. At least when my mind strayed to the phone conversation with my friend Sono that morning, I was still on track. Sono's truck had been stolen from her West Oakland studio, and she told me that though everyone responded to it as a disaster, she wasn't all that sorry it was gone, or in a hurry to replace it. There was a joy, she said, to finding that her body was adequate to get her where she was going, and it was a gift to develop a

---

—LUCY LIPPARD, OVERLAY   *We learn a place and how to visualize spatial relationships, as children, on*

---

more tangible, concrete relationship to her neighborhood and its residents. We talked about the more stately sense of time one has afoot and on public transit, where things must be planned and scheduled beforehand, rather than rushed through at the last minute, and about the sense of place that can only be gained on foot. Many people nowadays live in a series of interiors—home, car, gym, office, shops—disconnected from each other. On foot everything stays connected, for while walking one occupies the spaces between those interiors in the same way one occupies those interiors. One lives in the whole world rather than in interiors built up against it.

The narrow trail I had been following came to an end as it rose to meet the old gray asphalt road that runs up to the missile guidance station. Stepping from path to road means stepping up to see the whole expanse of the ocean, spreading uninterrupted to Japan. The same shock of pleasure comes every time I cross this boundary to discover the ocean again, an ocean shining like beaten silver on the brightest days, green on the overcast ones, brown with the muddy runoff of the streams and rivers washing far out to sea during winter floods, an opalescent mottling of blues on days of scattered clouds, only invisible on the foggiest days, when the salt smell alone announces the change. This day the sea was a solid blue running toward an indistinct horizon where white mist blurred the transition to cloudless sky. From here on, my route was downhill. I had told Sono about an ad I found in the *Los Angeles Times* a few months ago that I'd been thinking about ever since. It was for a CD-ROM encyclopedia, and the text that occupied a whole page read, "You used to walk across town in the pouring rain to use our encyclopedias. We're pretty confident that we can get your kid to click and drag." I think it was the kid's walk in the rain that constituted the real education, at least of the senses and the imagination. Perhaps the child with the CD-ROM encyclopedia will stray from the task at hand, but wandering in a book or a computer takes place within more constricted and less sensual parameters. It's the unpredictable incidents between official events that add up to a life, the incalculable that gives it value. Both rural and urban walking have for two centuries been prime ways of exploring the unpredictable and the incalculable, but they are now under assault on many fronts.

---

foot and with imagination. Place and the scale of place must be measured against our bodies and their

---

The multiplication of technologies in the name of efficiency is actually eradicating free time by making it possible to maximize the time and place for production and minimize the unstructured travel time in between. New timesaving technologies make most workers more productive, not more free, in a world that seems to be accelerating around them. Too, the rhetoric of efficiency around these technologies suggests that what cannot be quantified cannot be valued—that that vast array of pleasures which fall into the category of doing nothing in particular, of woolgathering, cloud-gazing, wandering, window-shopping, are nothing but voids to be filled by something more definite, more productive, or faster paced. Even on this headland route going nowhere useful, this route that could only be walked for pleasure, people had trodden shortcuts between the switchbacks as though efficiency was a habit they couldn't shake. The indeterminacy of a ramble, on which much may be discovered, is being replaced by the determinate shortest distance to be traversed with all possible speed, as well as by the electronic transmissions that make real travel less necessary. As a member of the self-employed whose time saved by technology can be lavished on daydreams and meanders, I know these things have their uses, and use them—a truck, a computer, a modem—myself, but I fear their false urgency, their call to speed, their insistence that travel is less important than arrival. I like walking because it is slow, and I suspect that the mind, like the feet, works at about three miles an hour. If this is so, then modern life is moving faster than the speed of thought, or thoughtfulness.

Walking is about being outside, in public space, and public space is also being abandoned and eroded in older cities, eclipsed by technologies and services that don't require leaving home, and shadowed by fear in many places (and strange places are always more frightening than known ones, so the less one wanders the city the more alarming it seems, while the fewer the wanderers the more lonely and dangerous it really becomes). Meanwhile, in many new places, public space isn't even in the design: what was once public space is designed to accommodate the privacy of automobiles; malls replace main streets; streets have no sidewalks; buildings are entered through their garages; city halls have no plazas; and everything has walls, bars,

capabilities.—GARY SNYDER, "BLUE MOUNTAINS CONSTANTLY WALKING" *Then one day walking*

gates. Fear has created a whole style of architecture and urban design, notably in southern California, where to be a pedestrian is to be under suspicion in many of the subdivisions and gated "communities." At the same time, rural land and the once-inviting peripheries of towns are being swallowed up in car-commuter subdivisions and otherwise sequestered. In some places it is no longer possible to be out in public, a crisis both for the private epiphanies of the solitary stroller and for public space's democratic functions. It was this fragmentation of lives and landscapes that we were resisting long ago, in the expansive spaces of the desert that temporarily became as public as a plaza.

And when public space disappears, so does the body as, in Sono's fine term, adequate for getting around. Sono and I spoke of the discovery that our neighborhoods—which are some of the most feared places in the Bay Area—aren't all that hostile (though they aren't safe enough to let us forget about safety altogether). I have been threatened and mugged on the street, long ago, but I have a thousand times more often encountered friends passing by, a sought-for book in a store window, compliments and greetings from my loquacious neighbors, architectural delights, posters for music and ironic political commentary on walls and telephone poles, fortune-tellers, the moon coming up between buildings, glimpses of other lives and other homes, and street trees noisy with songbirds. The random, the unscreened, allows you to find what you don't know you are looking for, and you don't know a place until it surprises you. Walking is one way of maintaining a bulwark against this erosion of the mind, the body, the landscape, and the city, and every walker is a guard on patrol to protect the ineffable.

Perhaps a third of the way down the road that wandered to the beach, an orange net was spread. It looked like a tennis net, but when I reached it I saw that it fenced off a huge new gap in the road. This road has been crumbling since I began to walk on it a decade ago. It used to roll uninterruptedly from sea to ridgetop. Along the coastal reach of the road a little bite appeared in 1989 that one could edge around, then a little trail detoured around the growing gap. With every winter's rain, more and more red earth and road surface crumbled away, sliding into a heap at the ruinous bottom of the steep slope the

----

round Tavistock Square I made up, as I sometimes make up my books, To the Lighthouse, *in a great,*

----

road had once cut across. It was an astonishing sight at first, this road that dropped off into thin air, for one expects roads and paths to be continuous. Every year more of it has fallen. And I have walked this route so often that every part of it springs associations on me. I remember all the phases of the collapse and how different a person I was when the road was complete. I remember explaining to a friend on this route almost three years earlier why I liked walking the same way over and over. I joked, in a bad adaptation of Heraclitus's famous dictum about rivers, that you never step on the same trail twice; and soon afterward we came across the new staircase that cut down the steep hillside, built far enough inland that the erosion wouldn't reach it for many years to come. If there is a history of walking, then it too has come to a place where the road falls off, a place where there is no public space and the landscape is being paved over, where leisure is shrinking and being crushed under the anxiety to produce, where bodies are not in the world but only indoors in cars and buildings, and an apotheosis of speed makes those bodies seem anachronistic or feeble. In this context, walking is a subversive detour, the scenic route through a half-abandoned landscape of ideas and experiences.

I had to circumnavigate this new chunk bitten out of the actual landscape by going to a new detour on the right. There's always a moment on this circuit when the heat of climbing and the windblock the hills provide give way to the descent into ocean air, and this time it came at the staircase past the scree of a fresh cut into the green serpentine stone of the hill. From there it wasn't far to the switchback leading to the other half of the road, which winds closer and closer to the cliffs above the ocean, where waves shatter into white foam over the dark rocks with an audible roar. Soon I was at the beach, where surfers sleek as seals in their black wet suits were catching the point break at the northern edge of the cove, dogs chased sticks, people lolled on blankets, and the waves crashed, then sprawled into a shallow rush uphill to lap at the feet of those of us walking on the hard sand of high tide. Only a final stretch remained, up over a sandy crest and along the length of the murky lagoon full of water birds.

It was the snake that came as a surprise, a garter snake, so called because of the yellowish stripes running the length of its dark body, a snake tiny and enchanting as it writhed like waving water across the

apparently involuntary rush.—VIRGINIA WOOLF, *MOMENTS OF BEING*   *In my room, the world is beyond*

path and into the grasses on one side. It didn't alarm me so much as alert me. Suddenly I came out of my thoughts to notice everything around me again—the catkins on the willows, the lapping of the water, the leafy patterns of the shadows across the path. And then myself, walking with the alignment that only comes after miles, the loose diagonal rhythm of arms swinging in synchronization with legs in a body that felt long and stretched out, almost as sinuous as the snake. My circuit was almost finished, and at the end of it I knew what my subject was and how to address it in a way I had not six miles before. It had come to me not in a sudden epiphany but with a gradual sureness, a sense of meaning like a sense of place. When you give yourself to places, they give you yourself back; the more one comes to know them, the more one seeds them with the invisible crop of memories and associations that will be waiting for you when you come back, while new places offer up new thoughts, new possibilities. Exploring the world is one of the best ways of exploring the mind, and walking travels both terrains.

---

my understanding, / But when I walk I see that it consists of three or four hills and a cloud.—WALLACE

THE MIND AT THREE MILES

AN HOUR

I. PEDESTRIAN ARCHITECTURE

Jean-Jacques Rousseau remarked in his *Confessions*, "I can only meditate when I am walking. When I stop, I cease to think; my mind only works with my legs." The history of walking goes back further than the history of human beings, but the history of walking as a conscious cultural act rather than a means to an end is only a few centuries old in Europe, and Rousseau stands at its beginning. That history began with the walks of various characters in the eighteenth century, but the more literary among them strove to consecrate walking by tracing it to Greece, whose practices were so happily revered and misrepresented then. The eccentric English revolutionary and writer John Thelwall wrote a massive, turgid book, *The Peripatetic,* uniting Rousseauian romanticism with this spurious classical tradition. "In one respect, at least, I may boast of a resemblance to the simplicity of the ancient sages: I pursue my meditations on foot," he remarked. And after Thelwall's book appeared in 1793, many more would make the claim until it became an established idea that the ancients walked to think, so much so that the very picture seems part of cultural history: austerely draped men speaking gravely as they pace through a dry Mediterranean landscape punctuated with the occasional marble column.

This belief arose from a coincidence of architecture and language. When Aristotle was ready to set up a school in Athens, the city assigned him a plot of land. "In it," explains Felix Grayeff's history of this school, "stood shrines to Apollo and the Muses, and perhaps other smaller buildings. . . . A covered colonnade led to the temple of Apollo, or perhaps connected the temple with the shrine of the Muses; whether it had existed before or was only built now, is not known. This colonnade or walk (peripatos) gave the school its name; it seems that it was here, at least at the beginning, that the pupils assembled and the teachers gave their lectures. Here they wandered to and fro; for this reason it was later said that Aristotle himself lectured and taught while walking up and down." The philosophers who came from it were called the Peripatetic philosophers or the Peripatetic school, and in English the word *peripatetic* means "one who walks habitually and extensively." Thus their name links thinking with walking. There is something more to this than the coincidence that established a school of philosophy in a temple of Apollo with a long colonnade—slightly more.

The Sophists, the philosophers who dominated Athenian life before Socrates, Plato, and Aristotle, were famously wanderers who often taught in the grove where Aristotle's school would be located. Plato's assault on the Sophists was so furious that the words *sophist* and *sophistry* are still synonymous with deception and guile, though the root *sophia* has to do with wisdom. The Sophists, however, functioned something like the chautauquas and public lecturers in nineteenth-century America, who went from place to place delivering talks to audiences hungry for information and ideas. Though they taught rhetoric as a tool of political power, and the ability to persuade and argue was crucial to Greek democracy, the Sophists taught other things besides. Plato, whose half-fabricated character Socrates is one of the wiliest and most persuasive debaters of all times, is somewhat disingenuous when he attacks the Sophists.

Whether or not the Sophists were virtuous, they were often mobile, as are many of those whose first loyalty is to ideas. It may be that loyalty to something as immaterial as ideas sets thinkers apart from those whose loyalty is tied to people and locale, for the loyalty that ties down the latter will often drive the former from place to place. It is an attachment that requires detachment. Too, ideas are not as reliable or popular a crop as, say, corn, and those who cultivate them often must keep moving in pursuit of support as well as truth. Many

---

undergraduate, he recalls in his autobiography, Pilgrim's Way (1940), *that "the works of Aristotle are*

professions in many cultures, from musicians to medics, have been nomadic, possessed of a kind of diplomatic immunity to the strife between communities that keeps others local. Aristotle himself had at first intended to become a doctor, as his father had been, and doctors in that time were members of a secretive guild of travelers who claimed descent from the god of healing. Had he become a philosopher in the era of the Sophists, he might have been mobile anyway, for settled philosophy schools were first established in Athens in his time.

It is now impossible to say whether or not Aristotle and his Peripatetics habitually walked while they talked philosophy, but the link between thinking and walking recurs in ancient Greece, and Greek architecture accommodated walking as a social and conversational activity. Just as the Peripatetics took their name from the *peripatos* of their school, so the Stoics were named after the stoa, or colonnade, in Athens, a most unstoically painted walkway where they walked and talked. Long afterward, the association between walking and philosophizing became so widespread that central Europe has places named after it: the celebrated Philosophenweg in Heidelberg where Hegel is said to have walked, the Philosophen-damm in Königsberg that Kant passed on his daily stroll (now replaced by a railway station), and the Philosopher's Way Kierkegaard mentions in Copenhagen.

And philosophers who walked—well, walking is a universal human activity. Jeremy Bentham, John Stuart Mill, and many others walked far, and Thomas Hobbes even had a walking stick with an inkhorn built into it so he could jot down ideas as he went. Frail Immanuel Kant took a daily walk around Königsberg after dinner—but it was merely for exercise, because he did his thinking sitting by the stove and staring at the church tower out the window. The young Friedrich Nietzsche declares with superb conventionality, "For recreation I turn to three things, and a wonderful recreation they provide!—my Schopenhauer, Schumann's music, and, finally, solitary walks." In the twentieth century, Bertrand Russell recounts of his friend Ludwig Wittgenstein, "He used to come to my rooms at midnight, and for hours he would walk backwards and forwards like a caged tiger. On arrival, he would announce that when he left my rooms he would commit suicide. So, in spite of getting sleepy, I did not like to turn him out. One such evening after an hour or two of dead silence, I said to him, 'Wittgenstein, are you

---

forever bound up with me with the smell of peat and certain stretches of granite and heather."—ON JOHN

thinking about logic or about your sins?' 'Both' he said, and then reverted to silence." Philosophers walked. But philosophers who thought about walking are rarer.

II. CONSECRATING WALKING

It is Rousseau who laid the groundwork for the ideological edifice within which walking itself would be enshrined—not the walking that took Wittgenstein back and forth in Russell's room, but the walking that took Nietzsche out into the landscape. In 1749 the writer and encylopedist Denis Diderot was thrown into jail for writing an essay questioning the goodness of God. Rousseau, a close friend of Diderot's at the time, took to visiting him in jail, walking the six miles from his home in Paris to the dungeon of the Château de Vincennes. Though that summer was extremely hot, says Rousseau in his not entirely reliable *Confessions* (1781–88), he walked because he was too poor to travel by other means. "In order to slacken my pace," writes Rousseau, "I thought of taking a book with me. One day I took the *Mercure de France* and, glancing through it as I walked, I came upon this question propounded by the Dijon Academy for the next year's prize: Has the progress of the sciences and arts done more to corrupt morals or improve them? The moment I read that I beheld another universe and became another man." In this other universe, this other man won the prize, and the published essay became famous for its furious condemnation of such progress.

Rousseau was less an original thinker than a daring one; he gave the boldest articulation to existing tensions and the most fervent praise to emerging sensibilities. The assertion that God, monarchical government, and nature were all harmoniously aligned was becoming untenable. Rousseau, with his lower-middle-class resentments, his Calvinist Swiss suspicion of kings and Catholicism, his desire to shock, and his unshakable self-confidence, was the person to make specific and political those distant rumblings of discord. In the *Discourse on the Arts and Letters*, he declared that learning and even printing corrupt and weaken both the individual and the culture. "Behold how luxury, licentiousness, and slavery have in all periods been punishment for the arrogant attempts we have made to emerge from the happy ignorance in which eternal wisdom had placed us." The arts and sciences, he asserted, lead not to happiness nor to self-knowledge, but to distraction and corruption.

Now the assumption that the natural, the good, and the simple are all aligned seems commonplace at best; then, it was incendiary. In Christian theology, nature and humanity had both fallen from grace after Eden; it was Christian civilization that redeemed them, so that goodness was a cultural rather than a natural state. This Rousseauian reversal that insists that men and nature are better in their original condition is, among other things, an attack on cities, aristocrats, technology, sophistication, and sometimes theology, and it is still going on today (though curiously the French, who were his primary audience and whose revolution he contributed to, have in the long run been less responsive to these ideas than the British, the Germans, and the Americans). Rousseau developed these ideas further in his *Discourse on the Origin and Foundation of Inequality* (1754) and in his novels *Julie* (1761) and *Emile* (1762). Both novels portray, in various ways, a simpler, more rural life—though none of them acknowledge the hard manual labor of most rural people. His fictional characters lived, as he himself did at his happiest, in unostentatious ease, supported by invisible toilers. The inconsistencies in Rousseau's work don't matter, for it is less a cogent analysis than the expression of a new sensibility and its new enthusiasms. That Rousseau wrote with great elegance is one of those inconsistencies and one of the reasons he was so widely read.

In the *Discourse on Inequality*, Rousseau portrays man in his natural condition "wandering in the forests, without industry, without speech, without domicile, without war and without liaisons, with no need of his fellow-men, likewise with no desire to harm them," even while he admits that we cannot know what this condition was. The treatise offhandedly ignores Christian narratives of human origins and reaches toward a prescient comparative anthropology of social evolution instead (and though it reiterates Christianity's theme of the fall from grace, it reverses the direction of this fall: it is no longer into nature but into culture). In this ideology, walking functions as an emblem of the simple man and as, when the walk is solitary and rural, a means of being in nature and outside society. The walker has the detachment

---

MOUNTAINEERING   ... *while he himself began to walk around at a lively pace in a "Keplerian ellipse," all*

---

of the traveler but travels unadorned and unaugmented, dependent on his or her own bodily strength rather than on conveniences that can be made and bought—horses, boats, carriages. Walking is, after all, an activity essentially unimproved since the dawn of time.

In portraying himself so often as a pedestrian, Rousseau claimed kin to this ideal walker before history, and he did walk extensively throughout his life. His wandering life began when he returned to Geneva from a Sunday stroll in the country, only to find that he had come back too late: the gates of the city were shut. Impulsively, the fifteen-year-old Rousseau decided to abandon his birthplace, his apprenticeship, and eventually his religion; he turned from the gates and walked out of Switzerland. In Italy and France he found and left many jobs, patrons, and friends during a life that seemed aimless until the day he read the *Mercure de France* and found his vocation. Ever after, he seemed to be trying to recover the carefree wandering of his youth. He writes of one episode, "I do not remember ever having had in all my life a spell of time so completely free from care and anxiety as those seven or eight days spent on the road. . . . This memory has left me the strongest taste for everything associated with it, for mountains especially and for travelling on foot. I have never travelled so except in my prime, and it has always been a delight to me. . . . For a long while I searched Paris for any two men sharing my tastes, each willing to contribute fifty louis from his purse and a year of his time for a joint tour of Italy on foot, with no other attendant than a lad to come with us and carry a knapsack."

Rousseau never found serious candidates for this early version of a walking tour (and never explained why companions were necessary to its execution, unless they were to pay the bills). But he continued to walk at every opportunity. Elsewhere he claimed, "Never did I think so much, exist so vividly, and experience so much, never have I been so much myself—if I may use that expression—as in the journeys I have taken alone and on foot. There is something about walking which stimulates and enlivens my thoughts. When I stay in one place I can hardly think at all; my body has to be on the move to set my mind going. The sight of the countryside, the succession of pleasant views, the open air, a sound appetite, and the good health I gain by walking, the easy atmosphere of an inn, the absence of everything

the time explaining in a low voice his thoughts on "complementarity." He walked with bent head and knit

that makes me feel my dependence, of everything that recalls me to my situation—all these serve to free my spirit, to lend a greater boldness to my thinking, so that I can combine them, select them, and make them mine as I will, without fear or restraint." It was, of course, an ideal walking that he described—chosen freely by a healthy person amid pleasant and safe circumstances—and it is this kind of walking that would be taken up by his countless heirs as an expression of well-being, harmony with nature, freedom, and virtue.

Rousseau portrays walking as both an exercise of simplicity and a means of contemplation. During the time he wrote the *Discourses*, he would walk alone in the Bois de Boulogne after dinner, "thinking over subjects for works to be written and not returning till night." The *Confessions*, from which these passages come, were not published until after Rousseau's death (in 1762 his books had been burned in Paris and Geneva, and his life as a wandering exile had begun). Even before the *Confessions* were finished, however, his readers already associated him with peripatetic excursions. When an adoring James Boswell came to visit Rousseau near Neuchâtel, Switzerland, in 1764, Boswell wrote, "To prepare myself for the great Interview, I walked out alone. I strolled pensive by the side of the River Ruse in a beautifull Wild Valley surrounded by immense Mountains, some covered with frowning rocks, others with glittering snow." Boswell, who was at twenty-four as self-conscious as Rousseau and more foppish about it, already knew that walking, solitude, and wilderness were Rousseauian and clad his mind in their effects as he might have adorned his body for a more conventional meeting.

Solitude is an ambiguous state throughout Rousseau's writings. In the *Discourse on the Origin and Foundations of Inequality*, he portrays human beings in their natural state as isolated dwellers in a hospitable forest. But in his more personal work, he often portrays solitude not as an ideal state but as a consolation and refuge for a man who has been betrayed and disappointed. In fact, much of his writing revolves around the question of whether and how one should relate to one's fellow humans. Hypersensitive almost to the point of paranoia and convinced of his own rightness under the most dubious circumstances, Rousseau overreacted to the judgments of others, yet never could or would subdue his unorthodox and often abrasive ideas and

brows: from time to time, he looked up at me and underlined some important point by a sober gesture. As he

acts. It is now popularly proposed that his writing universalizes his experience, and that his picture of man's fall from simplicity and grace into corruption is little more than a portrait of Rousseau's fall from Swiss simplicity and security or merely from childhood naïveté into his uncertain life abroad among aristocrats and intellectuals. Whether this is so or not, his version has been so influential that few are entirely beyond its reach nowadays.

Finally, at the end of his life, he wrote _Reveries of a Solitary Walker_ (_Les Rêveries du promeneur solitaire_ in the original, 1782), a book that is and is not about walking. Each of its chapters is called a walk, and in the Second Walk, he states his premise: "Having therefore decided to describe my habitual state of mind in this, the strangest situation which any mortal will ever know, I could think of no simpler or surer way of carrying out my plan than to keep a faithful record of my solitary walks and the reveries that occupy them." Each of these short personal essays resembles the sequence of thoughts or preoccupations one might entertain on a walk, though there is no evidence they are the fruit of specific walks. Several are meditations on a phrase, some are recollections, some are little more than aired grievances. Together the ten essays (the eighth and ninth were still drafts and the tenth was left unfinished at the time of his death in 1778) portray a man who has taken refuge in the thoughts and botanical pursuits of his walks, and who through them seeks and recalls a safer haven.

A solitary walker is in the world, but apart from it, with the detachment of the traveler rather than the ties of the worker, the dweller, the member of a group. Walking seems to have become Rousseau's chosen mode of being because within a walk he is able to live in thought and reverie, to be self-sufficient, and thus to survive the world he feels has betrayed him. It provides him with a literal position from which to speak. As a literary structure, the recounted walk encourages digression and association, in contrast to the stricter form of a discourse or the chronological progression of a biographical or historical narrative. A century and a half later, James Joyce and Virginia Woolf would, in trying to describe the workings of the mind, develop the style called stream of consciousness. In their novels _Ulysses_ and _Mrs. Dalloway_, the jumble of thoughts and recollections of their protagonists unfolds best during walks. This kind of unstructured, associative

spoke, the words and sentences which I had read before in his papers suddenly took life and became loaded

thinking is the kind most often connected to walking, and it suggests walking as not an analytical but an improvisational act. Rousseau's *Reveries* are one of the first portraits of this relationship between thinking and walking.

Rousseau walks alone, and the plants he gathers and strangers he encounters are the only beings toward whom he expresses tenderness. In the Ninth Walk, he reminisces about earlier walks, which seem to slide out from each other like sections of a telescope to focus on his distant past. He begins with a walk two days before to the Ecole Militaire, then moves to one outside Paris two years before, and then to a garden excursion with his wife four or five years before, and finally recounts an incident that pre-dated even this last recollection by many years, in which he bought all the apples a poor girl was selling and distributed them among hungry urchins loitering nearby. All these recollections were prompted by reading an obituary of an acquaintance that mentioned her love for children and triggered Rousseau's guilt about his own abandoned children (though modern scholars sometimes doubt he had any children to begin with, his *Confessions* say he had five by his common-law wife Thérèse and put all of them in orphanages). These recollections argue against a charge no one but he himself has leveled, and they do so by declaring his affection for children, as demonstrated in these casual encounters. The essay is a ruminative defense for an imagined trial. The conclusion shifts the subject to the tribulations his fame has brought him and his inability to walk unrecognized and in peace among people. The implication is that even in this most casual of social exchanges he has been thwarted, so that only the terrain of the reverie leaves him freedom to roam. Most of the book was written while he was living in Paris, isolated by his fame and his suspicion.

If the literature of philosophical walking begins with Rousseau, it is because he is one of the first who thought it worthwhile to record in detail the circumstances of his musing. If he was a radical, his most profoundly radical act was to revalue the personal and the private, for which walking, solitude, and wilderness provided favorable conditions. If he inspired revolutions, revolutions in imagination and culture as well as in political organization, they were for him only necessary to overthrow the impediments to such experience. The full force of

---

with meaning. It was one of the few solemn moments that count in an existence, the revelation of a world of

---

his intellect and his most compelling arguments had been made in the cause of recovering and perpetuating such states of mind and life as he describes in *Reveries of a Solitary Walker*.

In two walks, he recollects the interludes of rural peace he most treasured. In the famous Fifth Walk, he describes the happiness he found on the island of Saint-Pierre on Lake of Bienne, to which he fled after being stoned and driven out of Motiers, near Neuchâtel, where Boswell had visited him. "Wherein lay this great contentment?" he asks rhetorically, and goes on to describe a life in which he owned little and did little, save botanize and boat. It is the Rousseauian peaceable kingdom, privileged enough that no manual labor is required, but without the sophistication and socializing of an aristocratic retreat. The Tenth Walk is a paean to the similar rural happiness he had with his patroness and lover, Louise de Warens, when he was a teenager. It was written when he had finally found a replacement for Saint-Pierre, the estate of Ermenonville. He died at the age of seventy-five, leaving the Tenth Walk unfinished. The marquis de Girardin, Ermenonville's owner, buried Rousseau on an isle of poplars there and established a pilgrimage for the sentimental devotees who came to pay tribute. It included an itinerary that instructed the visitor not only how to walk through the garden toward the tomb but how to feel. Rousseau's private revolt was becoming public culture.

III. WALKING AND THINKING AND WALKING

Ssren Kierkegaard is the other philosopher who has much to say about walking and thinking. He chose cities—or one city, Copenhagen—as his place to walk and study his human subjects, though he compared his urban tours to rural botanizing: human beings were the specimens he gathered. Born a hundred years later in another Protestant city, he had a life in some ways utterly unlike Rousseau's: the harsh ascetic standards he set for himself could not be less like Rousseau's self-indulgence, and he kept to his birthplace, to his family, and to his religion throughout his life, though he quarreled with them all. In other respects—in his social isolation, his prolific writing of works both literary and philosophical, his chafing self-consciousness—the

dazzling thought.—LEON ROSENFELD, ON A 1929 ENCOUNTER WITH NIELS BOHR *Last Sunday I*

resemblance is strong. The son of a wealthy and grimly devout merchant, Kierkegaard lived off his inheritance and under his father's thumb for most of his life. In a memory he ascribes to one of his pseudonyms but which is almost certainly his own, he tells of how his father, rather than let him leave the house, would walk back and forth in a room with him, describing the world so vividly that the boy seemed to see all the variety evoked. As the boy grew older, his father let him join in: "What had at first been an epic now became a drama; they conversed in turn. If they were walking along well-known paths they watched one another sharply to make sure that nothing was overlooked; if the way was strange to Johannes, he invented something, whereas the father's almighty imagination was capable of shaping everything, of using every childish whim as an ingredient of the drama. To Johannes it seemed as if the world were coming into existence during the conversation, as if the father were the Lord God and he was his favorite."

The triangle between Kierkegaard, his father, and his God would consume his life, and sometimes it seems that he made his God in his father's image. With these walks in the room, his father seems to be consciously shaping the strange character Kierkegaard would become. He described himself as already an old man in childhood, as a ghost, as a wanderer, and this pacing back and forth seems to have been instruction in living in a disembodied magical realm of the imagination that had only one real inhabitant, himself. Even the myriad pseudonyms under which he published many of his best-known works seem devices to lose himself while revealing himself and to make a crowd out of his solitude. Throughout his adult life, Kierkegaard almost never received guests at home, and indeed, throughout his life he almost never had anyone he could call a friend, though he had a vast acquaintance. One of his nieces says that the streets of Copenhagen were his "reception room," and Kierkegaard's great daily pleasure seems to have been walking the streets of his city. It was a way to be among people for a man who could not be with them, a way to bask in the faint human warmth of brief encounters, acquaintances' greetings, and overheard conversations. A lone walker is both present and detached from the world around, more than an audience but less than a participant. Walking assuages or legitimizes this alienation:

---

took a Walk toward highgate and in the lane that winds by the side of Lord Mansfield's park I met Mr.

---

one is mildly disconnected because one is walking, not because one is incapable of connecting. Walking provided Kierkegaard, like Rousseau, with a wealth of casual contacts with his fellow humans, and it facilitated contemplation.

In 1837, just as his literary work was beginning, Kierkegaard wrote, "Strangely enough, my imagination works best when I am sitting alone in a large assemblage, when the tumult and noise require a substratum of will if the imagination is to hold on to its object; without this environment it bleeds to death in the exhausting embrace of an indefinite idea." He found the same tumult on the street. More than a decade later, he declared in another journal, "In order to bear mental tension such as mine, I need diversion, the diversion of chance contacts on the streets and alleys, because association with a few exclusive individuals is actually no diversion." In these and other statements, he proposes that the mind works best when surrounded by distraction, that it focuses in the act of withdrawing from surrounding bustle rather than in being isolated from it. He reveled in the turbulent variety of city life, saying elsewhere, "This very moment there is an organ-grinder down in the street playing and singing—it is wonderful, it is the accidental and insignificant things in life which are significant."

In his journals, he insists that he composed all his works afoot. "Most of *Either / Or* was written only twice (besides, of course, what I thought through while walking, but that is always the case); nowadays I like to write three times," says one passage, and there are many like it protesting that although his extensive walks were perceived as signs of idleness they were in fact the foundation of his prolific work. The recollections of others show him during his pedestrian encounters, but there must have been long solitary intervals in which he could compose his thoughts and rehearse the day's writing. Perhaps it was that the city strolls distracted him so that he could forget himself enough to think more productively, for his private thoughts are often convolutions of self-consciousness and despair. In a journal passage from 1848, he described how on his way home, "overwhelmed with ideas ready to be written down and in a sense so weak that I could scarcely walk," he would often encounter a poor man, and if he refused to speak with him, the ideas would flee "and I would sink into the most dreadful spiritual tribulation at the idea that God could do

---

Green our Demonstrator at Guy's in conversation with Coleridge—I joined them, after enquiring by a look

to me what I had done to that man. But if I took the time to talk with the poor man, things never went that way."

Being out in public gave him almost his only social role, and he fretted over how his performance on the stage of Copenhagen would be interpreted. In a way, his appearances on the street were like his appearances on the printed page: endeavors to be in touch, but not too closely and on his own terms. Like Rousseau, he had an exacting relationship with the public. He chose to publish many of his works under pseudonyms and then to complain that he was considered an idler, since none knew he went home from his roaming to write. After he broke off his engagement with Regine Olsen in what was to be the defining tragedy of his life, he continued to see her on the street but nowhere else. Years later, they would both appear repeatedly at the same time on a portside street, and he worried over what this meant. The street, which is the most casual arena for people with full private lives, was the most personal for him.

The other great crisis of his uneventful life came when he wrote a small attack on Denmark's scurrilous satirical magazine the *Corsair*. Though its editor admired him, the magazine began to publish mocking pictures and paragraphs about him, and the Copenhagen public took up the joke. Most of the jokes were mild enough—they depicted him as having trouser legs of uneven lengths, made fun of his elaborate pseudonyms and compositional style, published pictures of him as a spry figure in a frock coat that belled out around his spiky legs. But the parodies made him better known than he wished to be, achingly anxious about being mocked and seeing mockery everywhere. Kierkegaard seems to have exaggerated the effect of the *Corsair's* jabs and suffered horribly—not least because he no longer felt free to roam the city. "My atmosphere has been tainted for me. Because of my melancholy and my enormous work I needed a situation of solitude in the crowd in order to rest. So I despair. I can no longer find it. Curiosity surrounds me everywhere." One of his biographers says that it was this final crisis of his life, after those of his father and his fiancée, that pushed him into his last phase as a theological rather than a philosophical and aesthetic writer. Nevertheless, he continued to walk the streets of Copenhagen, and it was on one of those walks that he collapsed and was taken to the hospital, where he died some weeks later.

---

whether it would be agreeable—I walked with him at his alderman-after-dinner pace for nearly two miles I

Like Rousseau, Kierkegaard is a hybrid, a philosophical writer rather than a philosopher proper. Their work is often descriptive, evocative, personal, and poetically ambiguous, in sharp contrast to the closely reasoned argument central to the Western philosophical tradition. It has room for delight and personality and something as specific as the sound of an organ grinder in a street or rabbits on an island. Rousseau branched out into the novel, the autobiography, and the reverie, and play with forms was central to Kierkegaard's work: creating a massive postscript to a relatively short essay, layering pseudonymous authors like Chinese boxes within his texts. As a writer his heirs seem to be literary experimentalists like Italo Calvino and Jorge Luis Borges, who play with the way form, voice, reference, and other devices shape meaning.

Rousseau and Kierkegaard's walking is only accessible to us because they wrote about it in more personal, descriptive, and specific works—Rousseau's *Confessions* and *Reveries,* Kierkegaard's journals—rather than staying in the impersonal and universal realm of philosophy at its most pure. Perhaps it is because walking is itself a way of grounding one's thoughts in a personal and embodied experience of the world that it lends itself to this kind of writing. This is why the meaning of walking is mostly discussed elsewhere than in philosophy: in poetry, novels, letters, diaries, travelers' accounts, and first-person essays. Too, these eccentrics focus on walking as a means of modulating their alienation, and this kind of alienation was a new phenomenon in intellectual history. They were neither immersed in the society around them nor—save in Kierkegaard's later years, after the *Corsair* affair—withdrawn from it in the tradition of the religious contemplative. They were in the world but not of it. A solitary walker, however short his or her route, is unsettled, between places, drawn forth into action by desire and lack, having the detachment of the traveler rather than the ties of the worker, the dweller, the member of a group.

IV. The Missing Subject

In the early twentieth century, a philosopher actually addressed walking directly as something central to his intellectual project. Of course

---

suppose. In those two Miles he broached a thousand things—let me see if I can give you a list

---

walking had been an example earlier. Kierkegaard liked to cite Diogenes: "When the Eleatics denied motion, Diogenes, as everyone knows,
came forward as an opponent. He literally did come forward, because
he did not say a word but merely paced back and forth a few times,
thereby assuming he had sufficiently refuted them." The phenomenol-
ogist Edmund Husserl described walking as the experience by which we
understand our body in relationship to the world, in his 1931 essay, "The
World of the Living Present and the Constitution of the Surrounding
World External to the Organism." The body, he said, is our experience
of what is always here, and the body in motion experiences the unity of
all its parts as the continuous "here" that moves toward and through the
various "theres." That is to say, it is the body that moves but the world
that changes, which is how one distinguishes the one from the other:
travel can be a way to experience this continuity of self amid the flux of
the world and thus to begin to understand each and their relationship to
each other. Husserl's proposal differs from earlier speculations on how a
person experiences the world in its emphasis on the act of walking rather
than on the senses and the mind.

Still, this is slim pickings. One would expect that postmodern
theory would have much to say about walking, given that mobility and
corporeality have been among its major themes—and when corporeality gets mobile, it walks. Much contemporary theory was born out of
feminism's protest at the way earlier theory universalized the very specific experience of being male, and sometimes of being white and privileged. Feminism and postmodernism both emphasize that the specifics
of one's bodily experience and location shape one's intellectual perspective. The old idea of objectivity as speaking from nowhere—
speaking while transcending the particulars of body and place—was
laid to rest; everything came from a position, and every position was
political (and as George Orwell remarked much earlier, "The opinion
that art should not be political is itself a political opinion"). But while
dismantling this false universal by emphasizing the role of the ethnic
and gendered body in consciousness, these thinkers have apparently
generalized what it means to be corporeal and human from their own
specific experience—or inexperience—as bodies that, apparently, lead
a largely passive existence in highly insulated circumstances.

The body described again and again in postmodern theory does

—Nightingales, poetry—on Poetical Sensation—Metaphysics—Different genera and species of

not suffer under the elements, encounter other species, experience primal fear or much in the way of exhilaration, or strain its muscles to the utmost. In sum, it doesn't engage in physical endeavor or spend time out of doors. The very term "the body" so often used by post-modernists seems to speak of a passive object, and that body appears most often laid out upon the examining table or in bed. A medical and sexual phenomenon, it is a site of sensations, processes, and desires rather than a source of action and production. Having been liberated from manual labor and located in the sensory deprivation chambers of apartments and offices, this body has nothing left but the erotic as a residue of what it means to be embodied. Which is not to disparage sex and the erotic as fascinating and profound (and relevant to walk-ing's history, as we shall see), only to propose that they are so empha-sized because other aspects of being embodied have atrophied for many people. The body presented to us in these hundreds of volumes and essays, this passive body for which sexuality and biological function are the only signs of life, is in fact not the universal human body but the white-collar urban body, or rather a theoretical body that can't even be theirs, since even minor physical exertions never appear: this body described in theory never even aches from hauling the complete works of Kierkegaard across campus. "If the body is a metaphor for our locatedness in space and time and thus for the fin-itude of human perception and knowledge, then the postmodern body is no body at all," writes Susan Bordo, one feminist theorist at odds with this version of embodiment.

Travel, the other great theme of recent postmodern theory, is about being utterly mobile; the one has failed to modify the other, and we seem to be reading about the postmodern body shuttled around by airplanes and hurtling cars, or even moving around by no apparent means, muscular, mechanical, economic, or ecological. The body is nothing more than a parcel in transit, a chess piece dropped on an-other square; it does not move but is moved. In a sense, these are problems arising from the level of abstraction of contemporary theory. Much of the terminology of location and mobility—words like *nomad, decentered, marginalized, deterritorialized, border, migrant,* and *exile*—are not attached to specific places and people; they represent instead ideas of rootlessness and flux that seem as much the result of the ungrounded

---

Dreams—Nightmare—a dream accompanied by a sense of touch—single and double touch—A dream

theory as its putative subject. Even in these endeavors to come to terms with the tangible world of bodies and motion, abstraction dematerializes them again. The words themselves seem to move freely and creatively, unburdened by the responsibilities of specific description.

Only in maverick writings does the body become active. In Elaine Scarry's magisterial book *The Body in Pain: The Unmaking and Making of the World*, she considers first how torture destroys the conscious world of its subjects, then theorizes how creative efforts—making both stories and objects—construct that world. She describes tools and manufactured objects as extensions of the body into the world and thus ways of knowing it. Scarry documents how the tools become more and more detached from the body itself, until the digging stick that extends the arm becomes a backhoe that replaces the body. Though she never discusses walking directly, her work suggests philosophical approaches to the subject. Walking returns the body to its original limits again, to something supple, sensitive, and vulnerable, but walking itself extends into the world as do those tools that augment the body. The path is an extension of walking, the places set aside for walking are monuments to that pursuit, and walking is a mode of making the world as well as being in it. Thus the walking body can be traced in the places it has made; paths, parks, and sidewalks are traces of the acting out of imagination and desire; walking sticks, shoes, maps, canteens, and backpacks are further material results of that desire. Walking shares with making and working that crucial element of engagement of the body and the mind with the world, of knowing the world through the body and the body through the world.

Chapter 3

RISING AND FALLING:

The Theorists of Bipedalism

It was a place as blank as a sheet of paper. It was the place I had always been looking for. Out train and car windows, in my imagination, and on my walks through more complicated terrain, flat expanses would call to me, promising walking as I imagined it. And now I had arrived at the pure plane of a dry lake bed where I could walk uninterrupted and utterly free. The desert holds many of these dry lake beds or playas, washed long ago or annually to a surface as flat and inviting as a dance floor when dry. These are the places where the desert is most itself: stark, open, free, an invitation to wander, a laboratory of perception, scale, light, a place where loneliness has a luxurious flavor, like in the blues. This one, near Joshua Tree National Park, in southeastern California's Mojave Desert, was occasionally a lake bed but mostly a pure plain of cracked dust in which nothing grew. To me these big spaces mean freedom, freedom for the unconscious activity of the body and the conscious activity of the mind, places where walking hits a steady beat that seems to be the pulse of time itself. Pat, my companion on this walk across the lake bed, prefers rock climbing, in which every move is an isolated act that absorbs the whole of his attention and seldom rises to a rhythm. It's a difference of style that cuts deep in our lives: he is something of a Buddhist and conceives of spirituality as being conscious in the moment; while I am a sucker for symbolisms, interpretations, histories, and a Western kind of spirituality that is

many metaphysicians from a want of smoking the second consciousness—Monsters—the Kraken—

located less in the here than in the there. But both of us share the same notion of being out in the land as an ideal way to exist.

Walking, I realized long ago in another desert, is how the body measures itself against the earth. On this lake bed, each step brought us minutely closer to one of the ranges of mountains, blue in the late afternoon light, that circled our horizon like the bleachers rising above a field. Picture the lake bed as a pure geometric plane that our steps measured like the legs of a protractor swinging back and forth. The measurements recorded that the earth was large and we were not, the same good and terrifying news most walks in the desert provide. On this afternoon even the cracks in the ground cast long sharp shadows, and a shadow like a skyscraper stretched from Pat's van. Our shadows moved alongside us on our right, growing longer and longer, longer than I had ever seen them. I asked him how long he thought they were, and he told me to stand still and he'd pace it off. I faced east into my shadow, toward the closest mountains that all the shadows stretched toward, and he began to walk.

I stood alone, my shadow like a long road Pat traveled. He seemed, in that pellucid air, not to grow distant but only to grow smaller. When I could frame him between my thumb and forefinger held close together and his own shadow stretched almost to the mountains, he had reached the shadow of my head—but as he arrived, the sun suddenly slipped below the horizon. With that, the world changed: the plain lost its gilding, the mountains became a deeper blue, and our sharp shadows grew blurry. I called for him to stop at the now-vague shadow of my head, and when I had myself covered the distance between us, he told me he'd gone a hundred paces—250 or 300 feet—but what constituted my shadow had become harder and harder to distinguish as he went. We walked back to the van as night approached, the experiment concluded. But where did it begin?

Rousseau thought that humanity's true nature could be found in its origins, and that to understand those origins was to understand who we were and who we should be. The subject of human origins has itself evolved immensely since he cobbled together a few sketchy

---

Mermaids—Southey believes in them—Southey's belief too much diluted—A Ghost story—.... —JOHN

---

descriptions of non-European customs with some groundless specu-
lation on the "noble savage." But the argument that who we were
originally—whether _originally_ means 1940 or three million years
ago—is who we are or ought to be has only become stronger with
time. Popular books and scientific articles debate again and again
whether we are a bloodthirsty, violent species or a communitarian one
and what kind of differences between the genders are encoded in our
genes. Both are often just-so stories about who we are, could be, or
should be, told by everyone from conservatives arguing the adequacy
of tradition to health seekers arguing that we ought to eat some just-
discovered primordial diet. This, of course, makes who we were an
intensely political subject. The scientists researching human origins
have been contentious about these questions of human nature, and in
recent years walking has become a central part of their conversation.

While philosophers have had little to say about what walking
means, scientists have of late had a great deal to say. Paleontologists,
anthropologists, and anatomists have launched a passionate and often
partisan argument over when and why the ancestral ape got up on its
hind legs and walked so long that its body became our upright, two-
legged, striding body. They were the philosophers of walking I had
been looking for, speculating endlessly about what each bodily shape
says about function and about how those forms and functions eventually added up to our humanity—though what that humanity consists of is equally debatable. The only given is that upright walking is
the first hallmark of what became humanity. Whatever its causes, it
caused much more: it opened up vast new horizons of possibility, and
among other things, it created the spare pair of limbs dangling from
the upright body, seeking something to hold or make or destroy, the
arms freed to evolve into ever more sophisticated manipulators of the
material world. Some scholars see two-legged walking as the mechanism that set our brains expanding, others as the structure that established our sexuality. So, although the debate about the origins of
bipedalism is full of detailed descriptions of hip joints and foot bones
and geologic dating methods, it is ultimately about sex, landscape,
and thinking.

Usually the uniqueness of human beings is portrayed as a matter
of consciousness. Yet the human body is also unlike anything else on

earth, and in some ways has shaped that consciousness. The animal kingdom has nothing else like this column of flesh and bone always in danger of toppling, this proud unsteady tower. The few other truly two-legged species—birds, kangaroos—have tails and other features for balance, and most of these bipeds hop rather than walk. The alternating long stride that propels us is unique, perhaps because it is such a precarious arrangement. Four-legged animals are as stable as a table when all four feet are on the ground, but humans are already precariously balanced on two before they begin to move. Even standing still is a feat of balance, as anyone who has watched or been a drunk knows.

Reading the accounts of human walking, it is easy to begin to think of the Fall in terms of the falls, the innumerable spills, possible for a suddenly upright creature that must balance all its shifting weight on a single foot as it moves. John Napier, in an essay on the ancient origins of walking, wrote, "Human walking is a unique activity during which the body, step by step, teeters on the edge of catastrophe. . . . Man's bipedal mode of walking seems potentially catastrophic because only the rhythmic forward movement of first one leg and then the other keeps him from falling flat on his face." This is easiest to see in small children for whom the many aspects that will later unite seamlessly into walking are still distinct and awkward. They learn to walk by flirting with falling—they lean forward with their body and then rush to keep their legs under that body. Their plump bowed legs always seem to be lagging behind or catching up, and they often tumble into frustration before they master the art. Children begin to walk to chase desires no one will fulfill for them: the desire for that which is out of reach, for freedom, for independence from the secure confines of the maternal Eden. And so walking begins as delayed falling, and the fall meets with the Fall.

Genesis may seem out of place in a discussion of science, but it is often the scientists who have dragged it in with them, unwittingly or otherwise. The scientific stories are as much an attempt to account for who we are as any creation myth, and some of them seem to hark back to the central creation myth of Western culture, that business of Adam and Eve in the Garden. Many of the hypotheses have been wildly speculative, seemingly based less on the evidence than on

---

against the human race, and I thank you. . . . Never was so much intelligence used to make us stupid. While

---

modern desires or old social mores, particularly as they relate to the roles of the sexes. During the 1960s, the Man the Hunter story was widely accepted and made popular by such books as Robert Ardrey's *African Genesis*, with its famous opening line "Not in innocence, and not in Asia, was man born." It suggested that violence and aggression are ineradicable parts of human disposition, but redeemed them by proposing that they were the means by which we evolved (or males evolved; most of the mainstream theories have tended to leave females doing little but passing along the genes of their evolving mates). Early challengers of the Man the Hunter scenario, writes the feminist anthropologist Adrienne Zihlman, "point out parallels between the interpretation of hunting as propelling humankind into humanity, on the one hand, and the biblical myth of expulsion from Eden, after Eve's eating of the tree of knowledge, on the other. The authors argue that both fates—that of hunting and of the expulsion—were precipitated by an act of eating—meat in the first instance and forbidden fruit in the other." And they argue that the division of labor—men as hunters, women as gatherers—reflects the distinct division of roles given Adam and Eve in Genesis. Similarly, during the 1960s and 1970s, the theory went that human walking evolved during a time of radical climate change, when the species was transformed from an arboreal forest dweller to a creature of the savannah, another expulsion from Eden. Nowadays both the dominance of hunting and the residence on the savannah have fallen from favor as evolutionary explanations. But the language remains: scientists now pursuing human origins not in fossils but in genes describe our hypothetical common ancestor as "African Eve" or "Mitochondrial Eve."

These scientists have sometimes looked for what they wanted to find, or found what they were looking for. The Piltdown man hoax was believed from 1908 to its denouement in 1950 because British scientists were eager to believe the evidence of a large-brained creature with an animal jaw. The bones suggested that our intelligence was of great age and gratified them by showing up in England. Much was made of clever Piltdown man as an Englishman, until new technologies proved him a liar cobbled together from a modern ape's jaw and a human skull. When Raymond Dart found a child's skull in South Africa in 1924 that, unlike Piltdown man, turned out to be

reading it, one longs to go on all fours. —VOLTAIRE TO ROUSSEAU, ON THE DISCOURSE ON THE ORIGIN

genuine, it was widely discredited as a human ancestor by the British masters so pleased by Piltdown. It was discredited because the scientists of the era preferred not to come from Africa and because the skull of the Taung child, as it was called, had a small cranium but evidently walked upright, suggesting that our intelligence had come late rather than early in our evolution. At the base of the skull is an opening called the foramen magnum through which the spinal cord connects to the brain. The foramen magnum of the Taung child was in the center of the skull, as it is in us, rather than at the back, as it is in apes, and so it was evident that this creature had walked upright, its head poised atop the spine rather than hanging down from it. Like most of the skulls of the australopithecine hominids who would evolve into humans, this one looks to the modern eye like a house with odd proportions: the porch of the brow and jutting jaw is enormous, the attic where the modern brain rises is nonexistent. Most early evolutionists proposed that our human characteristics—walking, thinking, making—originated together, perhaps because they found it hard or unpleasant to imagine a creature who shared only a part of our humanity. Dart's counter-hypothesis was advanced by Louis and Mary Leakey's spectacular Kenyan finds in the 1950s, 1960s, and 1970s and all but confirmed by Donald Johanson's celebrated discovery of the "Lucy" skeleton and related fossils in Ethiopia in the 1970s. Walking came first.

Nowadays walking upright is considered to be the Rubicon the evolving species crossed to become hominid, distinct from all other primates and ancestral to human beings. The list of what we eventually got from bipedalism is long and alluring, full of all the gothic arches and elongations of the body. Start with the straight row of toes and high arch of the foot. Go up the long straight walker's legs to the buttocks, round and protuberant thanks to the massively developed gluteus maximus of walkers, a minor muscle in apes but the largest muscle in the human body. Then go on to the flat stomach, the flexible waist, the straight spine, the low shoulders, the erect head set atop a long neck. The upright body's various sections are balanced on top of each other like the sections of a pillar, while the weight of quadrupeds' heads and torsos hangs from their spines like the roadway from a suspension bridge, with a pair of pierlike legs toward

OF INEQUALITY   *The diminution of the olfactory stimuli seems itself to be a consequence of man's raising*

either end. The great apes are knuckle-walkers: creatures adapted to life in tropical forests who for the most part move only short distances on the ground between trees, on long forelimbs that give them a kind of diagonal posture. Apes have—when compared to humans— arched backs, no waists, short necks, chests shaped like inverted funnels, protuberant abdomens, scrawny hips and bottoms, bandy legs, and flat feet with opposable big toes.

When I think about this evolutionary history of walking, I see a small figure, like my companion on the lake bed, only this time it is dawn and the figure is moving toward me, an indecipherable dot in the distance that seems somehow unfamiliar as it becomes distinguishable as an upright figure and finally, when it draws close, is just another walker. But what was that casting a long shadow in the middle distance? Lucy—as they named the small 3.2-million-year-old *Australopithecus afarensis* skeleton found in Ethiopia in 1974, presuming from various details that it was female—was apelike in many respects; she had little in the way of a waist or neck, short legs, longish arms, and the funnel-like rib cage of an ape. Her pelvis, however, was wide and shallow, and so she had a stable gait with hip joints far apart tapering to close-together knees like humans and unlike chimps (whose narrow hips and far-apart knees make them lurch from side to side when they walk upright). Some say she would have been a terrible runner and not much of a walker. But she walked. This much is certain, and then come the arguments.

Dozens of scientists have interpreted her bones, reconstructed her flesh, her gait, her sex life, in dozens of different ways and argued over whether she walked well or poorly. Discovery conveys a certain privilege of interpretation, and so Johanson, who worked at the Cleveland Museum, took the bones he found in Hadar, Ethiopia, to his friend Owen Lovejoy, an anatomist at Cleveland State University and an expert on human locomotion. Lovejoy issued the orthodox verdict. In his book *Lucy,* Johanson reports that Lovejoy said of the *afarensis* knee joint he had brought in the year before,

"This is like a modern knee joint. This little midget was fully bipedal."
"But could he walk upright?" I persisted.

---

himself from the ground, of his assumption of an upright gait, this made his genitals, which were previously

---

"My friend, he could walk upright. Explain to him what a hamburger was and he'd beat you to the nearest McDonald's nine times out of ten."

Johanson's knee joint came along as the first material support for Lovejoy's bold theory that bipedalism had begun and been perfected far earlier than anyone else had assumed. The following year, the Lucy skeleton—or the 40 percent of it that was recovered—further confirmed his hypothesis about the antiquity of human walking, as did the 3.7-million-year-old footprints of a pair of walkers Mary Leakey's team found at Laetoli, Tanzania, in 1977. But why had these creatures become bipedal?

By 1981 Lovejoy had evolved a complicated explanation for why we got up and walked. His 1981 *Science* article "The Origin of Man" has become the focal point for the arguments in the field about the reasons why walking appeared 4 million or more years ago. Lovejoy evolved an elaborate thesis that decreasing the time between births would increase the survival rate of the species. "In most primate species," he wrote, "male fitness is largely determined by consort success of one sort or another"—that is, in the ability or opportunity to mate and thereby pass along their genes. He proposed that in the Miocene era, 5 million and more years ago, the human ancestor changed its—or rather his—behavior. Males, he proposed, began to bring back provisions for the females; the females thus provided for were able to bear more children as the challenge of feeding themselves and their young was lessened, and the male-headed nuclear family was born. In other words, male fitness had expanded to include provisioning, which would allow them to pass along those genes more frequently and certainly. "Bipedalism," he wrote in a 1988 summary, "figured in this new reproductive scheme because by freeing the hands it made it possible for the male to carry food gathered far from his mate." But, he added, such daily separation of the sexes would only genetically favor males if they could come home and propagate their own genes and no one else's—thus the behavior must have selected for monogamous females as well as responsible males. Lovejoy explained, "The highly unusual sexual behavior of man may now be brought into focus. Human females are continually sexually receptive

---

concealed, visible and in need of protection, and so provided feelings of shame in him.—FREUD,

and . . . male approach may be considered equally stable." Since unlike the females of most species this one no longer signaled her fertile times, they had sex a lot to procreate and to bond. If we regard it as a creation myth, it is one in which the two-parent family is far older than the human species, hominid males are mobile and responsible partners and parents, and females are needy, faithful, stay-at-home mates who are *not* the instruments of bipedal evolution.

The 1960s myth of Man the Hunter had been succeeded by two theories in the 1970s. One dubbed Woman the Gatherer proposed that the primordial diet was probably mostly vegetarian and was mostly collected then, as in hunter-gatherer societies today, by females. The other emphasized food sharing as instrumental in ensuring survival and generating a home base to which food was brought and shared, resulting in more complex social consciousness. In this theory, a communitarian First Supper takes the place of Ardrey's blood sports as the event that propelled us into humanity. Lovejoy combined aspects of both these new theories to create his Man the Gatherer, who brought food home and shared it, though only with his mate and offspring. His theory suggested not only that walking had been a male business, and that the males in question had been full of family virtues, but that the virtues in question had made us walkers. In fact, he said, Lucy and her ilk could walk better than we do, and further, the species had lost its ability to climb.

I was staying with Pat in his shack just outside Joshua Tree National Park while I wrote this chapter. Preoccupied and trying to sort out the sea of material before me, I kept recounting theories to him about why we became bipedal, about details of anatomy and function, and he laughed incredulously at the more outlandish ones. "People get grants and tenure for *that?*" he'd say. His favorite was R. D. Guthrie's 1974 proposal that when hominids became bipedal, the males used their now-exposed penises as a "threat display organ" to intimidate opponents, and we speculated on the origins of human laughter. The following day, after he came home from guiding clients up and down rock faces all day and was lounging with a drink, I read him anthropologist Dean Falk's attack on Lovejoy. Lovejoy's term "copulatory vigi-

---

CIVILIZATION AND ITS DISCONTENTS   *Hand in hand with equal plod they go. In the free hands—no. Free*

lance" caught his attention, and he laughed more at the strange stuff I was immersed in. Not that his world was exactly a bastion of seriousness: while he was climbing for pleasure the day before, I had lain in the shade idly flipping through his guidebook and been entertained by some of the names of the climbing routes up and down the park's myriad giant boulders: "Presbyterian Dental Floss" was right next to "Episcopalian Toothpick," while "Boogers on a Lampshade" mocked the climb genteelly called "Figures in a Landscape," and innumerable poodle, political, and anatomical jokes described other vertical routes up the rocks. That evening, as I read bipedal theory to him and the quail bobbed about the backyard and the setting sun pushed the shadows of the hills farther and farther across the valley, he swore he would get his friends who founded and named many of the park's climbing routes to name the next one "Copulatory Vigilance," an obscure monument to a theory that we had lost our ability to climb and to his opinion of the more far-flung theories of human origins. Lovejoy's theory has become famous, if only because no one can resist attacking it.

Among the earlier critics were the anatomists Jack Stern and Randall Sussman at the State University of New York at Stonybrook, who I visited. Two unathletic men with identical clipped gray mustaches, they looked something like the Walrus and the Carpenter, with Stern as the compact Carpenter and Sussman the expansive Walrus. They talked to me for hours in an office full of bones and books, and periodically one or the other would grab a chimp pelvis or a cast of a fossil femur to illustrate a point. Obviously enthusiastic about their work, they often went off on conversational tangents with each other that left me far behind, and they delighted too in dishing their colleagues in this contentious field. They had argued that Johanson's Lucy-era *Australopithecus afarensis* fossils were those of apprentice walkers who, based on the evidence—big arms and smallish legs, curved fingers and toes—continued climbing trees well and frequently for a long time afterward. Another feature of the *afarensis* fossils they took up was gender size: if the large and small skeletons Johanson and company had found in Ethiopia were the same species (which Richard Leakey and others contest), then the sizes must represent small females and large males, which made it unlikely they

empty hands. Backs turned both bowed with equal plod they go. The child hand raised to reach the holding

practiced Lovejoy's monogamous arrangement. Living primate species where the males are far larger—baboons, gorillas—are usually polygamous; only those without size differences, such as gibbons, are monogamous. So their version of Lucy was that she was a lousy walker with big floppy feet, a pretty good climber with long, strong arms, and probably part of a polygamous group in which small females spent more time in the trees than large males.

Sussman said, "Back when we started this work, and I don't think it's unhumble to say it, the majority of the people in the field would say we evolved in the savannah, in the open country of the veldt of South Africa or the savannah of East Africa. I think that's a load of crap. I think that what happened was that *afarensis* was living in forest and open country mosaics like you see today in places like the French Congo or along rivers where there's a lot of trees. I mean, that probably went on for a million years when you had an animal that was climbing and an apprentice biped." He added that in the old pictures re-creating this phase of evolution, the creatures were strolling across the grassland; newer ones showed them in much more mixed habitat, and the most recent *National Geographic* articles had paintings that placed these creatures in forests with some of them in the trees. That the creatures were forest dwellers and tree climbers had become, Stern said, so obvious that no one bothered to credit Stern and Sussman for pushing the idea early on.

The argument before had been circular: that hominids had learned to walk in order to venture onto the savannah, and that if they survived on the savannah, they must have been competent walkers. And the savannah seemed to be an image of freedom, of unlimited space in which the possibilities were likewise unlimited, a nobler space than the primeval forest that was less like the open forest of Rousseau's solitary wanderers and more like the jungles from which Jane Goodall and Dian Fossey sent back their primate reports. Stern said a little later on, "I worry most about the manner of their bipedal walking. I wrote a paper saying they could not have walked as we do. It's not fast, it's not energetically efficient. . . . Are we wrong? Was their method of bipedalism actually pretty good?" Sussman cut in, "Or did they combine very good tree-climbing with shitty bipedalism and gradually the proportions reversed . . ." Stern continued, "The argument that I sometimes soothe myself with is that chimpanzees are really pretty crappy quadrupeds themselves, as four-footed animals go. So if they can be pretty crappy quadrupeds for seven million years, then we could've been pretty crappy bipeds for a couple of million years."

At the 1991 Conference on the Origins of Bipedalism in Paris, three anthropologists had reviewed all the current theories on walking as a kind of academic stand-up comedy routine. They described the "schlepp hypothesis," which explained walking as an adaptation for carrying food, babies, and various other things; "the peek-a-boo hypothesis," which involved standing up to see over the grass of the savannah; "the trench coat hypothesis," which, like Guthrie's theory that so amused Pat, connected bipedalism to penile display, only this time to impress females rather than intimidate other males; "the all wet hypothesis," which involved wading and swimming during a proposed aquatic phase of evolution; "the tagalong hypothesis," which involved following migratory herds across that ever-popular savannah; "the hot to trot hypothesis," which was one of the more seriously reasoned theories, claiming that bipedalism limited solar exposure in the tropical midday sun and thus freed the species up to move into hot, open habitat; and the "two feet are better than four" hypothesis, which proposes that bipedalism was more energy-efficient than quadrupedalism, at least for the primates who would become humans.

It was quite a collection of theories, though since talking to Stern and Sussman I had grown accustomed to the fluctuating interpretations of what to a lay person exposed to only one source sounds like established fact. The bones unearthed in Africa in ever greater quantity remain enigmatic in crucial ways, and the business of their interpretation recalls the ancient Greeks reading the entrails of animals to divine the future, or the Chinese throwing I Ching sticks to understand the world. They are constantly being rearranged to correspond to a new evolutionary family tree, a new set of measurements. Two Zurich anthropologists, for example, recently declared that the famous Lucy skeleton is actually that of a male, while Falk argues that she is not a human ancestor. Paleontology sometimes seems like a courtroom full of lawyers, each waving around evidence that confirms their hypotheses and ignoring the evidence that contradicts it (though Stern and Sussman impressed me as being exceptionally

---

plod on and never recede. Backs turned. Both bowed. Joined by held holding hands. Plod on as one. One

---

committed to evidence rather than ideology). Only one thing seemed agreed upon in all these competing stories of the bones, the thing that Mary Leakey had said when she wrote about the footprints her team had found in Laetoli: "One cannot overemphasize the role of biped-alism in hominid development. It stands as perhaps the salient point that differentiates the forebears of man from other primates. This unique ability freed the hands for myriad possibilities—carrying, tool-making, intricate manipulation. From this single development, in fact, stems all modern technology. Somewhat oversimplified, the formula holds that this new freedom of forelimbs posed a challenge. The brain expanded to meet it. And mankind was formed."

Falk wrote the most devastating reply to Lovejoy's hypothesis in a 1997 essay titled "Brain Evolution in Human Females: An Answer to Mr. Lovejoy." She declared, "According to this view, early hominid females were left not only four-footed, pregnant, hungry and in fear of too much exercise in a central core area, they were also left 'waiting for their man.'" And she went on to say, after reviewing details such as the unlikelihood of monogamy between such differently scaled males and females, to comment, "The Lovejoy hypothesis may also be viewed at an entirely different level, i.e., as being preoccupied with questions/ anxieties about male sexuality. At its most basic level, the hypothesis focuses on the evolution of how men got/get sex." She goes on to point out that the behavior of terrestrial female primates suggests that female ancestral hominids chose multiple partners for reproductive and recreational sex, and "much of the world appears to fear that this might still be the case as indicated by the universal close observation and control of sexual conduct in human communities, not to mention all those male insecurities simmering beneath the surface of Lovejoy's hypothesis."

Having dismissed the notion that a providing male brought home the bacon to a monogamous, immobilized mate, Falk took up the al-ternate and much simpler theory that walking upright minimized the amount of direct sun the earliest hominids received as they moved in the open spaces between patches of trees, thereby freeing them to move farther and farther out from the shade of the forest. Falk ex-plains that Peter Wheeler, whose hypothesis it was, proposed that "these features led to 'whole-body cooling' that regulated temperature

---

shade. Another shade.—SAMUEL BECKETT   *John and the Austrian walked one way along the shore*

of blood circulating to (among other regions) the brain, helped prevent heatstroke, and thereby released a physiological constraint on brain size in *Homo*." Thus the changes freed the species to grow larger and larger brains, as well as to wander farther and farther. She buttresses Wheeler's theory with information drawn from her own research into brain evolution and structure and concludes, as Mary Leakey did, though for a different reason, that becoming upright walkers didn't create but did make possible the rise of intelligence.

Intelligence may be located in the brain, but it affects other parts of the anatomy. Consider the pelvis as a secret theater where thinking and walking meet and, according to some anatomists, conflict. One of the most elegant and complicated parts of the skeleton, it is also one of the hardest to perceive, shrouded as it is in flesh, orifices, and preoccupations. The pelvis of all other primates is a long vertical structure that rises nearly to the rib cage and is flattish from back to front. The hip joints are close together, the birth canal opens backward, and the whole bony slab faces down when the ape is in its usual posture, as do the pelvises of most quadrupeds. The human pelvis has tilted up to cradle the viscera and support the weight of the upright body, becoming a shallow vase from which the stem of the waist rises. It is comparatively short and broad, with wide-set hip joints. This width and the abductor muscles that extend from the iliac crests—the bone on each side that sweeps around toward the front of the body just below the navel—steady the body as it walks. The birth canal points downward, and the whole pelvis is, from the obstetrical point of view, a kind of funnel through which babies fall—though this fall is one of the most difficult of human falls. If there is a part of anatomical evolution that recalls Genesis, it is the pelvis and the curse, "In sorrow thou shalt bring forth children."

Giving birth for apes, as for most mammals, is a relatively simple process, but for humans it is difficult and occasionally fatal for mother and child. As hominids evolved, their birth canals became smaller, but as humans have evolved, their brains have grown larger and larger. At birth the human infant's head, already containing a brain as big as that of an adult chimpanzee, strains the capacity of this bony theater. To exit, it must corkscrew down the birth canal, now facing forward, now sideways, now backward. The pregnant woman's body has already

discussing the formation of sand banks and the theories of the tides, and Charlotte & I went in the opposite

increased pelvic capacity by manufacturing hormones that soften the ligaments binding the pelvis together, and toward the end of pregnancy the cartilage of the pubic bone separates. Often these transformations make walking more difficult during and after giving birth. It has been argued that the limitation on our intelligence is the capacity of the pelvis to accommodate the infant's head, or contrarily that the limitation on our mobility is the need for the pelvis to accommodate birth. Some go further to say that the adaptation of the female pelvis to large-headed babies makes women worse walkers than men, or makes all of us worse walkers than our small-brained ancestors. The belief that women walk worse is widespread throughout the literature of human evolution. It seems to be another hangover from Genesis, the idea that women brought a fatal curse to the species, or that they were mere helpmeets along the evolutionary route, or that if walking is related to both thinking and freedom, they have or deserve less of each. If learning to walk freed the species—to travel to new places, to take up new practices, to think—then the freedom of women has often been associated with sexuality, a sexuality that needs to be controlled and contained. But this is morality, not physiology.

I got so annoyed by the ambiguous record on gender and walking that early one fine morning in Joshua Tree, while the cottontails were hopping in the yard, I called up Owen Lovejoy. He pointed to some differences between male and female anatomy that, he said, *ought* to make women's pelvises less well adapted to walking. "Mechanically," he said, "women are less advantaged." Well, I pressed, do these differences actually make a practical difference? No, he conceded, "it has no effect on their walking ability at all," and I walked back out into the sunshine to admire a huge desert tortoise munching on the prickly pear in the driveway. Stern and Sussman had laughed when I asked them whether women were indeed worse walkers and said that as far as they knew, no one had ever done the scientific experiments that would back up this assertion. Great runners tend to converge in certain body types, whichever gender they are, they ruminated, but walking is not running, and the question of what constitutes greatness there is more ambiguous. What, they asked, does better mean? Faster? More efficient? Humans are slow animals, they said, and what we excel at is distance, sustaining a pace for hours or days.

----

direction for above two hours and lastly lay down among the long grass and gathered shells until our

----

Those in other disciplines who speculate about walking speculate about what meanings it can be invested with—how it can be made an instrument of contemplation, of prayer, of competition. What makes these scientists, for all their squabbles, significant is their attempt to discuss what meanings are intrinsic to walking—not what we make it but how it made us. Walking is an odd fulcrum in human evolutionary theory. It is the anatomical transformation that propelled us out of the animal kingdom to eventually occupy our own solitary position of dominion over the earth. Now it remains as a limitation, no longer leading us into a fantastic future but linking us to an ancient past as the same gait of a hundred thousand or a million, or if you go with Lovejoy, three million years ago. It may have made possible the work of the hands and the expansion of the mind, but it remains as something not particularly powerful or fast. If it once separated us from the rest of the animals, it now—like sex and birth, like breathing and eating—connects us to the limits of the biological.

The morning before I left I went walking in the national park, starting out from the rocks where Pat was teaching climbing, pacing myself to stay cool and hydrated. His father had told him, and he had told me, that the landscape never looks the same coming and going, so turn around periodically and look at the view you'll see coming back. It's good advice for this confusing landscape. I started amid a big cluster of rocks, an archipelago or a neighborhood of rock piles each the size of a huge building; like buildings they cut off views, so you have to know the lay of the land and local landmarks rather than counting on distant sights to steer by as in other deserts. With the morning sun at my left, I went south along a path that crossed a road to become a fainter road itself, with tufts of grass in the center; it curved along to the southwest and ended in another, much-used road. Small lizards darted into the bushes as I went by, and a faint flush of tender green grass was everywhere in the shade, spears an inch or two high from the downpour a few weeks ago. Drifting across the vast space, silent except for wind and footsteps, I felt uncluttered and unhurried for the first time in a while, already on desert time. My road reached the dead end of a private property boundary, so I circled around, guessing I

Handkerchiefs were quite full.—EFFIE GRAY RUSKIN   *You've got to walk / that lonesome valley / Walk it*

could find another path back to the rock cluster, flirting with being lost. Mountain ranges appeared and disappeared on the horizon as I rotated around the plain and returned to the rocks. Eventually I met the point where my trail crossed the disused road, found my own footprints going the other way, printed crisply atop the softer footprints of people who'd passed this way on previous days, and followed that trace of my own passing an hour or so ago back to where I started.

Chapter 4

THE UPHILL ROAD TO GRACE:
Some Pilgrimages

Walking came from Africa, from evolution, and from necessity, and it went everywhere, usually looking for something. The pilgrimage is one of the basic modes of walking, walking in search of something intangible, and we were on pilgrimage. The red earth between the piñon and juniper trees was covered with a shining mix of quartz pebbles, chips of mica, and the cast-off skins of cicadas who had gone underground again for another seventeen years. It was a strange pavement to be walking on, both lavish and impoverished, like much of New Mexico. We were walking to Chimayó, and it was Good Friday. I was the youngest of the six people setting out cross country for Chimayó that day, and the only nonlocal. The group had coalesced a few days before, when various characters, myself included, asked Greg if he would mind company. Two of the others were members of Greg's cancer survivors' group, a surveyor and a nurse, and my friend Meridel had brought her neighbor David, a carpenter.

Although we were on our own route—or rather Greg's route—we had joined the great annual pilgrimage to the Santuario de Chimayó and thus were walking as pilgrims. Pilgrimage is one of the fundamental structures a journey can take—the quest in search of something, if only one's own transformation, the journey toward a goal—and for pilgrims, walking is work. Secular walking is often imagined as play, however competitive and rigorous that play, and uses gear and techniques to make the body more comfortable and more efficient. Pilgrims, on the other hand, often try to make their journey harder, recalling the origin of the word travel in *travail*, which also means work, suffering, and the pangs of childbirth. Since the Middle Ages, some pilgrims have traveled barefoot or with stones in their shoes, or fasting, or in special penitential garments. Irish pilgrims at Croagh Patrick still climb that stony mountain barefoot on the last Sunday of every July, and pilgrims in other places finish the journey on their knees. An early Everest mountaineer noted a still more arduous mode of pilgrimage in Tibet. "These devout and simple people travel sometimes two thousand miles, from China and Mongolia, and cover every inch of the way by measuring their length on the ground," wrote Captain John Noel. "They prostrate themselves on their faces, marking the soil with their fingers a little beyond their heads, arise and bring their toes to the mark they have made and fall again, stretched full length on the ground, their arms extended, muttering an already million-times-repeated prayer."

In Chimayó, a few pilgrims every year come carrying crosses, from lightweight and relatively portable models to huge ones that must be dragged step by weary step. Inside the chapel that is their destination one such cross is preserved to the right of the altar, and a small metal plaque by its carrier declares, "This cross is a symbol in thanking God for the safe return of my son Ronald E. Cabrera from combat duty in Viet-Nam. I Ralph A. Cabrera promised to make a pilgrimage, which consisted of walking 150 miles from Grants New Mexico to Chimayó. This pilgrimage was finished on the 28th day of November 1986." Cabrera's plaque and knobby wooden crucifix, about six feet high with a folkloric carved Christ attached to it, make it clear that a pilgrimage is work, or rather labor in a spiritual economy in which effort and privation are rewarded. Nobody has ever quite articulated whether this economy is one in which benefits are incurred for labor expended or the self is refined into something more worthy of such benefit—and nobody needs to; pilgrimage is almost universally embedded in human culture as a literal means of spiritual journey, and asceticism and physical exertion are almost universally understood as means of spiritual development.

Some pilgrimages, such as that to Santiago de Compostela in

----

because there is no light in him.—JOHN 11:10   *But as for me, I will walk in mine integrity: redeem me, and*

----

northwest Spain, are entirely on foot from beginning to end; the pilgrimage begins with the first step, and the journey itself is the most important part. Others, such as the Islamic hajj in Mecca or various denominations' visits to Jerusalem, nowadays are as likely to begin with airplanes, and the walking only begins upon arrival (though West African Muslims may spend a lifetime or generations slowly walking toward Saudi Arabia, and a whole culture of nomads has grown up whose eventual goal is Mecca). Chimayó is still a walking pilgrimage, though most walkers have a driver who dropped them off and will pick them up. It's a pilgrimage in an intensely automotive culture, alongside the highway north from Santa Fe and then on the shoulder of the smaller road northeast to Chimayó. The roadside for the last several miles is studded with cars whose drivers are keeping track of family or friends, and in town the air can be noxious with carbon monoxide from the traffic jam; from Santa Fe onward, it's also studded with signs to drive slowly and watch for pilgrims.

Greg's route began about twelve miles north of Santa Fe and cut across country to join up with the rest of the pilgrims only a few miles from Chimayó. We had arrived at eight in the morning at the land Greg and his wife MaLin had bought long ago, and for him the walk tied their land to the holy land due north some sixteen or so miles. It made sense for the rest of us too; none of us were Catholics or even Christians, and walking cross-country let us be in that nonbeliever's paradise, nature, before we arrived at this most traditional of religious destinations. I kept having to remind myself it wasn't a hike and get over my desire to move at my own speed and make good time. As it turned out, it was slowness that would make this walk hard.

Like much of northern New Mexico, the town of Chimayó exudes a sense of ancientness that sets it apart from the rest of the forgetful United States. The Indians here embedded the landscape with stone buildings, potsherds, and petroglyphs, and Pueblo, Navajo, and Hopi people have remained a very visible portion of the population. The Hispanic population is also large and old, and their ancestors established Santa Fe as the first European-inhabited town in what would become the United States. Neither of these peoples has

---

be merciful unto me. My foot standeth in an even place.—PSALMS 26:1–12   The farther pilgrims move

been forgotten or eradicated as they have in other parts of the country; nobody imagines that this landscape was uninhabited wilderness before the Yankees came. And in fact the Yankees who come tend to borrow and revel in the cultures, becoming connoisseurs of adobe architecture and Indian silver work, of Pueblo dances and Hispanic crafts and everyone's customs, including the pilgrimage.

Before the Conquistadors came, Chimayó had been inhabited by ancestors of the contemporary Tewa Pueblo people, and they named the hill above the Santuario Tsi Mayo, "the place of good flaking stone." Records of Spanish settlement in the Chimayó valley date back to 1714, and the plaza at the north end of this narrow, well-watered agricultural valley is said to be one of the best remaining examples of colonial architecture in the region. Like much of New Mexico, it is insular; one of its children, Don Usner, says in his history of the place that those of the plaza didn't intermarry with people at the Potrero in the southern end of the valley. In colonial times the Spanish settlers were forbidden to travel without permission, and an extremely local, land-based identity evolved. In another northern New Mexico village I had lived in the year before this pilgrimage, someone once tartly remarked of a neighbor, "They're not from here. We remember when their great-grandfather moved here." The Spanish spoken here is old-fashioned, and it is often noted that the culture derives from pre-Enlightenment Spain. In its strong agricultural and local ties and traditions, its widespread poverty, its conservative social views, and its devout, magical Catholicism, this culture often seems like a last outpost of the Middle Ages.

The Santuario is in the southern end of Chimayó, on its own little unpaved plaza past a street of crumbling adobe houses and shops with hand-lettered signs and chile *ristras*. Graves fill the courtyard of this small, sturdily built adobe church. Inside it's covered in faded murals depicting the saints and Christ hung on a green cross in a style reminiscent of both Byzantine and Pennsylvania Dutch painting. The northern chapels are what make the church exceptional, though. The first is full of pictures of Jesus, Mary, and the saints brought in by devotees, and hand-painted images mingle with 3D and decoupage icons, a silver-glitter Virgin of Guadalupe, and a printed, varnished, cracked Last Supper. The outer wall of this chapel is covered with

---

from their common world, the closer they come to the realm of the divine. We might mention that in Japanese

crucifixes, in front of which hang a solid row of crutches, their silvery aluminum forming a surface as regular as prison bars through which many Christs peer. Through a low doorway to the west is the most important part of the church, a little chapel where the hole in the unpaved floor yields up the dirt pilgrims take home. This year it had in it a small green plastic scoop from a detergent carton with which to take up the moistly crumbling sandy earth. People used to drink this earth dissolved in water, and they still collect it to apply to diseased and injured areas and write to the church of miraculous cures. The crutches here testify, as they do in many pilgrimage sites, to cures of lameness.

When I first came here several years before, I had heard of many holy wells of water, but I was astonished to find a holy well of dirt. The Catholic church doesn't generally consider dirt much of a medium for holiness, but the dirt well in Chimayó is exceptional. The anthropologists Victor and Edith Turner use the term "baptizing the customs" to describe how the Catholic church assimilated local practices as it spread across Europe and the Americas—which is why, for example, so many of Ireland's holy wells were holy before they were Christian. It is now thought that the Tewa considered the earth here sacred or at least of medicinal virtue before the Spanish came, and that in the smallpox plague of the 1780s the Spanish women acquired some of their customs. To consider earth holy is to connect the lowest and most material to the most high and ethereal, to close the breach between matter and spirit. It subversively suggests that the whole world might potentially be holy and that the sacred can be underfoot rather than above. On earlier visits, I was given to understand that the well was supposed to replenish itself magically, and such inexhaustibility has been the stuff of miracles since the bottomless drinking horns of Celtic literature and Jesus' own multiplying loaves and fishes. Certainly the hole in the dirt floor of the chapel is still only about the size of a bucket after nearly two centuries of devotees scooping out soil to take home. But the religious literature I bought next door made it clear that the priests add earth from elsewhere that has been blessed, and on Good Friday a large box of such earth rests on the altar.

The story goes that during Holy Week early in the nineteenth century a local landowner, Don Bernardo Abeyta, was performing the

---

the word for "walk" is the same word which is used to refer to Buddhist practice; the practitioner (gyōja) is

customary penances of his religious society in the hills. He saw a light shining from a hole in the ground and found in it a silver crucifix that, when brought to other churches, would be found again in the hole in Chimayó. After the crucifix returned to the hole three times, Don Bernardo understood that the miracle was tied to the site, and he built a private chapel there in 1814–16. The curative properties of the earth were already known in 1813—a pinch of it in the fire was said to abate storms. The miracle story fits the pattern for many pilgrimage sites, notably the medieval "cycle of the shepherds" in which a cowherd, shepherd, or farmer discovers a holy image in the earth or some other humble place amid miraculous light or music or homage by the beasts, an image that cannot be relocated, for the miracle and the place are one. The Turners write of Christian pilgrimage, "All sites of pilgrimage have this in common: they are believed to be places where miracles once happened, still happen, and may happen again."

Pilgrimage is premised on the idea that the sacred is not entirely im- material, but that there is a geography of spiritual power. Pilgrimage walks a delicate line between the spiritual and the material in its em- phasis on the story and its setting: though the search is for spirituality, it is pursued in terms of the most material details—of where the Buddha was born or where Christ died, where the relics are or the holy water flows. Or perhaps it reconciles the spiritual and the ma- terial, for to go on pilgrimage is to make the body and its actions ex- press the desires and beliefs of the soul. Pilgrimage unites belief with action, thinking with doing, and it makes sense that this harmony is achieved when the sacred has material presence and location. Protes- tants, as well as the occasional Buddhist and Jew, have objected to pilgrimages as a kind of icon worship and asserted that the spiritual should be sought within as something wholly immaterial, rather than out in the world.

There is a symbiosis between journey and arrival in Christian pilgrimage, as there is in mountaineering. To travel without arriving would be as incomplete as to arrive without having traveled. To walk there is to earn it, through laboriousness and through the transfor- mation that comes during a journey. Pilgrimages make it possible to

then also the walker, one who does not reside anywhere, who abides in emptiness. All of this is of course move physically, through the exertions of one's body, step by step, toward those intangible spiritual goals that are otherwise so hard to grasp. We are eternally perplexed by how to move toward forgiveness or healing or truth, but we know how to walk from here to there, however arduous the journey. Too, we tend to imagine life as a journey, and going on an actual expedition takes hold of that image and makes it concrete, acts it out with the body and the imagination in a world whose geography has become spiritualized. The walker toiling along a road toward some distant place is one of the most compelling and universal images of what it means to be human, depicting the individual as small and solitary in a large world, reliant on the strength of body and will. In pilgrimage, the journey is radiant with hope that arrival at the tangible destination will bring spiritual benefits with it. The pilgrim has achieved a story of his or her own and in this way too becomes part of the religion made up of stories of travel and transformation.

Tolstoy captures this in a longing that comes to Princess Marya in *War and Peace* as she feeds the myriad Russian pilgrims that pass by her home: "Often as she listened to the pilgrims' tales she was so fired by their simple speech, natural to them but to her full of deep meaning, that several times she was on the point of abandoning everything and running away from home. In imagination she already pictured herself dressed in coarse rags and with her wallet and staff, walking along a dusty road." She has imagined her life of genteel seclusion become clear, sparse, and intense with a purpose she can move toward. Walking expresses both the simplicity and the purposefulness of the pilgrim. As Nancy Frey writes of the long-distance pilgrimage to Santiago de Compostela in Spain, "When pilgrims begin to walk several things usually begin to happen to their perceptions of the world which continue over the course of the journey: they develop a changing sense of time, a heightening of the senses, and a new awareness of their bodies and the landscape. . . . A young German man expressed it this way: 'In the experience of walking, each step is a thought. You can't escape yourself.'"

In going on pilgrimage, one has left behind the complications of one's place in the world—family, attachments, rank, duties—and become a walker among walkers, for there is no aristocracy among

---

related to the notion of Buddhism as a path: practice is a concrete approach to Buddhahood.—ALLAN G.

---

pilgrims save that of achievement and dedication. The Turners talk about pilgrimage as a liminal state—a state of being between one's past and future identities and thus outside the established order, in a state of possibility. Liminality comes from the Latin *limin*, a threshold, and a pilgrim has both symbolically and physically stepped over such a line: "Liminars are stripped of status and authority, removed from a social structure maintained and sanctioned by power and force, and leveled to a homogeneous social state through discipline and ordeal. Their secular powerlessness may be compensated for by a sacred power, however—the power of the weak, derived on the one hand from the resurgence of nature when structural power is removed, and on the other from the reception of sacred knowledge. Much of what has been bound by social structure is liberated, notably the sense of comradeship and communion, or communitas."

We started easily enough, on a flat wooden bridge across a stream that watered the banks around it into rare lushness, then up through Greg and MaLin's dogleg cornfield bordered by oaks. From there we went over an irrigation ditch and through the fence that divided their land from the Nambe reservation, the first of many fences we would crawl under, scramble over, or unlatch a wire-fastened gate and pass through. On the Nambe reservation, we passed Nambe Falls, which we could hear roaring in its gorge but not quite see. I liked its invisibility as a reminder that we were not on a scenic walk or the territory of people imbued with the mainstream European tradition of such walks. We could hear it as we approached and, by going to a promontory point and craning, could see part of it, but the only possible clear view on our route would be the quick one during the plummet from the cliff into the deep channel below. So we glimpsed the foaming white edges and lower streambed and went on. We all kept pace with each other for the first half of the expedition, and though the way utterly failed to resemble the route that had looked so coherent when Greg had shown it to us on the topographical maps, the roads and irrigation ditches and landmarks made it clear enough to him.

"Wherever you go, there you are," he said whenever someone

asked him if we were lost yet. We had a cheerful morning of it. Sue said that she had expected us to proceed in somber silence, but everyone told stories and made observations. We ate a first snack under a roadside cottonwood tree past the San Juan Reservoir on the Nambe reservation, which adjoins Greg's land, then walked through the outskirts of the reservation town with its horses, fruit trees, sweat lodge, buffalo pasture, and many scattered houses. For the whole length of that road into Nambe, Meridel told us about her first New Age experience in Santa Fe, having her aura balanced in the 1970s, and we variously inquired and wisecracked about the notion. Sue taught us the acronym AFGO, "another fucking growth opportunity," for the plethora of spiritual opportunities (and opportunists) in Santa Fe. Three in our party had had Christian upbringings, and I had come out partly to help Meridel celebrate her fiftieth birthday with a revisionist Passover dinner the day after our walk (she was raised as a nonreligious Jew, and I was raised as nothing in particular by a lapsed Catholic and a nonpracticing Jew). Since the Last Supper was a Passover seder, even Good Friday and Easter are overlaid on the Jewish holiday celebrating the flight from Egypt, and this pilgrimage was built on top of all those layers of meeting, suffering, moving, dying.

We began to drift apart north of the Nambe settlement when we reached the rough sandstone expanse of the badlands, with wind-carved pillars of red stone studding a hot, airless expanse of sand and gravel and ruddy dirt stretching to the red cliffs in the distance. The two other women began to trail, and the two men I didn't know went on ahead. We all met up at the windmill, which marked a turn in terrain and in direction, and lounged around the shade of its waterless tank. Afterward, Greg and Sue decided to go around a hill the rest of us were going to go straight over, because she was wearing out. The badlands had given way to more of that intricate terrain of hillocks so hard to navigate in, and rather than going over the single hill I had expected, we found ourselves surmounting and descending innumerable tree-studded red-soil rises. We shouted, but we couldn't find them, so we kept walking. One of the other men had gone on far ahead; the other was walking faster than Meridel could. She is an athletic woman, but she is small and had pulled something in her knee, and her steps had grown short.

SPACE IN JAPANESE RELIGIONS"   *At the same time continue to count inhalations and exhalations as you*

This drifting apart was dispiriting. When I think about what we were doing, it seems as if it ought to have been an experience of paradise attained—dear friends and amiable new acquaintances moving across a varied landscape toward a remarkable goal under an azure sky. But, alas, we had various bodies and various styles. I had been frustrated for the last few hours by the pace. Someone would stop to pull out binoculars or to confer, and everyone would come to a halt that would grow protracted. Standing or wandering slowly makes my feet hurt; it's why museums and malls are more painful than mountains. And if the devil is in the details, mine was in the heavy-duty boots I thought I had broken in but which had begun to break my feet in all over again. So I oscillated between the man ahead and the woman behind until we finally reached the open grassland. Three of us arrived at the road on the far side of the grassland together. A steady stream of walkers and cars was going by—the former all uphill, the latter in both directions—and Meridel and I joined it. We were now part of the much larger community spread out for dozens of miles along the highway that is the main pilgrimage route. The trail of empty water bottles and orange peels bore evidence to the volunteers farther down the road, people who came every year and set up tables bearing slices of oranges, water, soft drinks, cookies, and occasionally Easter candy that everyone was welcome to take. This was one of the most moving parts of the pilgrimage to me, these people who were out not to earn their own salvation but to support others doing so.

On Good Friday of the year before, I had been struck by how little preparation most of the pilgrims seemed to have made for a long walk. Their everyday clothes had been something of a rebuke to me that this was not a hike, and many stout people who looked as if they never walked much otherwise persevered. This year the day was much warmer, and everything seemed different: with our aching feet and our packs, *we* looked more serious, more dogged, than the jaunty young pilgrims in their colorful shorts and jeans and T-shirts (though Meridel's husband Jerry told us when he met us in Chimayó that he had seen a woman from a very small town walking in a fancy white dress—"the kind of dress you would get married in, or buried in"— and two days earlier and thirty miles west I had seen two men in

walk slowly around the room. Begin walking with the left foot and walk in such a way that the foot sinks

fatigues walking eastward, one of them carrying a large cross). Both times I joined this pilgrimage I had the strange sense that I was walking alongside people in another world, the world of believers, people for whom the Santuario up ahead contained a definite power in a cosmos organized around the Trinity, the mother of God, the saints, and the geography of churches, shrines, altars, and sacraments. But I had suffered like a pilgrim; my feet were killing me.

Pilgrimages are not athletic events, not only because they often punish the body but because they are so often gone on by those who are seeking the restoration of their own or a loved one's health. They are for the least equipped rather than the most. Greg told me, when I called him up to ask if I could join in, that when he had leukemia he made a deal with the gods. Framed in the same easygoing humor he brought to other subjects, the deal's terms were flexible: that if he lived, he would try to go on the pilgrimage when he could. This was his third year of walking it, and it got easier every year. Four years before, when he was deathly ill, Jerry and Meridel walked for him and brought him back some dirt from the Santuario.

This Easter week in which we were walking to Chimayó, a similar pilgrimage from Paris to Chartres would be taking place again, and far larger crowds of Christians would be gathering in Rome and Jerusalem. In the last half century or so, a wide variety of secular and nontraditional pilgrimages have evolved that extend the notion of the pilgrimage into political and economic spheres. Not long before I had set out, a march in San Francisco commemorated the farmworker organizer César Chávez's birthday with a crosstown "Walk for Justice"; and in Memphis, Tennessee, civil rights activists commemorated the thirtieth anniversary of Martin Luther King's assassination there with another march. In the southwest in April, I could have instead joined the Franciscan-led Nevada Desert Experience on their annual peace walk from Las Vegas to the Nevada Test Site (akin to another pilgrimage route from Chimayó to Los Alamos, birthplace of the atomic bomb, thirty miles west). Then there was the Muscular Dystrophy Association's annual walkathon on the first week of April and the March of Dimes's WalkAmerica the last weekend of that month. I had

---

into the floor, first the heel and then the toes. Walk calmly and steadily, with poise and dignity. The walking

---

come across a flyer in Gallup, New Mexico, for "Native Americans for Community Action, Inc. 15th Annual Sacred Mountain 10k Prayer Run and 2k Fun Run/Walk" to be held in Flagstaff in June, which sounded like the Spirit Runs held by the five tribes fighting the proposed Ward Valley nuclear waste dump in southeastern California, and I knew that the annual breast cancer and AIDS walks were coming up in San Francisco's Golden Gate Park and other locations around the country. And no doubt somewhere somebody was walking across the continent for some other good cause. All these were outgrowths of the pilgrimage, or adaptations of its terms.

Imagine all those revisionist versions of pilgrimage as a mighty river of walkers flowing from many sources. The first small trickle comes, like March ice melt from a high glacier, from a single woman almost half a century ago. On January 1, 1953, a woman known to the world only as Peace Pilgrim set out, vowing "to remain a wanderer until mankind has learned the way of peace." She had found her vocation years before when she walked all night through the woods and felt, in her words, "a complete willingness, without any reservations, to give my life to God and to service," and she prepared for her vocation by walking 2,000 miles on the Appalachian Trail. Raised on a farm and active in peace politics before she abandoned her name and began her pilgrimage, she was a peculiarly American figure, plainspoken and confident that the simplicity of life and thought that worked for her could work for everyone. Her cheery accounts of her long years of walking the roads and talking to the people she met are unburdened by complexity, dogma, or doubt and rife with exclamation marks.

She started her pilgrimage by joining the Rose Bowl Parade in Pasadena, and something about setting out on her long odyssey from this corny festivity recalls Dorothy in *The Wizard of Oz*, with her own farmgirl can-do determination, starting down the Yellow Brick Road amid dancing munchkins. Peace Pilgrim kept walking for twenty-eight years through all kinds of weather and every state and Canadian province as well as parts of Mexico. An older woman at the time she first set out, she wore navy blue pants and shirt, tennis shoes, and a navy blue tunic whose front was stenciled with the words "Peace Pilgrim" and whose back text changed over the years from "walking

---

must not be done absentmindedly, and the mind must be taut as you concentrate on the counting.

coast to coast for peace" to "walking 10,000 miles for world disarmament" to "25,000 miles on foot for peace." Something of her brisk, practical piety comes across in her explanation of the choice of dark blue—"it doesn't show dirt," she wrote, and "does represent peace and spirituality." Though she attributes her extraordinary health and stamina to her spirituality, it is hard not to wonder if it was the other way around. She continued her pilgrimage in her simple outfit through snowstorms, rain, a harsh dust storm, and heat, sleeping in cemeteries, in Grand Central Station, on floors, and on an endless succession of the couches of new acquaintances.

Though most of her writings are nonpartisan, she took a strong stand on national and global politics, arguing against the Korean War, the cold war, the arms race, and war in general. The war in Korea was still going on when she set out from Pasadena, as was Senator Joe McCarthy's anticommunist intimidation. It was one of the bleakest periods in American history, with fear of nuclear war and communism driving most Americans into the bunkers of conformity and repression. Even to argue for peace took heroic courage. To set out, as Peace Pilgrim did on the first day of 1953, with nothing more than her single outfit, whose pockets contained "a comb, a folding toothbrush, a ballpoint pen, copies of her message and her current correspondence," was astonishing. While the economy was booming and capitalism was becoming enshrined as a sacrament of freedom, she had dropped out of the money economy—she never carried or used money for the rest of her life. She says of her lack of material possessions, "Think of how free I am! If I want to travel, I just stand up and walk away. There is nothing to tie me down." Though her models were largely Christian, her pilgrimage seems to have arisen from the same 1950s crisis of culture and spirituality that pushed John Cage, Gary Snyder, and many other artists and poets into investigations of Zen Buddhism and other nonwestern traditions and sent Martin Luther King to India to study Gandhi's teachings on nonviolence and *satyagraha*, or soul-force.

Most people who diverge from the mainstream withdraw from its spaces, but Peace Pilgrim had withdrawn from the former to enter the latter, where she would be most required to mediate the gap between her beliefs and national ideology—she was as much an evangelist as a

---

—INSTRUCTIONS ON WALKING MEDITATION IN *THREE PILLARS OF ZEN*   *Sigmund Freud believed,*

pilgrim. She had set out to walk 25,000 miles for peace, and it took her nine years to do so. Afterward, she continued walking for peace but stopped counting the miles. As she put it, "I walk until given shelter, fast until given food. I don't ask—it's given without asking. Aren't people good!. . . I usually average twenty-five miles a day walking, depending on how many people stop to talk to me along the way. I have gone up to fifty miles in one day to keep an appointment or because there was no shelter available. On very cold nights I walk through the night to keep warm. Like the birds, I migrate north in the summer and south in the winter." Later she became a widely recognized public speaker and occasionally accepted a ride to get her to her speaking engagements. She died, ironically, in a head-on car crash in July 1981.

Like a pilgrim, she had entered the liminal condition the Turners would later describe, leaving behind an ordinary identity and the goods and circumstances that bolster such identities to achieve that state of anonymous simplicity and clear purpose Tolstoy's Princess Marya longed for. Her walking became a testament to the strength of her convictions and suggests several things. One is that the world was in such trouble that she herself had to drop her ordinary name and ordinary life to try to heal it. Another is that if she could break with the ordinary and go forth unprotected by money, by buildings, and by a place in the world, then perhaps profound change and profound trust were possible on a larger scale. A third is that of the carrier: like Christ taking on the sins of all his followers or the Hebrew scapegoat driven out into the wilderness, burdened with the sins of the community, she had taken personal responsibility for the state of the world, and her life was testimony and expiation as well as example. But what makes her unorthodox is that she adapted a religious form, the pilgrimage, to carry political content. The pilgrimage traditionally dealt with disease and healing of self or loved ones, but she had taken on war, violence, and hate as plagues ravaging the world. The political content that motivated her and the way in which she endeavored to achieve change through influencing her fellow human beings rather than through divine intervention make her the first of a horde of modern political pilgrims.

She foreshadowed this shift in the nature of the pilgrimage, from appealing for divine intervention or holy miracle to demanding po-

---

for example, that the psychical foundation of all travel was the first separation and the various other

litical change, making the audience no longer God or the gods, but the public. Perhaps the postwar era marked the end of belief that divine intervention alone was adequate; God had failed to prevent the Jewish Holocaust, and the Jews had seized their promised land through political and military means. African Americans, who had long used metaphors of the Promised Land, stopped waiting too. At the height of the civil rights movement, Martin Luther King said that he was going to Birmingham to lead demonstrations until "Pharaoh lets God's people go." The collective walk brings together the iconography of the pilgrimage with that of the military march and the labor strike and demonstration: it is a show of strength as well as conviction, and an appeal to temporal rather than spiritual powers—or perhaps, in the case of the civil rights movement, both.

Because of the involvement of so many ministers, the practice of nonviolence, and the language of religious redemption and, occasionally, martyrdom, the civil rights movement was more saturated with the temperament and imagery of pilgrimage than most struggles. It was in large part about the rights of access of black people, and it was first fought on the contested sites: sitting down in and then boycotting buses, bringing children into schools, sitting in at lunch counters. But it found its momentum in events that united the protest or the strike with the pilgrimage: the march from Selma to Montgomery to petition for voting rights, the many marches in Birmingham and throughout the country, the culminating March on Washington. In fact, the first major event organized by the newly founded Southern Christian Leadership Conference (SCLC) was the "prayer pilgrimage" at the Lincoln Memorial in Washington, D.C., on May 17, 1957, the third anniversary of the Supreme Court ruling in favor of desegregating schools. It was so called to make it sound less threatening; a pilgrimage makes an appeal while a march makes a demand. King was profoundly influenced by the writings and actions of Mahatma Gandhi, and he adapted from Gandhi both the general principle of nonviolence and the specifics of marches and boycotts that had hastened India's liberation from British rule. Perhaps Gandhi was the founder of the political pilgrimage with his famous 200-mile-long Salt March in 1930, in which he and many people living inland walked to the sea to make their own salt in violation of British law and British

---

departures from one's mother, including the final journey into death. Journeying is therefore an activity

taxes. Nonviolence means that activists are asking their oppressors for change rather than forcing it, and it can be an extraordinary tool for the less powerful to wring change out of the more powerful.

Six years after the founding of the SCLC, Martin Luther King decided that nonviolent resistance by itself was inadequate, and the violence the southern segregationists inflicted on blacks should be made as public as possible. The audience would no longer be merely the oppressors, but the world. This was the strategy of the Birmingham struggle, perhaps the central episode of the civil rights movement, which began on Good Friday of 1963 with the first of many marches, or processions. It is from these protests that the most famous images come, of people being blasted by high-pressure fire hoses and savaged by police dogs, images that provoked worldwide indignation. King and hundreds of others were arrested for marching in Birmingham, and after the supply of willing adults began to run out, high school students were recruited, and their younger siblings volunteered. They marched for freedom with bold jubilance, and on May 2 of that year 900 of these children were arrested. To go out onto the streets knowing they risked attack, injury, arrest, and death took an extraordinary resolve, and the religious ardor of Southern Baptists as well as the Christian iconography of martyrdom seems to have strengthened them. A month after the Birmingham campaign had begun, writes one of King's biographers, "Reverend Charles Billups and other Birmingham ministers led more than 3,000 young people on a prayer pilgrimage to Birmingham jail singing 'I Want Jesus to Walk with Me' as they moved."

A photograph of the 1965 Selma-to-Montgomery march has been on my refrigerator for months, and it speaks of this inspired walking. Taken by Matt Heron, it show a steady stream of marchers three or four wide moving from right to left across the photograph. He must have lain low to take it, for it raises its subjects up high against a pale, clouded sky. They seem to know they are walking toward transformation and into history, and their wide steps, upraised hands, the confidence of their posture, express the will with which they go to meet it. They have found in this walk a way to make their history rather than suffer it, to measure their strength and test their

----

related to a larger feminine realm, so that it is not surprising that Freud himself was ambivalent about it. Of

----

freedom, and their movement expresses the same sense of destiny and meaning that resonates in King's deep-voiced, indomitable oratory.

In 1970 the form of the pilgrimage was moved yet further from its origins when the first Walkathon was held by the March of Dimes. Tony Choppa, who has been working on these walks since 1975 and is their unofficial historian, says it was risky at the time, since walking the streets en masse was associated with more radical demonstrations. The first walkers were high school students in San Antonio, Texas, and Columbus, Ohio, and this first "walkathon" was modeled after a fund-raiser for a hospital in Canada. It rained on both walks, he says, and there was "no money but great potential. People did actually come out and walk." Over the years the route was trimmed from the initial twenty-five miles to ten kilometers, and participation mush-roomed. The year we walked to Chimayó from Greg's land, nearly a million people were expected to join what the March of Dimes now calls WalkAmerica, and they would raise about $74 million for infant and prenatal health care and supporting research. The walk was co-sponsored by K-Mart and Kellogg's, among others. This walkathon structure, with corporations sponsoring the event in return for pro-motional opportunities and walkers raising the money for the charity, has been adopted by hundreds of organizations, the great majority of them dealing with disease and health care.

The summer before I had accidentally run into the eleventh annual AIDS Walk San Francisco in Golden Gate Park. A huge throng of people in shorts and caps milled around the starting area that sunny day, holding various free beverages, advertisements, and product samples. The hundred-page booklet for the walk consisted almost entirely of advertisements for the dozens of corporate sponsors—clothing companies, brokerage houses—who also had tables set up around the lawn. It was a strange atmosphere, a cross between a gym and a convention, crawling with logos and ads. Yet it must have been profound for some of its participants. The next day the paper said 25,000 walkers had raised $3.5 million for local AIDS organizations and described a walker who wore a T-shirt printed with photographs

---
the landscape he said, "All of these dark woods, narrow defiles, high grounds and deep penetrations are

of his two sons who had died of AIDS and said, "You never get over it. The walk is a way to cope with it."

These fund-raising walks have become the mainstream American version of the pilgrimage. In many ways they have traveled far from its original nature, notably in the evolution from devoutly appealing for divine intervention to pragmatically asking friends and family for money. And yet, however banal these walks are, they retain much of the content of the pilgrimage: the subject of health and healing, the community of pilgrims, and the earning through suffering or at least exertion. Walking is crucial to these events, or at least it has been. Bikeathons have come into being, and the last indignity dealt to this highly mutated form of pilgrimage came with the virtual walk, including the San Francisco Art Institute's "nonwalk," in which people were asked to give money and were given a T-shirt but weren't obliged to show up, and AIDS Action's "Until It's Over e-March," which proposed that participants electronically sign their names to a letter on the Internet as a substitute for marching or walking.

Fortunately, walkathons are not the end of the story. Though mutant forms of the pilgrimage keep springing up, the older ones thrive, from religious pilgrimages to long political walks. A month after 25,000 people walked ten kilometers to raise money for AIDS organizations in San Francisco, gang counselor Jim Hernandez and antiviolence organizer Heather Taekman finished a 500-mile walk from East Los Angeles to Richmond, California, carrying more than 150 photographs of young murder victims and meeting with teenagers along the way. In 1986 hundreds of people joined together to form the Great Peace March. They walked across the United States together to ask for disarmament in a mass pilgrimage that created its own culture and support structure and had a large impact in some of the small towns through which they trekked. The walk began as a sort of publicity event, but somewhere along the long way the walking itself took over, and the walkers became less concerned with media and message and more with what was happening within themselves. In 1992 two more cross-continental peace walks did much the same thing, and like the walkers of the Great Peace March they drew inspiration from Peace Pilgrim. Similar walks went across the Soviet Union and Europe during the early 1990s, and in 1993 strawberry pickers and other United Farm Workers (UFW) supporters reenacted the great three-hundred-mile Delano-to-Sacramento march César Chávez had organized in 1966 and called a pilgrimage.

Even the most sophisticated yield to the pilgrim's impulse, and even without the superstructure of religion, the ordeal of walking makes sense. The filmmaker Werner Herzog writes, "At the end of November, 1974, a friend from Paris called and told me that Lotte Eisner [a film historian] was seriously ill and would probably die. I said that this must not be, not at this time, German cinema could not do without her now, we would not permit her death. I took a jacket, a compass and a duffel bag with the necessities. My boots were so solid and new that I had confidence in them. I set off on the most direct route to Paris, in full faith, believing that she would stay alive if I came on foot. Besides, I wanted to be alone with myself." He walked the several hundred miles from Munich in winter weather, often wet, often smelly, often thirsty, and usually suffering from great pain in some part of his feet and legs.

Herzog, as anyone who has seen his films knows, is fond of deep passions and extreme behavior, however obtuse, and in his journals of his long walk to Paris he took on the qualities of one of the obsessives in his films. He walked in all weather, though he occasionally accepted a lift, and he slept in barns and a display mobile home he broke into as well as in strangers' homes and inns. The sparse prose describes walking, suffering, minor encounters, and fragments of scenery. Elaborate fantasies that themselves sound like outlines for Herzog movies are woven into the description of his ordeal. On the fourth day, he writes, "While I was taking a shit, a hare came by at arm's length without noticing me. Pale brandy on my left thigh which hurts from my groin downwards with every step. Why is walking so full of woe?" On the twenty-first day, he put his feet up in Eisner's room, and she smiled at him. "For one splendid fleeting moment something mellow flowed through my deadly tired body. I said to her, open the window, from these last days onward I can fly."

We had arrived too, along the curving road into Chimayó. Sal and I sat down and waited for Meridel on a sidewalk. Cars, policemen, and

---

MADNESS   *The geographical pilgrimage is the symbolic acting out of an inner journey. The inner journey*

children carrying Sno-Cones passed by in front of us; behind us bloomed a few stunted fruit trees in a knobby pasture. Afterward, Sal went to stand in the long line in front of the Santuario, and I went off to buy us some lemonade at a little mobile food-stand around the corner, near the Santo Niño Chapel, where people used to offer up children's shoes because the Santo Niño, a version of the Christ child, is said to have worn out his own running errands of mercy around the countryside at night. It was nice to be back on familiar ground. I knew what was inside the Santuario and thought of the thousands of crosses woven into the cyclone fence behind the outdoor chapel below, crosses made of grapevine and cottonwood twigs and larger sticks, and then of the irrigation ditch that flowed just the other side of the fence, of the swift shallow river that runs through the town, of the burrito stand that sold meatless alternatives for Lent, of the old adobe houses and the trailer homes that are beginning to look old, and of the many unwelcoming signs: "Notice: Please Don't Leave Your Belongings Unattended at Any Time," "Not Responsible for Theft," "Beware of Dog." Chimayó is a desperately poor town, known for drugs, violence, and crime as well as for sanctity. Jerry West was waiting for his wife, Meridel, in front of that chapel, and I made my last foot journey back with the lemonade, bade Sal farewell, and went off to my own culminating destination. About ten thousand pilgrims would come into town and stand in line to go into the chapel that day, and Jerry found Greg and Sue standing in line to go in too. When we left after the moon had risen, there were still more figures walking along the narrow shoulder of the road in the night, shadowy groups that no longer looked festive, but dedicated and fragile in the dark.

is the interpolation of the meanings and signs of the outer pilgrimage. One can have one without the other. It

Chapter 5

L ABYRINTHS AND C ADILLACS:

Walking into the Realm

of the Symbolic

I didn't mind not getting into the church at Chimayó along with all the long patient line of pilgrims, because I had another destination. The year before I'd walked the last six or so miles of the pilgrimage, and later, trying to catch up with my friend who'd driven in, I walked past the Cadillac with the stations of the cross painted on it. I kept going after a cursory inspection, and then I did the world's slowest double take. A Cadillac with the fourteen stations began, during the interim between those two Good Fridays, to seem more and more extraordinary, a gorgeous compression of many symbolic languages and desires into one divinely strange chariot. Jerry said, in front of the Santo Niño Chapel, that it was just up the road a hundred yards, and so I limped off to see it again.

Long and pale blue and somehow soft-looking, as though the metal body were dissolving into velvet or veils, this 1976 Cadillac was a contrary thing. The stations of the cross were wrapped around its long lean body, below the chrome line that bisected it horizontally. Jesus was condemned at the rear end of the driver's side and carried the cross, stumbled, and encountered his way around the car to be crucified in the middle of the passenger's side, next to the door handle, and he was buried at the back end of that side. All along those

----
is best to have both.—THOMAS MERTON   *I was the first in six generations to leave the Valley, the only*
----

sides was painted a dark gray sky full of lightning that made the place of his suffering into New Mexico, with its volatile thunderclouds. There was Jesus again on the trunk as a big soft-focus head with a crown of thorns, flanked by angels, thorny roses, and the same kind of undulating ribbons that bear inscriptions in medieval and old Mexican religious paintings. The thorns everywhere seemed like further reminders that Chimayó and Jerusalem were both arid landscapes, and the same thorny roses adorned the hood, where Mary, the Sacred Heart, an angel, and a centurion were.

This car was designed to be looked at standing still, but it retained the possibility of moving. It didn't matter if the car ever went far, just that it could, that these images could hurtle down the highway, whipped by wind and drops of rain running sideways. Imagine it doing seventy on the interstate, passing mesas and crumbling adobes and cattle and maybe some billboards for fake Indian trading posts, Dairy Queens, and cheap motels, an eight-cylinder Sistine Chapel turned inside out and speeding toward a stark horizon under changing skies. The artist, Arthur Medina, a slender, restless-looking man with wavy black hair, showed up while I was admiring it and leaned against the adjacent wooden shed to receive compliments and questions. Why a Cadillac? I asked, and he didn't seem to understand my premise that a luxury car is not the most natural and neutral thing on which to paint holy pictures. So I asked him why he painted the car with this subject matter, and he said, "To give the people something for Lent," and indeed he displayed it here every year.

He had, he said, painted other cars and had an Elvis car, and then he darkly intimated that other local artists were imitating him. It was true that another long 1970s car was parked nearer the Santuario, in front of a white-painted adobe shop, and that very shop was painted with perfect accuracy on the side of the car facing the street, while a radiant image of the Santuario itself covered the hood. This made it almost as dizzying a vehicle of meaning as Medina's car, a transformation of immobile place into speeding representations. But the tradition of customized low-rider cars goes back more than a quarter of a century in northern New Mexico, and this other car was painted much more professionally—which is not to say that Medina was a lesser artist, only that most such cars have an orthodox aesthetic that

----

one in my family to ever leave home. But I didn't leave all the parts of me: I kept the ground of my own

----

comes from a particular way of handling the airbrush, and Medina had made his figures simpler and flatter and created a much more lushly misty atmosphere. You could say that most low-rider cars are baroque, with a slightly cynical hyperreality of form, while Medina's had something of the flat devout force of medieval painting about it.

It was an extraordinarily quixotic object, a car about walking, a luxury item about suffering, sacrifice, and humiliation. And the car united two radically different walking traditions, one erotic and one religious. Customized cars exist both as art objects and as the vehicles for an updated version of an old Spanish and now Latin American custom, the paseo or *corso*. For hundreds of years, promenading the plaza in the center of town has been a social custom in these places, one that allows young people to meet, flirt, and stroll together and dictates that villages and cities from Antigua, Guatemala, to Sonoma, California, have a central plaza in which to do so (the more casual promenades of northern Europe take place in parks, quays, and boulevards). In some parts of Mexico and elsewhere the custom was once so formal that the men strolled in one direction and the women in the other, like the indefinitely extended steps of a line dance, but in most nowadays the plaza is the site of less structured promenading. The promenade is a special subset of walking with an emphasis on slow stately movement, socializing, and display. It is not a way of getting anywhere, but a way of being somewhere, and its movements are essentially circular, whether on foot or by car.

During the days I was writing this, I ran into my brother Steve's friend José in Dolores Park after San Francisco's May Day Parade and asked him about the custom. At first he said he knew nothing about it, but as we talked, more and more came back to him, and his eyes shone with the old memories flooding back in a new light. In his hometown in El Salvador, the custom was called "going around the park," though *park* meant the plaza at the center of town. Mostly teenagers used the park for this socializing, in part because the small houses and warm weather made it uncomfortable to socialize at home, at least at that age. Girls didn't go to the park alone, so he was much in request as a sort of midget chaperone by his older sister and his three beautiful cousins. Many Saturday and Sunday evenings of his childhood were spent licking an ice cream cone and ignoring their

being. On it I walked away, taking with me the land, the Valley, Texas.—GLORIA ANZALDUA *An*

conversations with boys. The paseo, like less structured courtship walks in other places, allows people to remain visually in public but verbally in private, giving them enough room to talk and enough supervision to do little more. Nobody could afford to stay in the village, he said, and so the romances kindled during strolls in the parks rarely led to marriage. But when people came back home, they would go around the park again, not to meet people but in reminiscence of this part of their life. Every small town and village in El Salvador and, he ventured, Guatemala had some form of this custom, and "the smaller the town the more important it was for keeping people's sanity." Other versions of the pedestrian paseo exist in Spain, southern Italy, and much of Latin America; the custom turns the world into a kind of ballroom and walking into a slow waltz.

It is hard to say how the customized car and the cruise came together, but the cruise is very much the successor to the paseo or _corso_, with the cars moving at promenade speed and the young people within flirting with and challenging each other. Meridel, my companion on the Chimayó pilgrimage, had in 1980 made one of her earliest series of photographs about New Mexico, a documentary project on low riders. At that time the subculture was booming, and the cars would slowly cruise the old plaza at the center of Santa Fe. Like low riders in most places, these ones met with the hostility of the civic authorities, who turned the four streets around the plaza into a one-way roundabout and took other steps to ban the practice. But when Meridel's series was complete, she organized a show of her work in the plaza, to which the low riders were invited and at which many of their cars were on display. By resituating them within the context of high art, she had reopened the space to them and introduced their work and world to the others in the region. It was the biggest art opening in Santa Fe history, with all kinds of people milling around the plaza to look at the cars, the photographs, and each other, an art paseo.

Though cruising came from the paseo, the cars' imagery sometimes spoke of a very different tradition. In devout New Mexico they bore far more religious imagery than, for example, low-rider cars in California, and Meridel came to see many of them as chapels, reliquaries, and, because of the plush velvet upholstery, even caskets.

---

active line on a walk moving freely, without goal. A walk for a walk's sake.—PAUL KLEE,

They express the culture of young people who are both devout and hard-partying as an indivisible whole, not a set of contradictions. And they express something of the centrality of the car in New Mexico, where sidewalks and roadside trails are often hard to find and both rural and urban life are built around the car (even on the pilgrimage, low riders cruised the road and did the occasional doughnut for us pedestrians). Still I find it strange that the paseo should have ceased to be a pedestrian event and become a vehicular one. Cars function best as exclusionary devices, as mobile private space. Even driven as slowly as possible, they still don't allow for the directness of encounter and fluidity of contact that walking does. Medina's car, however, was no longer a vehicle but an object. He stood beside it to receive compliments, and we walked around it less as devotees would walk the stations of the cross than as connoisseurs would tour a gallery.

The stations of the cross are themselves one of those cultural things made up of many strata laid down upon each other. The first layer is the presumed course of events from Jesus's condemnation to the laying of his dead body in the tomb in the cave, a walk from Pilate's house to Golgotha, the walk that the pilgrims dragging crosses to Chimayó imitate. During the Crusades pilgrims in Jerusalem would tour the sites of these events, praying as they went, laying down a second layer, a layer of devout retracing that brought pilgrimage close to tourism. In the fourteenth and fifteenth centuries Franciscan friars created the third layer by formalizing the route as a series of fixed events—the fourteen stages—and abstracting them from their site. From this tradition come the stations of the cross artworks—usually fourteen small paintings or prints running up and down the nave of the church—that adorn nearly all Catholic churches, and it is an amazing abstraction. No longer is it necessary to be in Jerusalem to trace these events two millennia ago. The time is past, the place is elsewhere, but walking and imagining are adequate means to enter into the spirit of those events. (Most of the recommendations on praying the stations emphasize reliving the events of the crucifixion, so that it is an act not merely of prayer but of identification and imagination.) Christianity is a portable religion, and even this route once so specific to Jerusalem was exported around the world.

A path is a prior interpretation of the best way to traverse a

---

ALLEGORIZING DRAWING   *Trebuchant sur les mots comme sur les paves (stumbling against words*

---

landscape, and to follow a route is to accept an interpretation, or to stalk your predecessors on it as scholars and trackers and pilgrims do. To walk the same way is to reiterate something deep; to move through the same space the same way is a means of becoming the same person, thinking the same thoughts. It's a form of spatial theater, but also spiritual theater, since one is emulating saints and gods in the hope of coming closer to them oneself, not just impersonating them for others. It's this that makes pilgrimage, with its emphasis on repetition and imitation, distinct amid all the modes of walking. If in no other way one can resemble a god, one can at least walk like one. And indeed, in the stations of the cross, Jesus appears at his most human, stumbling, sweating, suffering, falling three times, and dying in the course of redeeming the Fall. But by the time the stations of the cross had become a sequence of pictures in any church, anywhere, devotees were tracing a path that was no longer through a place but through a story. The stations are set up all along the nave of churches so that worshipers can walk themselves into Jerusalem, into the central story of Christianity.

There are many other devices besides the stations of the cross that let people bodily enter a story. I found one last summer. I had a date to meet some friends for drinks at the famously kitschy old mock-Polynesian bar the Tonga Room in the Fairmont Hotel atop Nob Hill. After walking over Nob Hill, past a grocery store advertising caviar, past a Chinese boy skipping with joy, past the less joyful adults in this posh neighborhood, and around the back of Grace Cathedral, I walked through a courtyard where a fountain was playing and a young man was waving a Bible around and mumbling something. At the far side of the space I saw, to my delight, something new there, a labyrinth. In pale and dark cement it repeated the same pattern made of stone in Chartres Cathedral: eleven concentric circles divided into quadrants through which the path winds until it ends at the six-petaled flower of the center. I was early for my rendezvous, and so I stepped onto the path. The circuit was so absorbing I lost sight of the people nearby and hardly heard the sound of the traffic and the bells for six o'clock.

---

as against cobblestones)—CHARLES BAUDELAIRE, "LE SOLEIL." *At the other extreme is a group of*

Inside the labyrinth the two-dimensional surface ceased to be open space one could move across anyhow. Keeping to the winding path became important, and with one's eyes fixed upon it, the space of the labyrinth became large and compelling. The very first length of path after the entrance almost reaches the center of the eleven rings, then turns away to snake round and round, nearer and farther, never so close as that initial promise until long afterward, when the walker has slowed down and become absorbed in the journey—which even on a maze forty feet in diameter like this can take a quarter hour or more. That circle became a world whose rules I lived by, and I understood the moral of mazes: sometimes you have to turn your back on your goal to get there, sometimes you're farthest away when you're closest, sometimes the only way is the long one. After that careful walking and looking down, the stillness of arrival was deeply moving. I looked up at last to see that white clouds like talons and feathers were tumbling east in a blue sky. It was breathtaking to realize that in the labyrinth, metaphors and meanings could be conveyed spatially. That when you seem farthest from your destination is when you suddenly arrive is a very pat truth in words, but a profound one to find with your feet.

The poet Marianne Moore famously wrote of "real toads in imaginary gardens," and the labyrinth offers us the possibility of being real creatures in symbolic space. I had thought of a children's story as I walked, and the children's books that I loved best were full of characters falling into books and pictures that became real, wandering through gardens where the statues came to life and, most famously, crossing over to the other side of the mirror, where chess pieces, flowers, and animals all were alive and temperamental. These books suggested that the boundaries between the real and the represented were not particularly fixed, and magic happened when one crossed over. In such spaces as the labyrinth, we cross over; we are really traveling, even if the destination is only symbolic, and this is in an entirely different register than is thinking about traveling or looking at a picture of a place we might wish to travel to. For the real is in this context nothing more or less than what we inhabit bodily. A labyrinth is a symbolic journey or a map of the route to salvation, but it is a map we can really walk on, blurring the difference between map and

figurative monuments in Kelly Ingram Park in Birmingham, which try to draw the viewer back into the world. If the body is the register of the real, then reading with one's feet is real in a way reading with one's eyes alone is not. And sometimes the map *is* the territory.

In medieval churches these labyrinths—once common, but now existing only in a few churches—were sometimes called *chemins à Jerusalem,* "roads to Jerusalem," and the center was Jerusalem or heaven itself. Though the historian of mazes and labyrinths W. H. Matthews cautions that there is no written evidence on their intended use, it is widely thought that they offered the possibility of compressing a pilgrimage into the compact space of a church floor, with the difficulties of spiritual progress represented by the twists and turns. At Grace Cathedral in San Francisco, the labyrinth was commissioned by cathedral canon Lauren Artress in 1991. "Labyrinths," she writes, "are usually in the form of a circle with a meandering but purposeful path, from the edge to the center and back out again. Each one has only one path, and once we make the choice to enter it, the path becomes a metaphor for our journey through life." Since then Artress has started something of a labyrinth cult, which has trained nearly 130 people to present labyrinth workshops and programs called "the theater of enlightenment," even publishing a quarterly newsletter on the labyrinth project (including a few pages hawking labyrinth tote bags, jewelry, and other items). Labyrinths as spiritual devices are proliferating around the country, and garden mazes are also undergoing a revival. In the 1960s and 1970s a very different kind of labyrinth proliferation took place, in the work of artists such as Terry Fox, and in the late 1980s Adrian Fisher became a wildly successful maze designer in Britain, designing and building garden mazes at Blenheim Palace and dozens of other locations.

Labyrinths are not merely Christian devices, though they always represent some kind of journey, sometimes one of initiation, death and rebirth, or salvation, sometimes of courtship. Some seem merely to signify the complexity of any journey, the difficulty of finding or knowing one's way. They were much mentioned by the ancient Greeks, and although the legendary labyrinth of Crete in which the minotaur was imprisoned has never been found and probably never existed, the shape now called the Cretan labyrinth appeared on its coins. Other labyrinths have been found: carved in the rock in Sardinia;

---

tumult of the past. Several works designed by James Drake along a path named Freedom Walk cleared in the stony desert surface in southern Arizona and California; made of mosaic by the Romans. In Scandinavia there are almost five hundred known labyrinths made of stones laid out upon the earth; until the twentieth century, fishermen would walk them before putting out to sea to ensure good catches or favorable winds. In England turf mazes—mazes cut into the earth—were used by young people for erotic games, often in which a boy ran toward a girl at the center, and the twists and turns of the maze seem to symbolize courtship's complexities. The much better known hedge mazes of that country are a later, more aristocratic innovation of the Renaissance garden. Many who've written about mazes and labyrinths distinguish between the two of them. Mazes, including most garden mazes, have many branchings and are made to perplex those who enter, whereas a labyrinth has only one route, and anyone who stays with it can find the paradise of the center and retrace the route to the exit. Another metaphorical moral seems built into these two structures, for the maze offers the confusions of free will without a clear destination, the labyrinth an inflexible route to salvation.

Like the stations of the cross, the labyrinth and maze offer up stories we can walk into to inhabit bodily, stories we trace with our feet as well as our eyes. There is a resemblance not only between these symbolically invested structures but between every path and every story. Part of what makes roads, trails, and paths so unique as built structures is that they cannot be perceived as a whole all at once by a sendentary onlooker. They unfold in time as one travels along them, just as a story does as one listens or reads, and a hairpin turn is like a plot twist, a steep ascent a building of suspense to the view at the summit, a fork in the road an introduction of a new storyline, arrival the end of the story. Just as writing allows one to read the words of someone who is absent, so roads make it possible to trace the route of the absent. Roads are a record of those who have gone before, and to follow them is to follow people who are no longer there—not saints and gods anymore, but shepherds, hunters, engineers, emigrants, peasants to market, or just commuters. Symbolic structures such as labyrinths call attention to the nature of all paths, all journeys.

----

commemorate the brutal police repression of the famous marches in the spring of 1963. In one work, the

----

This is what is behind the special relationship between tale and travel, and, perhaps, the reason why narrative writing is so closely bound up with walking. To write is to carve a new path through the terrain of the imagination, or to point out new features on a familiar route. To read is to travel through that terrain with the author as guide—a guide one may not always agree with or trust, but who can at least be counted upon to take one somewhere. I have often wished that my sentences could be written out as a single line running into the distance so that it would be clear that a sentence is likewise a road and reading is traveling (I did the math once and found the text of one of my books would be four miles long were it rolled out as a single line of words instead of being set in rows on pages, rolled up like thread on a spool). Perhaps those Chinese scrolls one unrolls as one reads preserve something of this sense. The songlines of Australia's aboriginal peoples are the most famous examples conflating landscape and narrative. The songlines are tools of navigation across the deep desert, while the landscape is a mnemonic device for remembering the stories: in other words, the story is a map, the landscape a narrative.

So stories are travels and travels are stories. It is because we imagine life itself as a journey that these symbolic walks and indeed all walks have such resonance. The workings of the mind and the spirit are hard to imagine, as is the nature of time—so we tend to metaphorize all these intangibles as physical objects located in space. Thus our relationship to them becomes physical and spatial: we move toward or away from them. And if time has become space, then the unfolding of time that constitutes a life becomes a journey too, however much or little one travels spatially. Walking and traveling have become central metaphors in thought and speech, so central we hardly notice them. Embedded in English are innumerable movement metaphors: steering straight, moving toward the goal, going for the distance, getting ahead. Things get in our way, set us back, help us find our way, give us a head start or the go-ahead as we approach milestones. We move up in the world, reach a fork in the road, hit our stride, take steps. A person in trouble is a lost soul, out of step, has lost her sense of direction, is facing an uphill struggle or going downhill, through a difficult phase, in circles, even nowhere. And

----

walkway passes between two vertical slabs, from which bronze attack dogs emerge on either side and lunge

----

there are the far more flowery phrases of sayings and songs—the primrose path, the road to ruin, the high road and the low road, easy street, lonely street, and the boulevard of broken dreams. Walking appears in many more common phrases: set the pace, make great strides, a great step forward, keep pace, hit one's stride, toe the line, follow in his footsteps. Psychic and political events are imagined as spatial ones: thus in his final speech Martin Luther King said, "I've been to the mountaintop," to describe a spiritual state, echoing the state Jesus attained after his literal mountain ascent. King's first book was called *Stride to Freedom*, a title echoed more than three decades later by Nelson Mandela's autobiography, *Long Walk to Freedom* (while his former countrywoman Doris Lessing called the second volume of her memoirs *Walking in the Shade*, and then there's Kierkegaard's *Steps on Life's Way* or the literary theorist Umberto Eco's *Six Walks in the Fictional Woods*, in which he describes reading a book as wandering in a forest).

If life itself, the passage of time allotted to us, is described as a journey, it's most often imagined as a journey on foot, a pilgrim's progress across the landscape of personal history. And often, when we imagine ourselves, we imagine ourselves walking; "when she walked the earth" is one way to describe someone's existence, her profession is her "walk of life," an expert is a "walking encyclopedia," and "he walked with God" is the Old Testament's way of describing a state of grace. The image of the walker, alone and active and passing through rather than settled in the world, is a powerful vision of what it means to be human, whether it's a hominid traversing grasslands or a Samuel Beckett character shuffling down a rural road. The metaphor of walking becomes literal again when we really walk. If life is a journey, then when we are actually journeying our lives have become tangible, with goals we can move toward, progress we can see, achievement we can understand, metaphors united with actions. Labyrinths, pilgrimages, mountain climbs, hikes with clear and desirable destinations, all allow us to take our allotted time as a literal journey with spiritual dimensions we can understand through the senses. If journeying and walking are central metaphors, then all journeys, all walks, let us enter the same symbolic space as mazes and rituals do, if not so compellingly.

---

into the pedestrian's space. In another the walkway leads through an opening in a metal wall faced

---

There are many other arenas in which walking and reading are conflated. Just as the church labyrinth had its secular sibling in the garden maze, so the reading of the stations of the cross has its secular equivalent in the sculpture garden. Premodern Europeans were expected to recognize a large cast of characters in painting, sculpture, and stained glass, from the saints—Saint Peter with his key, Saint Lucy with her eyes on a plate—to the graces, cardinal virtues, and deadly sins. Most churches would have some portion of the Bible translated into art; a particularly elaborate cathedral like Chartres would include such features as the Seven Liberal Arts and the Wise and Foolish Virgins as well as scenes from the life of Christ arranged symbolically. Though book literacy was far lower, image literacy was incomparably higher, and the more educated would be able to recognize the gods and mortals from classical mythology as well as Christian iconography. Because the sources were usually literary, each figure represented a story, and these stories could be arranged in various sequences and often were—sequences that could be "read" by strolling past (embodiments such as Liberty or Spring were not narrative, but they might be arranged in a sequence that was, while gods and heroes often appeared in some climactic moment from a familiar tale, making the sculpture equivalent to a film still). Many gardens were sculpture gardens, not in our modern sense of greenery as a sort of picture frame for various individual objects, but as whole spaces that could be read, making the garden as much an intellectual space as the library. Sculptures and, sometimes, architectural elements were arranged in sequences that the viewer-stroller interpreted as she passed, and part of the charm of these gardens is that walking and reading, body and mind, were harmoniously united there.

The cloisters that were part of every monastery and convent sometimes bore elaborate Christian stories. Usually a square arcade around a garden with a central well, pool, or fountain, the cloister was where monks or nuns could walk without leaving the contemplative space of the order. Renaissance gardens had elaborately arranged mythological and historical statues. Because the walker already knew the story, no words need be said, but in the space and time of the walk and its encounter with the statuary, the story was in a sense retold just by being called to mind. This makes the garden a poetic, literary,

by two water cannons, just off the wall, by the walk, are two bronze figures of African Americans,

mythological, and magical space. The great gardens of the Villa d'Este in Tivoli had a series of bas-reliefs that told the tales of Ovid's *Metamorphoses*. A more completely lost narrative was the labyrinth at Versailles, destroyed in 1775. In it were placed, along with a statue of Aesop, figure groupings from his fables, and "each of the speaking characters represented in the fable groups," writes W. H. Matthews, "emitted a jet of water, representing speech, and each group was accompanied by an engraved plate displaying more or less appropriate verses by the poet de Benserade." The labyrinth was thus a three-dimensional anthology in which walking, reading, and looking united into a journey into the fables' morals and meanings. Versailles, the largest of all Europe's formal gardens, had the most complex sculptural program, in which the Aesop maze was only a minor diversion. It organized nearly all its sculptures around the central image of Louis XIV as the Sun King (subsequent additions and subtractions make it hard to decipher now). Seventy sculptors labored that the sculptures, fountains, and very plants would speak to strollers of the power of the king, a power naturalized and endorsed by the imagery of the sun and the classical sun god Apollo, on a scale that made the symbolic not a scale model but a vast expanse of the world. A century later, the celebrated formal garden at Stowe in Buckingham, England, was transformed into a more naturalistic landscape, but its rolling hills and groves were studded with even more pointedly political architectural motifs. The Temple of Ancient Virtue was located near both the ruined Temple of Modern Virtue and, across a pool, the Temple of British Worthies, featuring the poets and statesmen most appealing to the garden's Whig owner. The conjunction deplored the state of the eighteenth-century world while setting up the Whigs as heirs to the noble ancients. Other elements at Stowe were more humorous for those who could read space and symbolism: the hermitage located near the Temple of Venus, for example, pitting asceticism against sensuality. If a narrative is a sequence of related events, then these sculpture gardens made the world into a book by situating these events in real space, far enough apart to be "read" by walking (and made Versailles and Stowe into books of political propaganda). Sometimes what is to be read in the garden is less literal. "A garden path," write the landscape architects Charles W. Moore, William J. Mitchell,

----

a man crumpled to the ground and a woman standing with her back against the imagined force of

----

and William Turnbull, "can become the thread of a plot, connecting moments and incidents into a narrative. The narrative structure might be a simple chain of events with a beginning, middle, and end. It might be embellished with diversions, digressions, and picaresque twists, be accompanied by parallel ways (subplots), or deceptively fork into blind alleys like the alternative scenarios explored in a detective novel." Los Angeles's contribution to this genre is the Walk of the Stars on Hollywood Boulevard, in which tourists read celebrity names as they tread them underfoot.

Sometimes walkers overlay their surroundings with their imaginings and tread truly invented terrain. The American minister and walking enthusiast John Finlay wrote a friend, "You may be interested to know that I have a little game that I play alone: namely, that of walking in some part of the world as many miles as I actually walk here day by day, with the result that I have walked nearly 20,000 miles here in the last six years, which means that I have covered the land part of the earth in a circuit of the globe. I finished last night 2,000 miles since the first of January 1934 and in doing so reached Vancouver from the north." The Nazi architect Albert Speer traversed the world in his imagination while pacing back and forth in his prison yard, like Kierkegaard and his father. The art critic Lucy Lippard found that after her return to Manhattan she could continue to take the daily walks that had been so important a part of her year's residence in rural England "in a kind of out-of-body form—step by step, weather, texture, views, seasons, wildlife encounters."

There is a very practical sense in which to trace even an imaginary route is to trace the spirit or thought of what passed there before. At its most casual, this retracing allows unsought memories of events to return as one encounters the sites of those events. At its most formal it is a means of memorizing. This is the technique of the memory palace, another inheritance from classical Greece widely used until the Renaissance. It was a means of committing quantities of information to memory, an important skill before paper and printing made the written word replace the memory for much storage of rote information. Frances Yates, whose magnificent *Art of Memory* re-

---

the water. Integrated into the pedestrian experience of the park, these monuments invite everyone—

covered the history of this strange technique for our time, describes the workings of the system in detail. "It is not difficult to get hold of the general principles of the mnemonic," she writes. "The first step was to imprint on the memory a series of loci or places. The commonest, though not the only, type of mnemonic place system used was the architectural type. The clearest description of the process is that given by Quintilian. In order to form a series of places in memory, he says, a building is to be remembered, as spacious and various a one as possible, the forecourt, the living room, bedrooms, and parlours, not omitting statues and other ornaments with which the rooms are decorated. The images by which the speech is to be remembered . . . are then placed in imagination in the places which have been memorized in the building. This done, as soon as the memory of the facts requires to be revived, all these places are visited in turn and the various deposits demanded of their custodians. We have to think of the ancient orator as moving in imagination through his memory building whilst he is making his speech, drawing from the memorized places the images that he has placed on them. The method ensures that the points are remembered in the right order, since the order is fixed by the sequence of places in the building."

Memory, like the mind and time, is unimaginable without physical dimensions; to imagine it as a physical place is to make it into a landscape in which its contents are located, and what has location can be approached. That is to say, if memory is imagined as a real space—a place, theater, library—then the act of remembering is imagined as a real act, that is, as a physical act: as walking. The scholarly emphasis is always on the device of the imaginary palace, in which the information was placed room by room, object by object, but the means of retrieving the stored information was walking through the rooms like a visitor in a museum, restoring the objects to consciousness. To walk the same route again can mean to think the same thoughts again, as though thoughts and ideas were indeed fixed objects in a landscape one need only know how to travel through. In this way, walking *is* reading, even when both the walking and reading are imaginary, and the landscape of the memory becomes a text as stable as that to be found in the garden, the labyrinth, or the stations.

But if the book has eclipsed the memory palace as a repository of

----

black or white, young or old—to step for a moment into someone else's shoes.—KIRK SAVAGE   *I*
----

information, it has retained some of its pattern. In other words, if there are walks that resemble books, there are also books that resemble walks and use the "reading" activity of walking to describe a world. The greatest example is Dante's *Divine Comedy*, in which the three realms of the soul after death are explored by Dante, guided by Virgil. It is an unearthly travelogue of sorts, moving past sights and characters steadily, always keeping the pace of a tour. The book is so specific about its geography that many editions contain maps, and Yates suggests that in fact this masterpiece was a memory palace of sorts. Like a vast number of stories before and after, it is a travel story, one in which the movement of the narrative is echoed by the movement of the characters across an imaginary landscape.

---

stride along with calm, with eyes, with shoes, / with fury, with forgetfulness—PABLO NERUDA

Part II

FROM THE GARDEN TO THE *Wild*

Chapter 6

THE PATH OUT OF THE GARDEN

I. TWO WALKERS AND THREE WATERFALLS

Two weeks before the end of the century, a brother and sister went walking across the snow. Both were dark-complexioned, and their friends remarked that you could see their bad posture when they walked, but the resemblance ended there. He was tall, Roman-nosed, calm, while she was small and had fiery eyes that everyone noticed. The first day of their journey, December 17, they had gone twenty-two miles on horseback before they parted with their friend, the horses' owner, and walked another twelve miles to their lodgings, "having walked the last three miles in the dark and two of them over hardfrozen road to the great annoyance of our feet and ancles. Next morning the earth was thinly covered with snow, enough to make the road soft and prevent its being slippery." As they had the day before, the travelers turned aside to see a waterfall amid this mountainous landscape. "Twas a keen frosty morning," the brother went on in his Christmas Eve letter, "showers of snow threatening us but the sun bright and active; we had a task of twenty one miles to perform in a short winter's day. . . . On a nearer approach the water seemed to fall down a tall arch or rather nitch which had shaped itself by insensible moulderings in the wall of an old castle. We left this spot with reluctance but highly exhilarated."

In the afternoon they came upon another waterfall, whose water seemed to turn to snow as it fell amid the ice. He continued, "The stream shot from between the rows of icicles in irregular fits of strength and with a body of water that momently varied. Sometimes it threw itself into the basin in one continued curve, sometimes it was interrupted almost midway in its fall and, being blown towards us, part of the water fell at no great distance from our feet like the heaviest thunder shower. In such a situation you have at every moment a feeling of the presence of the sky. Above the highest point of the waterfall large fleecy clouds drove over our heads and the sky appeared of a blue more than usually brilliant." After the detour to the waterfall, they walked the next ten miles in two and a quarter hours "thanks to the wind that drove behind us and the good road," and he seemed to relish their prowess in walking almost as much as the scenery. Seven more miles took them to their next resting spot, and in the morning they walked into Kendal, the gateway to the Lake District, where they had come to live.

The century they were approaching as fast as their new home was the nineteenth century, and the home was a cottage on the outskirts of the small lakeside village of Grasmere; the two vigorous walkers themselves were, as many may have already guessed, William and Dorothy Wordsworth. What they did on those four days across the Pennine Mountains of northern England, what they had done and would do as walkers, was extraordinary. What exactly makes it so is hard to pin down. People had traveled by foot much farther and in far worse conditions before. People had begun, by the time of the poet's and his sister's birth nearly thirty years before, to admire some of the wildest features of the British countryside—mountains, cliffs, moors, storms, and the sea, as well as waterfalls. In France and Switzerland a few people had begun to climb mountains—the summit of Mont Blanc, Europe's highest peak, had first been reached fourteen years earlier. Wordsworth and his companions are said to have made walking into something else, something new, and thereby to have founded the whole lineage of those who walk for its own sake and for the pleasure of being in the landscape, from which so much has sprung. Most who have written about this first generation of Romantics propose that they themselves introduced walking as a cultural act, as a part of aesthetic experience.

Christopher Morley wrote in 1917, "I have always fancied that walking as a fine art was not much practiced before the Eighteenth Century. We know from Ambassador Jussurand's famous book how

---
went from opera, park, assembly, play / To morning walks, and prayers three hours a day . . .
---

many wayfarers were abroad on the roads in the Fourteenth Century, but none of these were abroad for the pleasures of moving meditation and scenery. . . . Generally speaking, it is true that cross-country walks for the pure delight of rhythmically placing one foot before the other were rare before Wordsworth. I always think of him as one of the first to employ his legs as an instrument of philosophy." Morley is not far off the mark, in his first sentence, though much of the eighteenth century had passed before Wordsworth was born in 1770. But then he conflates walking as a fine art with cross-country walking, which is where the confusion slips in. Since Morley, the subject of walking and English culture has been taken up in three books, all of which go further in proposing that it was in the late eighteenth century, when Wordsworth and his peers set out afoot, that this walking began.

Morris Marples's delightful 1959 *Shanks's Pony: A Study of Walking*, Anne D. Wallace's 1993 *Walking, Literature, and English Culture*, and Robin Jarvis's 1998 *Romantic Poetry and Pedestrian Travel* all use as their demonstration case the German minister Carl Moritz. During his walk across England in 1782, Moritz often found himself scorned and ejected by innkeepers and their employees, while coachmen and carters continually asked him if he wanted a ride. He concluded that it was his mode of travel that made him seem out of place to those he encountered: "A traveller on foot in this country seems to be considered as a sort of wild man, or an out-of-the-way-being, who is stared at, pitied, suspected, and shunned by everybody that meets him." Reading his book, one is moved to speculate on whether his dress, manner, or accent disconcerted the people he encountered, rather than his walking. But his explanation is largely accepted by the three who cite him.

Travel itself was enormously difficult until the late eighteenth century in England. The roads were atrocious and plagued by highwaymen and their pedestrian equivalents, footpads. Those who could afford to went by horse or by coach, carriage, or at worst, wagon, sometimes with weapons; walking along the public roads often signified that one was either a pauper or a footpad, at least until the 1770s, when various intellectuals and eccentrics began to walk there for pleasure. By the late eighteenth century, the roads were improving in both quality and safety, and walking was becoming a more

---

—ALEXANDER POPE, "EPISTLE TO MISS BLOUNT"   *To Hyde-park they repaired, sir Philip boasting*

---

respectable mode of travel. On the cusp of the next century, the Wordsworths were having a splendid time walking not only roads but fells and byways; fear of crime and denigration seem to be the furthest things from their mind as they admired the view and enjoyed their own powers of walking in weather that would keep most people huddled indoors.

They had visited the Lake District six years before their midwinter walk. "I walked with my brother at my side, from Kendal to Grasmere, eighteen miles, and afterwards from Grasmere to Keswick, fifteen miles, through the most delightful country that ever was seen," wrote Dorothy in the initial flush of pleasure after that excursion in 1794, and then she wrote defensively to her aunt, "I cannot pass unnoticed that part of your letter in which you speak of my 'rambling about the country on foot.' So far from considering this as a matter of condemnation, I rather thought it would have given my friends pleasure to hear that I had courage to make use of the strength with which nature has endowed me, when it not only procured me infinitely more pleasure than I should have received from sitting in a post chaise—but was also the means of saving me at least thirty shillings." If we take Dorothy Wordsworth in 1794 rather than Carl Moritz in 1782 as our witness, we find walking cross-country was nothing worse than unladylike and unconventional.

Though Wordsworth is in some sense the founding father (and therefore Dorothy the aunt) of a modern taste that has done much to shape the more pleasant parts of our world and the imaginations of those in it, he himself was heir to a long tradition, and so it is more accurate to see him as a transformer, a fulcrum, a catalyst for the history of walking in the landscape. His precursors, it is true, had not walked much on the public roads (and for the most part, neither do his modern descendants, since cars have made roads dangerous and miserable again). Though many traveled on foot out of necessity before him, few did so for pleasure, and these historians therefore conclude that walking for pleasure was a new phenomenon. In fact, walking had already become an important activity, though not as travel. Few of Wordsworth's pedestrian antecedents are to be found traveling along the public roads, but many of them were strolling in gardens and parks.

---

all the way they walked, of the superior strength of his head. Clarence protested that his own was stronger

---

II. THE GARDEN PATH

Halfway through the nineteenth century, Thoreau wrote, "When we walk, we naturally go to the fields and woods: what would become of us, if we walked only in a garden or a mall?" For Thoreau, the desire to walk in the unaltered landscape no longer seemed to have a history, but to be natural—if nature means the timeless truth we have found, not the historic specific we have made. Though many nowadays go to the fields and woods to walk, the desire to do so is largely the result of three centuries of cultivating certain beliefs, tastes, and values. Before that, the privileged seeking pleasure and aesthetic experience did indeed walk only in a garden or a mall. The taste for nature already entrenched in Thoreau's time and magnified in our own has a peculiar history, one that has made nature itself cultural. To understand why people chose to walk out in certain landscapes with certain agendas, one must first understand how that taste was formed in and launched from English gardens.

We tend to consider the foundations of our culture to be natural, but every foundation had builders and an origin—which is to say that it was a creative construction, not a biological inevitability. Just as a twelfth-century cultural revolution ushered in romantic love as first a literary subject and then a way of experiencing the world, so the eighteenth century created a taste for nature without which William and Dorothy Wordsworth would not have chosen to walk long distances in midwinter and to detour from their already arduous course to admire waterfalls. This is not to say that no one felt a tender passion or admired a body of water before these successive revolutions; it is instead to say that a cultural framework arose that would inculcate such tendencies in the wider public, give them certain conventional avenues of expression, attribute to them certain redemptive values, and alter the surrounding world to enhance those tendencies. It is impossible to overemphasize how profound is the effect of this revolution on the taste for nature and practice of walking. It reshaped both the intellectual world and the physical one, sending populations of travelers to hitherto obscure destinations, creating innumerable parks, preserves, trails, guides, clubs, and organizations and a vast body of art and literature with almost no precedent before the eighteenth century.

---

than any man's in England, and observed, that at this instant he walked better than any person in

---

Some influences stand out like a landmark and leave a traceable legacy with evident heirs. But the most profound influences soak into the cultural landscape like rain and nourish everyday consciousness. Such an influence is likely to go undetected, for it comes to seem the way things have always been, the natural or even the only way to look at the world. This is the influence Shelley had in mind when he wrote, "Poets are the unacknowledged legislators of the world." Such an influence is the Romantic taste for landscape, for wild places, for simplicity, for nature as an ideal, for walking in the landscape as the consummation of a relationship with such places and an expression of the desire for simplicity, purity, solitude. Which is to say, walking is natural, or rather part of natural history, but choosing to walk in the landscape as a contemplative, spiritual, or aesthetic experience has a specific cultural ancestry. This is the history that had already become naturalized for Thoreau and that took walkers farther and farther afield—for the changing history of walking is inseparable from the changing taste in places in which to walk.

The real reason Wordsworth and his peers seem to be the founders rather than the transformers of a tradition of walking for aesthetic reasons is because the walks that preceded theirs are so unremarkable. In fact these short walks in safe places are only incidental to the histories of architecture and gardens; they have no literature of their own, only mentions in novels, journals, and letters. The core of their history is concealed within another history, of making places to walk, places that became larger and more culturally significant as the eighteenth century wore on. It is also the history of a radical transformation of taste, from the formal and highly structured to the informal and naturalistic. It seems, in its origins, a trivial history of the idle aristocracy and their architecture, but in its results it created some of the most subversive and delightful places and practices in the contemporary world. The taste for walking and landscape became a kind of Trojan horse that would eventually democratize many arenas and in the twentieth century literally bring down the barriers around aristocratic estates.

The practice of walking can be traced through places. By the sixteenth century, as castles were beginning to turn into palaces and mansions, galleries—long narrow rooms like corridors, though often

---

company, sir Philip Baddely not excepted. Now, sir Philip Baddely was a noted pedestrian, and he

leading nowhere—often began to be part of the design. They were used for exercise indoors. "Sixteenth-century doctors stressed the importance of daily walking to preserve health, and galleries made exercise possible when the weather would otherwise have prevented it," writes Mark Girouard in his history of the country house. (The gallery eventually became a place for displaying paintings, and though museum galleries are still a place where people stroll, the strolling is no longer the point.) Queen Elizabeth added a raised terrace walk to Windsor Castle and walked there for an hour before dinner on every day that was not too windy. Walking was still more for health than for pleasure, though gardens were also being used for walking, and some kind of pleasure must have accrued there. But the taste for landscape was still fairly limited. On October 11, 1660, Samuel Pepys went walking in St. James's Park after dinner, but he only notes the water pumps at work there. Two years later, on May 21, 1662, he writes that he and his wife went walking in White Hall Garden, but he seemed most interested in the lingerie of the king's mistress in the privy garden, evidently hung there to dry. It was society that interested him, not nature, and landscape was not yet a significant subject for British painting and poetry, as it was to become. Until the surroundings became important, the walk was just movement, not experience.

A revolution was under way, however, in gardens. The medieval garden had been surrounded by high walls, in part for security in unstable times. In pictures of these gardens, the occupants most often sit or recline, listening to music or conversing (the enclosed garden had been, since the Song of Songs, a metaphor for the female body, and at least since the rise of the courtly love tradition, the site of much courtship and flirtation). Flowers, herbs, fruit-bearing trees, fountains, and musical instruments made them places that speak to all the senses, and the world outside this voluptuous sanctum seemed to provide more than enough exercise, since medieval nobles were still bodily involved in military and household matters. As the world became safer and the aristocratic residence became more a palace than a fortress, the gardens of Europe began to expand. Flowers and fruit were disappearing from the gardens; it was the eye to which these expanded realms appealed. The Renaissance garden was a place in which one could take a walk as well as sit, and the Baroque garden

---

immediately challenged our hero to walk with him for any money he pleased.—MARIA EDGEWORTH,

grew vast. Just as walking was exercise for those who need no longer work, so these vast gardens were cultivated landscapes that need no longer produce anything more than mental, physical, and social stimulation for walkers.

Were the Baroque garden not so ostentatious a display of wealth and power, its abstractness could be called austere. Trees and hedges were forced into squares and cones; paths, avenues, and walks were laid out as straight lines; water was pumped into fountains or poured into geometrical pools. A platonic order, a superimposition of the ideal on the messy material of the real, triumphed. Such gardens extended the geometry and symmetry of architecture into the organic world. But they still provided opportunities for informal and private behavior: throughout their history, one of the major functions of aristocratic gardens was to give people a place to retreat from the household into contemplation or private conversation. In England, William and Mary added new gardens to Hampton Court in 1699, gardens in which one could walk for a mile before reaching the wall. Walks, or paths, were becoming increasingly important parts of gardens, and they are indirect evidence of the increasing popularity of walking (in this context, "a walk" meant a path broad enough for two to walk abreast; it could be called a conversational route). English traveler and chronicler Celia Fiennes wrote of a garden she visited near the beginning of the eighteenth century, "There is gravel walks and grass and close walks, there is one walk all the length of the Garden called the Crooked Walke, of grass well cut and rowled, it is indented in and out in corners and so is the wall, which makes you thinke you are at the end of the walke several times before you are." But the walls of the garden were disappearing, and the distinction between it and the landscape beyond became harder and harder to find. Renaissance Italian gardens had been built by preference upon slopes that gave views of the countryside beyond, connecting the garden to the world, but French and English gardens seldom had such settings. The line of sight only extended to the garden wall, then eventually through a variety of openings in the garden wall.

When the ha-ha came into being in the early decades of the eighteenth century, the walls came down in Britain. A ditch relatively invisible from any distance, the ha-ha—so named because strollers

---

BELINDA   *In the evening walked alone down to the Lake by the side of Crow-Park after sun-set and saw*

were said to exclaim "Ha ha!" in surprise when they came upon it—
provided an invisible barrier that allowed the garden's inhabitants to
gaze into the distance uninterrupted. Where the eye went, the walker
would soon follow. Most English estates consisted of a series of increasingly controlled spaces: the park, the garden, and the house.
Originally hunting preserves, parks remained as a kind of buffer zone
between the leisure classes and the agricultural land and workers
around them and often provided timber and grazing space. The
garden was typically a much smaller space surrounding the house.
Susan Lasdun, in her history of these parks, writes of the straight avenues of trees planted in parks and gardens in the seventeenth
century: "These avenues provided the shade and shelter for walks
which, having been made fashionable by Charles II, were now becoming de rigueur in parks. . . . Certainly the liking for air and exercise was already considered an 'English' taste. Walks were now laid
out by private owners in their country parks, and walking became as
much a part of the pleasure of a park as hunting, driving and riding.
The walks themselves were made increasingly interesting, with aesthetic considerations developing from the simple static vista from a
window or terrace, to something that took account of a more mobile
point of view . . . The walker in fact made a circuit, and in the eighteenth century this was to become the standard manner for viewing
gardens and parks. The days when it was only safe to walk on the
castle terrace—the allure—had long since passed."

The formal garden, with its patterns made of clipped hedges,
geometric pools, and trees in orderly ranks, had suggested that nature
was a chaos on which men imposed order (though starting in Italy in
the Renaissance, paintings of unaltered landscapes, if not the unaltered land itself, were appreciated). In England the garden would
become less and less formal as the eighteenth century progressed, and
this idea of naturalistic landscaping that would be called the *jardin
anglais*, the English garden, or the landscape garden is one of the great
English contributions to Western culture. As the visual barrier that
separated it from its surroundings vanished, the design of the garden
became less distinctly separate too. In 1709 Anthony Ashley Cooper,
earl of Shaftesbury, had effused, "O glorious Nature! Supremely Fair,
and sovereignly Good!. . . I shall no longer resist the Passion growing

---

the solemn colouring of night draw on, the last gleam of sunshine fading away on the hilltops, the deep

in me for Things of a natural kind where neither Art nor the Conceit or Caprice of Man has spoiled their genuine Order, by breaking in upon that primitive State. Even the rude Rocks, the mossy Caverns, the irregular unwrought Grottos and broken Falls of waters, with all the horrid Graces of the Wilderness itself, as representing Nature more, will be the more engaging, and appear with a Magnificence beyond the formal Mockery of Princely Gardens." Rhetoric raced ahead of practice. It would be many more decades before princely gardens gave way to wilderness. But Shaftesbury's optimistic view of nature as inherently good joined with the optimism that men could appropriate, improve upon, or invent nature in their gardens.

"Poetry, Painting, and Gardening, or the science of Landscape, will forever by men of Taste be deemed Three Sisters, or the Three New Graces who dress and adorn nature," famously declared Horace Walpole, the wealthy aesthete who did much to inculcate romantic tastes in his peers. One of the premises of this declaration is that gardening is as much an art as the more traditionally respected practices of poets and painters, and the period was a sort of golden age for attention to gardens—or a kind of age of incubation, in which the taste for nature was hatched out of those gardens, poems, and paintings. Another premise is that nature needs to be dressed and adorned, at least in the garden, and paintings suggested some of the ways in which this could be done. Among the influences on the emerging English landscape garden were the seventeenth-century Italian landscapes of Claude Lorrain, Nicholas Poussin, and Salvator Rosa, with their rolling ground stretching toward far horizons, their clumps of feathery trees framing the distance, their serene bodies of water and their classical buildings and ruins (and, in Rosa's case, the cliffs, torrents, and bandits that made him the most gothic of the three). Pillared temples and Palladian bridges were added to make English gardens resemble the Italian campagna of these paintings and to suggest that England was heir to Rome's virtues and beauties. "All gardening is landscape painting," said Alexander Pope, and people were learning to look at landscape in gardens as they had learned to look at landscape in paintings.

And though architectural items—ruins, temples, bridges, obelisks—

---

serene of the asters, and the long shadows of the mountains thrown across them, till they nearly touched

continued to be sprinkled over gardens for many decades, the subject of gardens was becoming nature itself—but a very specific version of nature, nature as a visual spectacle of plants and water and space, a serene thing to be contemplated serenely. Unlike the formal garden and the painting, which had a single ideal point of view from which they could be regarded, the English landscape garden "asked to be explored, its surprises and unsuspected corners to be discovered on foot," writes garden historian John Dixon Hunt. Carolyn Bermingham adds, in her history of the relations between class and landscape, "Whereas the French formal garden was based on a single axial view from the house, the English garden was a series of multiple oblique views that were meant to be experienced while one walked through it." To use anachronistic terms, the garden was becoming more cinematic than pictorial; it was designed to be experienced in motion as a series of compositions dissolving into each other rather than as a static picture. It was now designed aesthetically as well as practically for walkers, and walking and looking were beginning to become linked pleasures.

There were other factors in the increasing naturalization of the garden. Perhaps the most important was the equation of the landscape garden with English liberty. The English aristocrats cultivating a taste for nature were, in a sense, politically positioning themselves and their social order as natural, in contrast to French artifice. Thus their pursuit of country pastimes, their penchant for portraits of themselves in the landscape, their creation of naturalistic gardens, their cultivation of a taste for landscape, all had a political subtext, as Bermingham has so brilliantly pointed out. Yet other influences include reports of Chinese gardens, in which the paths and waterways were sinuous and winding, the overall effect celebrating rather than subduing natural complexity. Neither the early chinoiseries nor the imitations of nature bore much resemblance to their originals, but the intent was there, and evolving. Finally, this changing taste manifests an extraordinary confidence. The formal, enclosed garden and the castle are corollaries to a dangerous world from which one needs to be protected literally and aesthetically. As the walls come down, the garden proposes that there is already an order in nature and that it is

---

the bithermost shore. At distance heard the murmur of many waterfalls not audible in the day-time.

---

in harmony with the "natural" society enjoying such gardens. The growing taste for ruins, mountains, torrents, for situations provoking fear and melancholy, and for artwork about all these things suggests that life had become so placidly pleasant for England's privileged that they could bring back as entertainment the terrors people had once strived so hard to banish. Too, private experience and informal art were blooming elsewhere, notably in the rise of the novel.

The exemplary garden for this evolution is Stowe in Buckingham. Stowe itself went through most of the phases of the English garden in the eighteenth century and stands now as a kind of lexicon of eighteenth-century gardening, from its temples, grotto, hermitage, and bridges to its lake and landscaping. It had some of the earliest chinoiserie and Gothic-revival architecture in England. Its owner, Viscount Cobham, had replaced the formal "parlour garden" made in 1680 with a far larger formal garden that he slowly revised and erased as the new century advanced. First the Elysian Fields, with their Temples of British Worthies and of Ancient and Modern Virtue, mentioned in the preceding chapter, were transformed into something with softer, more undulating lines, and the rest of the garden eventually followed. Straight walks became serpentine, and their walkers no longer promenaded but wandered. Christopher Hussey describes Stowe, the political capital of the Whigs, as transforming politics into garden architecture, loosening the formal landscape design "into harmony with the age's humanism, its faith in disciplined freedom, its respect for natural qualities, its belief in the individual, whether man or tree, and its hatred of tyranny whether in politics or plantations." Most of the great landscape architects of the age worked for Cobham, and many of the great poets and writers were among his guests. And the gardens continually expanded, annexing dozens of acres at a time. "Within thirty years," summarizes one of the garden's historians, "his taste had moved from a preoccupation with regular arrangements of terraced lawns, statues and straight paths . . . to an essay in three-dimensional landscape painting, the creation of an ideal landscape."

Celebrated in many poems, pictures, and journals, Stowe was a central site in the cultural foment of the era, both as a subject and a retreat. "O lead me to the wide-extended walks / The fair majestic paradise of Stowe . . . While there with th' enchanted round I walk /

Wished for the moon, but she was dark to me and silent, hid in her vacant interlunar cave.—THOMAS

The regulated wild," wrote James Thomson, a guest during most of 1734 and 1735, in the "Autumn" section of his poem *The Seasons*. This enormously successful poem with its blank verse describing the changing year and minor dramas in the landscape probably did more than any other literary work to inculcate a taste for scenery; in the nineteenth century J.M.W. Turner was still appending big chunks of the poem to his paintings. Pope wrote at length of Stowe's glories too, and in a letter described a typical day at Stowe in the 1730s: "Everyone takes a different way, and wanders about till we meet at noon." Walpole visited Cobham's heir at Stowe in 1770. After breakfast, the party spent the day walking in the gardens "or drove about it in cabriolets, till it was time to dress" for dinner. It had become enormous, a place it takes a whole day to explore on foot, and no clear boundary but a ha-ha separates it from the surrounding countryside.

That year, the Gothic architect Sanderson Miller walked there with various people, including Lancelot "Capability" Brown, the landscape architect who was to complete the revolution in garden design with his unadorned expanses of water, trees, and grass. Brown created the Grecian Valley in Stowe, the largest and plainest stretch of the garden (and though it looks wholly natural, the valley itself was dug by hand by many laborers, whose views of landscape gardening do not survive). The Brownian garden, having largely banished sculpture and architecture, no longer commemorated human history and politics. Nature was no longer a setting, but the subject. And the walkers in such a garden were no longer being steered toward ready-made reflections on virtue or Virgil; they were free to think their own thoughts as they followed the meandering paths (though those thoughts might well be about nature, or rather Nature, as taught by myriad texts). From being an authoritarian, public, and essentially architectural space, the garden was becoming a private and solitary wilderness.

Not everyone was ready to accept the landscape garden as realized by Brown. Sir Joshua Reynolds, president of the Royal Academy, wrote, "Gardening, as far as Gardening is an Art, or entitled to the appellation, is a deviation from nature; for if the true taste consists, as many hold, in banishing every appearance of Art, or any traces of the footsteps of man, it would then no longer be a garden." Reynolds was

---

GRAY, "JOURNAL IN THE LAKES"   *On the day after Mary Wollstonecraft first made love to William*

onto something. The garden, in the course of becoming more and more indistinguishable from the surrounding landscape, had become unnecessary—Walpole had said of the landscape architect William Kent that he had "leapt the fence and saw all nature was a garden." If a garden was nothing more than a visually pleasing space in which to wander, then gardens could be found rather than made, and the tradition of the garden walk could expand to become the tourist's excursion. Rather than looking at the work of man, the scenic stroller could look at the works of nature, and to look at nature as a work of art completed a momentous revolution. In Shaftesbury's terms, princely gardens had finally given way to wilderness; the nonhuman world had become a fit subject for aesthetic contemplation.

The aristocratic garden had begun as part of the fortified castle, and slowly its boundaries had melted away; the melting of the garden into the world is a mark of how much safer England had become (and to a lesser degree, much of western Europe, where the fashion for the English garden soon caught on). Since about 1770, England had undergone a "transportation revolution" of improved roads, decreased roadside crime, and cheaper fares. The very nature of travel changed. Before the mid-eighteenth century, travel accounts have little to say about the land between religious or cultural landmarks. Afterward, an entirely new way of travel arose. In pilgrimage and practical travel, the space between home and destination had been an inconvenience or an ordeal. When this space became scenery, travel became an end in itself, an expansion of the garden stroll. That is to say, the experiences along the way could replace destinations as the purpose of travel. And if the whole landscape was the destination, one arrived as soon as one set out in this world that could be looked at as a garden or a painting. Walking had long been recreational, but travel had joined it, and it was only a matter of time before traveling on foot would itself become a widespread part of the pleasures of scenic travel, its slowness finally a virtue. The point at which a poor poet and his sister might travel across a snowy countryside for the pleasure of looking and walking was drawing near.

Afterward, Wordsworth himself was moved to write a guidebook to the Lake District, in which he summarized the history traced here.

Godwin she retreated in concern and self-doubt: "Consider what has passed as a fever of your

---

"Within the last sixty years," he wrote in 1810, "a practice, denominated Ornamental Gardening, was at that time becoming prevalent over England. In union with an admiration of this art, and in some instances in opposition to it, had been generated a relish for select parts of natural scenery: and Travellers, instead of confining their observations to Towns, Manufactories, or Mines, began (a thing till then unheard of) to wander over the island in search of sequestered spots, distinguished . . . for the sublimity or beauty of the forms of Nature there to be seen."

III. Inventing Scenic Tourism

The unhappy German traveler Carl Moritz, who felt rejected at many points on his pedestrian journey, in fact encountered a plethora of walkers, though neither he nor his modern readers made much of them. He takes little note of the many people he saw walking from Greenwich to London, but he does say of London's St. James's Park that what "greatly compensates for the mediocrity of this park, is the astonishing number of people who, towards evening, in fine weather, resort here; our finest walks are never so full even in the midst of summer. The exquisite pleasure of mixing freely with such a concourse of people, who are for the most part well dressed and handsome, I have experienced this evening for the first time." Moritz, in fact, is suggesting that walking is *more* genteel a pastime, or more public a genteel pastime, in England than in Germany, even if traveling on foot is not (he is also revealing that he is a snob, which may be why he resented road-walking's plebeian status). During his time in London, he also visited Ranelagh and Vauxhall Gardens. Cousins of the country fair and the modern amusement park, these popular sites offered music, spectacles, strolls, and refreshments in a garden atmosphere, and both the gentry and the middle class flocked there for evening amusements. Like modern strollers in Latin American plazas and parks or any carnival or mall now, they were there to look at each other as well as the scenery, and the scenery was often augmented with orchestras, pantomimes, refreshments, and other diversions. Social promenading was another aspect of a thriving culture of walking, whose more solitary moments developed in the private garden and park. "The People of London are as fond of walking as our friends at Pekin of riding," wrote Oliver Goldsmith in the guise of a Chinese visitor describing Vauxhall. "One of the principal entertainments of the citizens here in summer is to repair about nightfall to a garden not far from town, where they walk about, shew their best cloaths and best faces, and listen to a concert provided for the occasion."

Another significant aspect of Moritz's travels was his visit to the famous cavern in the Peak District of Derbyshire in northern England, not far from the Lakes. Significantly, there was already a guide in place to collect a fee and show him its marvels. Scenic tourism was coming into existence in the Peak District, the Lake District, Wales, and Scotland. And just as the development of the English landscape garden had been surrounded by a flurry of descriptive poems and epistles, so the growth of tourism was encouraged and informed by guidebooks. Like modern guidebooks and travel narratives, these tell of what is to be seen and where to find it. Some of them—notably the work of the clergyman William Gilpin—also tell *how* to see. A taste for landscape was a sign of refinement, and those wishing to become refined took instruction in landscape connoisseurship. One suspects that the contemporaries who made Gilpin so influential a writer consulted him much the way later generations consulted guides on which fork to use or how to thank a hostess, for Gilpin wrote when the middle class was acquiring the hitherto aristocratic taste for landscape. A landscape garden was a luxury that only a few could create or use, but the unaltered landscape was available to virtually everyone, and more and more middle-class people could travel to enjoy it as the roads became safer and smoother and transportation became cheaper. A taste for landscape was something to be learned, and Gilpin was many people's guide.

"She would have every book that tells her how to admire an old twisted tree," remarks Edward of the romantic Marianne in Jane Austen's *Sense and Sensibility*. Critic John Barrell writes that "there is a sense in which, in late eighteenth-century England, one can say that the simple contemplation of landscape, quite apart from its expression in painting, writing or whatever, came to be regarded as an important

----

day roaming through the valley. I stood beside the sources of the Arveiron, which take their rise in a

----

pursuit for the cultivated and almost in itself the practice of an art. To display a correct taste in landscape was a valuable social accomplishment quite as much as to sing well, or to compose a polite letter. The heroines of a number of late eighteenth-century novels are made to display this taste with an almost ostentatious virtuosity, and not only the simple fact of having a taste for landscape or not, but also variations of taste within the general one, are regarded by some novelists as legitimate indications of differences in character." Marianne Dashwood is asserting her romanticism with her taste for old twisted trees, though she apologizes for the fashionability of the taste: "It is very true . . . that admiration of landscape scenery is become a mere jargon. Everybody pretends to feel and tries to describe with the taste and elegance of him who first defined what picturesque beauty was." She too is speaking of Gilpin, who brought into common usage the word *picturesque*, which originally meant any landscape that resembled or could be perceived as a picture and eventually came to mean a wild, gnarled, rough, intricate kind of landscape.

For Gilpin was teaching people to look at landscape as pictures. Nowadays, his books give a sense of what a heady new pastime looking at landscape was and how much assistance was required. Gilpin tells his readers what to look for and how to frame it in the imagination. Of Scotland, for example, he declares, "Were it not for this general deficiency of objects, particularly of wood, in the Scotch views, I have no doubt but that they would rival those of Italy. The grand outlines are all laid in; a little finishing is all we want"—which is to say that the new subject of Scottish terrain can be understood through comparison to both art and the already hallowed landscapes of Italy. He wrote guidebooks to many parts of England, notably the Lake District, as well as to Wales and Scotland, enumerating the proper sites to visit. Others joined in; Richard Payne Knight wrote, in his abominable but influential 1794 *Landscape: A Didactic Poem in Three Books*, "Let us learn, in real scenes, to trace / The true ingredients of the painter's grace."

Like the taste for landscape, the emphasis on the pictorial and the existence of scenic tourism seem unremarkable things to present-day readers, and yet they were all invented in the eighteenth century. The poet Thomas Gray's celebrated Lake District tour of 1769 came two

---

glacier, that with slow pace is advancing down from the summit of the hills to barricade the valley. . . .

years after the first tourist came specifically to admire the scenery and write about it, and Gray wrote about it too. By the end of the century the Lakes were an established tourist destination, as they have remained since, thanks to Gilpin, Wordsworth—and Napoleon: English travelers who once might have gone abroad began to travel around their own island during the turmoil of the French Revolution and Napoleonic wars. Tourists traveled by coach, then train (and eventually car and plane). They read their guidebooks. They looked at their landscapes. They bought souvenirs. And when they arrived, they walked. Originally, the walking seems to have been incidental, part of the process of moving around to find the best view. But by the turn of the century walking was a central part of some touristic ventures, and walking tours and mountain climbing were coming into being.

IV. Mud on Her Petticoat

Though Jane Austen famously ignored the Napoleonic Wars in her novels, she pointedly addressed other topical subjects. In *Northanger Abbey* she mocked the current taste for the gothic novel, with its macabre and unlikely thrills, and in *Sense and Sensibility* she was almost as sarcastic about Marianne Dashwood's romantic views on love and landscape. Later in her life, she seems to have accepted the cult of landscape far more, and in her late novel, *Mansfield Park*, she more than once equates the heroine's sensitivity to natural beauty with her moral virtues. Her novels with their genteel young women in rural circumstances are also a wonderful index of the uses of the walk at the end of the eighteenth century and beginning of the nineteenth, and none so much as *Pride and Prejudice*. A tour of Elizabeth Bennet's walks will close our inspection of the circumstances in which William and Dorothy Wordsworth set out for Grasmere in December of 1799 (and here it should be noted that though *Pride and Prejudice* was published in 1813, the first version was composed in 1799). Austen was their peer, and she lets us glimpse the staid world they walked out on.

Walks are everywhere in *Pride and Prejudice*. The heroine walks on every possible occasion and in every location, and many of the crucial

----

These sublime and magnificent scenes afforded me the greatest consolation that I was capable of receiving.

----

encounters and conversations in the book take place while two characters are walking together. The very incidental role of the walks in-
dicates how much a part of the fabric of everyday life walking was for
such people as Austen's genteel characters. Throughout the eighteenth century and into the nineteenth in England, walking was a
particularly feminine pursuit—"They were country ladies, and of
course fond of the country lady's amusement, walking," wrote Dorothy
Wordsworth in a letter in 1792. It was something to do. In the writings
of men we find much about designing and admiring gardens, but it is
in the letters and novels of women that we most often find people
actually walking in them, perhaps because they address more minute
daily life, or perhaps because Englishwomen—or, rather, ladies—had
so few other activities open to them. Between social functions Elizabeth Bennet, the heroine of *Pride and Prejudice*, reads copiously, writes
letters, sews a little, plays the piano passably, and walks.

Not long after the novel opens, Jane Bennet catches a cold riding
to Netherfields, the house of her suitor Mr. Bingley, and her sister
Elizabeth walks over to nurse her. Going by foot is in part an act of
necessity, since she is "no horse-woman," and only one horse, rather
than a pair to pull the carriage, is available. But the bold verve that
makes her so charming a heroine also makes her an avid walker—"I
do not wish to avoid the walk. The distance is nothing when one has
a motive; only three miles"—and the walk is the first major demonstration of her unconventionality. Though not going nearly as far as
did Dorothy Wordsworth when reprimanded by her aunt, Elizabeth
is likewise walking beyond the bounds of propriety for women of her
class, and the characters at Mr. Bingley's house have much to say
about it. The transgression seems to be both that she went out into
the world alone, and that she turned the idyll of the genteel walk into
something utilitarian. "That she should have walked three miles so
early in the day, in such dirty weather, and by herself, was almost in-
credible to Mrs. Hurst and Miss Bingley; and Elizabeth was convinced
that they held her in contempt for it." When she is out of earshot
caring for her sister, who has become seriously ill, they expatiate on
the mud on her petticoat and her "abominable sort of conceited independence, a most country town indifference to decorum." Mr. Bingley,
on the other hand, remarks that the unorthodox excursion "shews an

---

They elevated me from all littleness of feeling, and although they did not remove my grief, they subdued

---

affection for her sister that is very pleasing," and Mr. Darcy notes that it has "brightened" her eyes.

Soon afterward, while Jane and Elizabeth are marooned at this worldly house, its inhabitants demonstrate the correct sort of walking— within the bounds of both garden shrubbery and society. Miss Bingley is still railing to Mr. Darcy about Elizabeth. "At that moment they were met from another walk, by Mrs. Hurst and Elizabeth herself." Mrs. Hurst takes Mr. Darcy's disengaged arm and leaves Elizabeth to walk alone. "Mr. Darcy felt their rudeness and immediately said,—

"'This walk is not wide enough for our party. We had better go into the avenue.'

"But Elizabeth, who had not the least inclination to remain with them, laughingly answered—

"'No, no; stay where you are.—You are charmingly group'd, and appear to uncommon advantage. The picturesque would be spoilt by admitting a fourth. Good bye.'"

They have castigated her cross-country walk across the boundaries of decorum; she is mocking their garden propriety by suggesting that they have become part of the garden's array of aesthetic objects, objects that she can contemplate as impersonally as trees and water. That evening Miss Bingley strolls about the narrower confines of the drawing room, where all the Netherfields characters but Jane are gathered. "Her figure was elegant, and she walked well," says Austen. The acuity of idle people about each other's conduct extended to critiques of movement and posture, and a person's walk was considered an important part of his or her appearance. When she invites Elizabeth to join her, Mr. Darcy remarks that they walk either to discuss things privately or because "you are conscious that your figures appear to the greatest advantage in walking." Walking can be for display, withdrawal, or both.

This novel and other novels of the time suggest that walking provided a shared seclusion for crucial conversations. Etiquette at the time required residents and guests of the country house to pass their day in the main rooms together, and the garden walk provided relief from the group, either in solitude or in tête-à-têtes (in a twist on this practice, modern political figures have often held crucial conversations on walks in order to avoid being bugged). Soon after Jane's

---

and tranquilized it.—MARY SHELLEY, *FRANKENSTEIN*   *He who does not know these sensations has*

---

recovery, she and Elizabeth gossip while walking in their own family's shrubbery. For *Pride and Prejudice* is also an incidental inventory of the types of landscape available to walk in. Toward the end of the book, further features of the Bennet gardens appear when Lady Catherine storms in to harangue Elizabeth about her intentions toward Mr. Darcy: "'Miss Bennet, there seemed to be a prettyish kind of a little wilderness on one side of your lawn. I should be glad to take a turn in it, if you will favour me with your company,'" she dissembles, seeking private conversation. "'Go, my dear,' cried her mother, 'and shew her ladyship about the different walks. I think she will be pleased with the hermitage.'" This lets us know that it is a mid-eighteenth-century garden of some size, with at least one architectural adornment in it.

What exactly Lady Catherine's park included we never learn, only that during her stay nearby Elizabeth's "favourite walk . . . was along the open grove which edged that side of the park, where there was a nice sheltered path, which no one seemed to value but herself, and where she felt beyond the reach of Lady Catherine's curiosity." Not Mr. Darcy's, however: "More than once did Elizabeth in her ramble within the Park, unexpectedly meet Mr. Darcy.—She felt all the perverseness of the mischance." She tells him it is her "favorite haunt," for she still wishes to avoid him. He, of course, is in love with her and repeatedly joins her in the park seeking private conversation: "it struck her in the course of their third rencontre that he was asking some odd unconnected questions—about her pleasure in being at Hunsford, her love of solitary walks . . ."

For the author and her readers, as for Mr. Darcy, these solitary walks express the independence that literally takes the heroine out of the social sphere of the houses and their inhabitants, into a larger, lonelier world where she is free to think: walking articulates both physical and mental freedom. Though Austen has not nearly as much to say about scenery in this novel as in *Mansfield Park*, Elizabeth's sensitivity to landscape is another of the features that signifies her refined intelligence. It is not Mr. Darcy but Pemberly, his estate, that begins to change her mind about him, and walking in his park becomes a peculiarly intimate act. She "had never seen a place for which nature had done more, or where natural beauty had been so little counteracted by an awkward taste. . . . At that moment she felt, that to be

---

never enjoyed a cool rest at the side of a spring after the hard walk of a summer's day. . . . If in such

mistress of Pemberley might be something!" Evidently a student of
Gilpin, she inspects the view from each window of the house, and
after they have left it to walk toward the river, the owner of both
house and river appears. Her uncle "expressed a wish of going round
the whole Park, but feared it might be beyond a walk. With a triumphant
smile, they were told, that it was ten miles round." Like Elizabeth's
fondness for solitary walks, Mr. Darcy's possession of a magnificent
naturalistic landscape evidently in the modern style of Capability Brown
is a sign of character. When they unexpectedly meet in this landscape, a
more civil and conscious relationship begins, as they "pursued the accustomed circuit; which brought them again, after some time, in a descent among hanging woods, to the edge of the water, in one of its
narrowest parts. They crossed it by a simple bridge, in character with
the general air of the scene . . . and the valley, here contracted into a
glen, allowed room only for the stream, and a narrow walk amidst the
rough coppice-wood which bordered it. Elizabeth longed to explore its
windings. . . ."

It is this shared taste for scenery that finally provides the literal
common ground on which they resolve their differences. Of course,
the hero and heroine of the novel have been brought together in the
glories of Pemberley because her aunt and uncle had offered to take
her to the Lake District (to which Elizabeth "rapturously cried, 'what
delight! what felicity! You give me fresh life and vigour. Adieu to disappointment and spleen. What are men to rocks and mountains?'").
Though Miss Bingley despises this aunt and uncle for being in trade
and residing in an unfashionable part of London, they have demonstrated *their* refinement by taking up this moderately avant-garde form
of scenic tourism. The trip has been cut short, bringing them to Derbyshire and Pemberley, not far south of the Lakes, and gathering together all the most admirably conscious characters in the book.
Austen interrupts the high abstract plane of her narration, in which
only the most necessary details of the material world briefly intrude,
to offer us luscious descriptions of Pemberley. Of the country in
which it is located, she remarks, however, "It is not the object of this
work to give a description of Derbyshire." Still, she lists the magnificent estate of Chatsworth and the natural wonders of Dove Dale,

---

moments I find no sympathy, and Charlotte does not allow me to enjoy the melancholy consolation of

Matlock, and the Peak as among the attractions these tourists have visited.

The multiplicity of walking's uses are notable in this novel. Elizabeth walks to escape society and to converse privately with her sister and, at the end of *Pride and Prejudice,* with her suitor. The landscapes she enjoys include old-fashioned and new gardens, wild landscapes of the north and Kentish countryside. She walks for exercise, as did Queen Elizabeth, for conversation, as did Samuel Pepys, and walks in gardens, as did Walpole and Pope. She walks in scenic spots as did Gray and Gilpin, and even walks for transportation, as did Moritz and the Wordsworths, and like them she meets with disapproval for it. Once or twice she promenades, as did they all. New purposes keep being added to the pedestrian repertoire, but none are dropped, so that the walk constantly increases in meanings and uses. It has become an expressive medium. It is also both socially and spatially the widest latitude available to the women contained within these social strictures, the activity in which they find a chance to exert body and imagination. On a walk where they manage to lose all their companions and "she went boldly on with him alone," Elizabeth and Darcy finally come to an understanding, and their communications and newfound happiness take up so much time that "'My dear Lizzy, where can you have been walking to?' was a question which Elizabeth received from Jane as soon as she entered the room, and from all the others when they sat down to table. She had only to say in reply, that they had wandered about, till she was beyond her own knowledge." Consciousness and landscape have merged, so that Elizabeth has literally gone "beyond her own knowledge" into new possibilities. It is the last service the walk performs for the restless heroine of this novel.

Notable too are the many times in which walking appears as a noun rather than a verb in this book and in this era: "Within a short walk of Longbourn lived a family"; "a walk to Meryton was necessary to amuse their morning hours"; "they had a pleasant walk of about half a mile across the park"; "her favourite walk . . . was along the open grove" and so forth. These uses of the word express that the walk is a set piece with known qualities, like a song or a dinner, and that in going on such a walk one does not merely move one's legs alternately

---

bathing her hand in my tears, I tear myself from her and roam through the country, climb some

but does so for a certain duration neither too long nor too short, for purposes sufficiently unproductive of anything but health and pleasure, in pleasing surroundings. The language implies a conscious attention to the refinement of everyday acts. People had always walked, but they had not always invested it with these formal meanings, meanings about to expand further.

V. Out of the Gate

The Romantic poets are popularly portrayed as revolutionaries breaking with everything that had come before. The young Wordsworth was radical in politics, as well as poetic style and subject matter, but he carried much polite eighteenth-century convention forward with him. Still in his mother's womb when Gray arrived in the Lake District, he helped further popularize the region's beauties, and though he was born on the edge of its steep, stony expanses, it was conventional aesthetics as well as personal associations that brought him back to live out the last fifty years of his life there. From Wales to Scotland to the Alps, Wordsworth chose already-celebrated landscapes to walk in and write about. He was, in some ways, the ideal tourist, a tourist with a unique gift for remembering and describing what he saw, and his relationship to the Lake District is an odd balancing act between the clear-eyed intimacy of the local and the enthusiasm of the tourist. He and his sister were consciously steeping themselves in the existing literature on landscape, educating themselves to see the same way Marianne Dashwood or Elizabeth Bennet might have, and bringing that sight into their everyday excursions. In 1794 Wordsworth asked his brother in London to send his books to him and singled out the volumes of Gilpin's Scottish and northern English tours as important to include. And in 1800, seven months after that long walk across the snow, Dorothy wrote in her journal, "In the morning, I read Mr Knight's _Landscape_ [_The Landscape: A Didactic Poem_, quoted earlier]. After tea we rowed down to Loughrigg Fell, visited the white foxglove, gathered wild strawberries and walked up to view Rydale. We lay a long time looking at the lake: the shores all embrowned with the scorching sun. The ferns were turning yellow,

---

precipitous cliff, or make a path through a trackless wood, where I am wounded and torn by thorns and

---

that is, here and there one was quite turned. We walked round by Benson's wood home. The lake was now most still, and reflected the beautiful yellow and blue and purple and grey colours of the sky." The passage reads as though she took instruction in landscape in the morning and carried it out in the afternoon. It also illustrates the Wordsworths' most common kind of walking—not as travel but as daily outings in the region around them, in some respects like the daily garden walks of the ladies and gentlemen whose traditions they were extending, and in some respects radically different.

---

briars, and there I find some relief. . . . Ossian has superseded Homer in my heart. What a world into

Chapter 7

THE LEGS OF

WILLIAM WORDSWORTH

"His legs were pointedly condemned by all the female connoisseurs in legs that I ever heard lecture on that topic," wrote Thomas De Quincey of William Wordsworth, with the mixture of admiration and animosity most of the next generation of poets brought to that looming presence. "There was no absolute deformity about them; and undoubtedly they had been serviceable legs beyond the average standard of human requisition; for I calculate, upon good data, that with these identical legs Wordsworth must have traversed a distance of 175 to 180,000 English miles—a mode of exertion, which to him, stood in the stead of wine, spirits, and all other stimulants whatsoever to the animal spirits; to which he has been indebted for a life of un- clouded happiness, and we for much of what is most excellent in his writings." While others walked before and after him, and many other Romantic poets went on walking tours, Wordsworth made walking central to his life and art to a degree almost unparalleled before or since. He seems to have gone walking nearly every day of his very long life, and walking was both how he encountered the world and how he composed his poetry.

To understand his walking, it is important to break away from the idea of "the walk" as meaning a brief stroll about a pleasant place and from that other definition of the recent writers on Romantic walking, of walking as long-distance travel. For Wordsworth walking was a mode not of traveling, but of being. At twenty-one, he set off on a two-thousand-mile journey on foot, but during the last fifty years of his life, he paced back and forth on a small garden terrace to compose his poetry, and both kinds of walking were important to him, as was cruising about the streets of Paris and of London, climbing mountains, and walking with sister and friends. All this walking found a way into his poetry. I could have written about his walking earlier, with the philosophical writers who made walking part of their thinking process, or later, when I turn to the histories of walking in the city. But he himself linked walking with nature, poetry, poverty, and vagrancy in a wholly new and compelling way. And of course Wordsworth himself emphatically valued the rural over the urban:

> Happy in this, that I with nature walked
> Not having a too early intercourse
> With the deformities of crowded life. . . .

Too, he is the figure to which posterity looks in tracing the history of walking in the landscape: he has become a trailside god.

Born in 1770 in Cockermouth, just north of the more wild and steep scenery of the Lake District, Wordsworth liked in later years to portray himself as a simple man born amid a kind of pastoral republic of lakeland freeholders and shepherds. In fact, his father was the agent of Lord Lowther, an immensely wealthy despot who owned much of the region. The future poet was not yet eight when his mother died; Dorothy was sent away to be raised by relatives, and he himself was sent to school in Hawkshead, in the heart of the Lakes. The death of their father when Wordsworth was thirteen left the children dependent on the goodwill of unenthused relations, for Lord Lowther managed to deprive the five Wordsworth children of their legacy from this successful father for nearly twenty years. But the years at Hawkshead's excellent school were idyllic despite or perhaps because of all the family turmoil. There he set snares, ice-skated, climbed the cliffs for birds' eggs, boated, and walked incessantly, at night and often in the morning before school, when he and a friend would go the five miles around the nearby lake. Or so says *The Prelude*, his great autobiographical poem of several thousand lines, which even with its scrambled chronologies and

---

conjure up by the feeble light of the moon the spirit of our ancestors. . . . At noon I went to walk by the

deleted facts provides a spectacular portrait of the poet's early life. Called by his family "The Poem to Coleridge," to whom it is addressed, it is also subtitled "The Growth of a Poet's Mind," signifying exactly what kind of an autobiography it is, and it was meant to be a prelude to a monumental philosophical poem *The Recluse*, of which only *The Prelude* and "The Excursion" were completed.

The Prelude reads almost as a single long walk that, though interrupted, never altogether stops, and this recurrent image of the walker gives it continuity amid all its digressions and detours. One pictures Wordsworth like Christian in *Pilgrim's Progress* or Dante in the *Divine Comedy*, a small figure touring the whole world on foot, only this time around it is a world of lakes, dances, dreams, books, friendships, and many many places. The poem is also a kind of atlas of the making of a poet, showing us the role of this city and that mountain, for places loom larger than people. In the same respectfully spiteful vein as De Quincey remarking on Wordsworth's legs, the essayist William Hazlitt once quipped, "He sees nothing but himself and the universe." In the history of English literature, the rise of the novel is often linked to the rise of awareness and interest about personal life—personal life as private thoughts, emotions and relations between people. Wordsworth went much further than the novels of his time in charting his own thoughts, emotions, memories, and relations to place, but his seems a curiously impersonal life, since he remains reticent on his personal relationships—thus Hazlitt's quip.

His passion for walking and for landscape seems to have originated in childhood, or been that curiosity so many children have, salvaged and refined into art in his later years, but the passion begins too early and goes too far to be merely the fashionable taste for admiring and describing landscapes. In the fourth of the *Prelude's* thirteen books, he describes walking home from an all-night dance somewhere in the Lakes, sometime in his late teens, to witness a dawn "more glorious than I had ever beheld." Early on this morning, while "The sea was laughing at a distance; all / The solid mountains were as bright as clouds" he committed to his vocation as poet—"I made no vows but vows / Were then made for me"—and he became a "dedicated spirit. On I walked / In blessedness, which even yet remains." In his early twenties, he seems to have set about to systematically fail at every

river—I had no appetite. Everything around seemed gloomy.—GOETHE, *THE SORROWS OF YOUNG*

alternative to being a poet and chosen wandering and musing as the
preliminaries for realizing his vocation.

> Should the guide I choose
> Be nothing better than a wandering cloud,
> I cannot miss my way,

he asserts amid the opening lines of this massive poem first finished in
1805, revised repeatedly during his lifetime, and only published after
his death in 1850.

The turning point in both his life and *The Prelude* is his amazing
1790 walk with his fellow student Robert Jones across France into
the Alps, when they should have been studying for their Cambridge
University exams. Wordsworth's most recent biographer, Kenneth
Johnston, dramatically declares, "With this act of disobedience his
career as a Romantic poet may be said to have begun." Travel has its
rogue and rebel aspects—straying, going out of bounds, escaping—
but this journey was as much a quest for an alternate identity as an
escapade. The Grand Tour had been a standard feature of English
gentlemen's educations; usually they went by coach to meet people of
their own class and see the artworks and monuments of France and
Italy. Those connoisseurs of gardens and landscapes Horace Walpole
and Thomas Gray went on such a tour in 1739, where they each
wrote excitedly of the Alps they crossed en route to Italy. To go on
foot and to make Switzerland, rather than Italy, the destination of the
trip expressed a radical shift in priorities, away from art and aristocracy toward nature and democracy. To go in 1790 meant joining
the flood of radicals converging on Paris to breathe the heady atmosphere of the early days of the French Revolution, before the blood
had begun to pour. The Alps themselves, already central objects in
the cult of the landscape sublime, were part of the attraction, but so
was Switzerland's republican government and its associations with
Rousseau. Their final destination before they boated back down the
Rhine was the island of Saint-Pierre, which Rousseau wrote about in
the *Confessions* and the *Reveries of a Solitary Walker* as a version of the
natural paradise. Rousseau is an obvious precursor for Wordsworth,
who walked as both a means and an end—to compose and to be.

WERTHER   *When his [Coleridge's] health was good, which it certainly was for long spells, he did the*

They had landed in Calais on July 13 and woke the next day to the joyous celebrations of the first anniversary of the storming of the Bastille, when France was "standing on top of golden hours / And human nature seeming born again." They walked through

> hamlets, towns,
> Gaudy with relics of that festival,
> Flowers left to wither on triumphal arcs, and window garlands. . . .
> Unhoused beneath the evening star we saw
> Dances of liberty, and in late hours
> Of darkness, dances in the open air.

Wordsworth and Jones had charted their journey with care, however, and walked about thirty miles a day in order to carry out their ambitious plans:

> A march it was of military speed
> And earth did change her images and form
> Before us fast as clouds are changed in heaven.
> Day after day, up early and down late,
> From vale to vale, from hill to hill we went,
> From province on to province did we pass,
> Keen hunters in a chase of fourteen weeks.

So vigorous were they that they crossed the Alps without realizing it, much to their disappointment. Already over the final pass and still thinking they had far higher to go, they had cut off on an uphill trail when a peasant set them straight and sent them to finish their descent into Italy, where they made a quick loop past Lake Como before re-entering Switzerland. Wordsworth breaks off this narrative at Lake Como, but *The Prelude* recounts his returns to France in 1791, where his politics continued to develop.

It is entirely Wordsworthian that he tried to understand the Revolution by walking the streets of Paris and visiting "each spot of old and recent fame" from the "dust of the Bastille" to the Champ de Mars and Montmartre. Among the Britons he may have met there are Colonel John Oswald and "Walking Stewart," two examples of a new

---

most amazing walks and mountain climbs all round the Lakes single-handed. He was in effect the first of

---

kind of pedestrian. Johnston writes, "Oswald had traveled to India, become a vegetarian and nature mystic, walked back to Europe overland, thrown himself into the French Revolution with the direct intent of carrying it back to England." He would later appear under his own surname in Wordsworth's early verse drama *The Borderers*. Stewart was a similar character whose nickname commemorated his remarkable walks—he too had walked back from India, as well as all over Europe and North America—but whose books were diatribes on other subjects. De Quincey wrote of Walking Stewart, "No region, pervious to human feet except, I think, China and Japan, but had been visited by Mr. Stewart in this philosophical style; a style which compels a man to move slowly through a country, and to fall in continually with the natives of that country." A third eccentric, John Thelwall (mentioned in chapter 2), suggests something of a pattern: autodidacts who took the trinity of radical politics, love of nature, and pedestrianism to extremes. Thelwall became well acquainted with Wordsworth and Coleridge in the early 1790s, and later in that decade, after he narrowly escaped hanging for his politics, sought refuge with them. Wordsworth owned a copy of Thelwall's *Peripatetic*, which amid its digressions on philosophy takes stock of the living and working conditions of the laborers being drawn into the beginnings of the industrial revolution. These characters suggest that traveling any distance on foot was the act of a political radical in England, expressing an unconventionality and a willingness to identify and be identified with the poor. Wordsworth himself wrote in a letter of 1795, "I have some thoughts of exploring the country westward of us, in the course of next summer, but in an humble evangelical way; to wit *à pied*," and in *The Prelude* he wrote, "So like a peasant I pursue my way."

To walk in this way summoned up Rousseau's complex equation of virtue with simplicity with childhood with nature. At the beginning of the eighteenth century, English aristocrats had linked nature with reason and the current social order, suggesting that things were as they should be. But nature was a dangerous goddess to enthrone. At the latter end of that century, Rousseau and romanticism equated nature, feeling, and democracy, portraying the social order as highly artificial and making revolt against class privilege "only natural." In his

---

the modern fell walkers, the first known outsider to set out to climb the mountain tops, just for the pleasure

---

history of eighteenth-century ideas of nature, Basil Willey remarks, "Throughout that turbulent time 'Nature' remained the dominant concept," but its meaning was protean. "The Revolution was made in the name of Nature, Burke attacked it in the name of Nature, and in *eodem nomine* Tom Paine, Mary Wollstonecraft, and [radical philosopher William] Godwin replied to Burke." To walk in the gracious and expensive confines of the garden was to associate walking, nature, the leisure classes, and the established order that secured that leisure. To walk in the world was to link walking with a nature aligned instead with the poor and whatever radicalism would defend their rights and interests. Too, if society deformed nature, then children and the uneducated were, in a radical reversal, the purest and the best. Wordsworth, perfect sponge of his age, soaks up these values and pours them forth as his extraordinary poetry of childhood—his own, and those of his many fictional characters—and of the poor. He took up Rousseau's task and improved upon it, portraying rather than arguing a relationship between childhood, nature, and democracy. Though only the first two of this trinity are remembered by the worshipers of the trailside god, the third is central to at least the early work. "You know perhaps already that I am of that odious class of men called democrats," he wrote a friend in 1794, continuing with a confidence that proved unwarranted, "and of that class I shall for ever continue."

Somewhere on these roads, among these people and these questions, Wordsworth met up with his style. His earliest poetry is lofty, vague, and studded with conventional images, in the mode of Thomson's *Seasons*, but it seems to be his revolutionary ardor and sympathetic identification with the poor that saved him from being a minor landscape poet (during the same decade of the 1790s, Dorothy's writing undergoes a similar transformation, from the aphoristic abstruseness of a Dr. Johnson or Jane Austen to something vividly descriptive and down-to-earth). It changed both subject matter and style. In his retroactive preface to the *Lyrical Ballads*, the epochal book of poems by Wordsworth and Coleridge published in 1798, he wrote, "The principal object, then, proposed in these poems, was to choose incidents and situations from common life, and to relate or describe them throughout, as far as was possible, in a selection of language really used by men, and, at

---

of doing so. His ascent in 1802 of Scafell Pike, the highest Lakeland peak, is the first recorded climb of

the same time, to throw over them a certain coloring of imagination. . . . Humble and rustic life was generally chosen, because in that condition the essential passions of the heart find a better soil . . . and speak a plainer and more emphatic language." He wrote about the poor as people rather than as figures in fables of virtue or pity, as he wrote about landscapes in their specific details rather than in high-flown generalizations and classical allusions. Choosing plainer language was a political act, with spectacular artistic results.

What is marvelous about Wordsworth's early poetry is its union of the radical walk for the sake of encounters with the scenic stroll of aesthetic connoisseurs. Looking back, it seems there should have been some tensions between scenery and poverty as subjects, but for the young Wordsworth in that exuberant moment there were none. The landscapes are the more incandescent for being populated by vagrants rather than nymphs, and that incandescence is the more necessary as the birthright and backdrop of the desperate. The recurrent structure of these early poems is a walk interrupted by an encounter with those displaced by the economic turbulence of the time into fellow wanderers. Earlier poets and artists had looked at the cottages and bodies of the poor and found them picturesque or pitiful, but no one with such a voice had found it worthwhile to talk to them before. "When we walk, we naturally go to the fields and woods," remarked Thoreau, but Wordsworth headed as eagerly to the public roads as to mountains and lakes. People walk streets for the sake of encounters and paths for solitude and scenery; on the road Wordsworth seems to have found an ideal intermediary, a space providing long quiet spells broken by the occasional meeting. He affirmed:

> I love a public road: few sights there are
> That please me more—such object has had power
> O'er my imagination since the dawn
> Of childhood, when its disappearing line
> Seen daily afar off, on one bare steep
> Beyond the limits which my feet had trod,
> Was like a guide into eternity,
> At least to things unknown and without bound.

---

that magnificent mountain. —HUNTER DAVIES, *WILLIAM WORDSWORTH: A BIOGRAPHY*   *I said he has*

Which is to say that the road had a kind of perspectival magic, an allure of the unknown. But it also had a populace:

> When I began to enquire,
> To watch and question those I met, and held
> Familiar talk with them, the lonely roads
> Were schools to me in which I daily read
> With most delight the passions of mankind,
> There saw into the depth of human souls
> Souls that appear to have no depth at all
> To vulgar eyes. . . .

This education had begun during his schooldays, when he boarded with a retired carpenter and his wife and met peddlers, shepherds, and similar characters. These early experiences seem to have set him at ease with people of another class and at least partially relieved him of that mental barrier that separates the English classes from each other. He once remarked, "Had I been born in a class which would have deprived me of what is called a liberal education, it is not unlikely that, being strong in body, I should have taken to a way of life such as that in which my Pedlar passed the greater part of his days." The terrible uncertainty of his own early life, with parents dead and relatives shuttling the children around, seems to have generated a sympathy for the displaced, while his passion for traveling made these mobile characters, in a word, romantic to him. The times themselves were uncertain; the old order had been shaken by the revolutions and insurrections in France, America, and Ireland, and the poor were being displaced by the changing rural scene and dawning industrial revolution. The modern world of people cast adrift, unanchored by the securities of place, work, family, had dawned.

The mobile figure recurs in the work of Wordsworth's contemporaries too, and walking seems to have provided literal common ground between those traveling to seek adventure and pleasure and those on the road to seek survival. Even now English people tell me that walking plays so profound a role in English culture in part because it is one of the rare classless arenas in which everyone is roughly equal and welcome. The young Wordsworth wrote about discharged

---

just finished an exquisite ode to Pan—and as he had not a copy I begged Keats to repeat it—which he

soldiers, tinkers, peddlers, shepherds, stray children, abandoned wives, "The Female Vagrant," "The Leech Gatherer," "The Old Cumberland Beggar," and others who tended to be nomadic or displaced; even the Wandering Jew made an appearance in his poetry and that of many other Romantics. Or as Hazlitt put it in describing the revolutionary transformation of English poetry at the hands of Coleridge, Wordsworth, and Robert Southey, "They were surrounded, in company with Muses, by a mixed rabble of idle apprentices and Botany Bay convicts, female vagrants, gipsies, meek daughters in the family of Christ, of ideot boys and mad mothers, and after them 'owls and night-ravens flew.'"

The peddler Wordsworth might have been is the principal narrator of his first long narrative poem, "The Ruined Cottage." It is typical of his early poetry in that in it a fortunate young man encounters, while walking, someone who tells him the tale that makes up the body of the poem, so that the young man and his saunters make a kind of frame around the sad picture, serving as frames do both to underscore the value and to isolate the work within. This time around, the Wordsworth figure arrives at a ruined cottage where the Pedlar tells him the pathos-drenched tale of the last residents of the place: a family torn apart into wanderers and lingerers by economic hardship. Everyone in the story is in some kind of pedestrian motion: the strolling narrator, the nomadic Pedlar, the husband enlisted and gone to a distant land, the heartbroken wife wearing a path into the grass by pacing back and forth, watching the road for his return.

The walkers in the garden had been anxious to distinguish their walking for pleasure from that of those who walked for necessity, which is why it was important to stay within the garden's bounds and not to walk as travel—but Wordsworth sought out meetings with those who represented this other kind of walking (or, frequently, borrowed those characters as met and vividly described by Dorothy in her journals, from which he gathered much). For all its meaty radical politics, *The Prelude* is a thirteen-book sandwich whose bread is landscape. The poem ends with a visionary experience atop Mount Snowdon in Wales that leads into another long soliloquy—but no further geographical details. A shepherd—shepherds were among the first mountain guides in Europe—leads him and an unnamed friend

---

did in his usual half chant, (most touchingly) walking up & down the room. . . . —BENJAMIN

up during the night so they can see the sunrise from the peak. Because the young men are so fit, they arrive early at their destination. The narrative leaves Wordsworth atop the mountain in a sudden flood of moonlight, scenery, and revelation. Climbing a mountain has become a way to understand self, world, and art. It is no longer a sortie from but an act of culture.

But walking wasn't only a subject for Wordsworth. It was his means of composition. Most of his poems seem to have been composed while he walked and spoke aloud, to a companion or to himself. The results were often comic; the Grasmere locals found him spooky, and one remarked, "He won't a man as said a deal to common fwoak, but he talked a deal to hiseen. I oftenn seead his lips a ganin," while another recalled, "He would set his head a bit forrad, and put his hands behint his back. And then he would start a bumming, and it was bum, bum, bum, stop; then bum, bum bum, reet down till t'other end; and then he'd set down and git a bit o'paper and write a bit." In *The Prelude* he describes a dog he used to walk with who would, when a stranger drew near, cue him to shut up and avoid being taken for a lunatic. He possessed a remarkable memory that allowed him to recollect with visual detail and emotional vividness scenes long past, to quote long passages of the poets he admired, and to compose afoot and write the result down later. Most modern writers are deskbound, indoor creatures when they write, and nothing more than outline and ideas can be achieved elsewhere; Wordsworth's method seemed a throwback to oral traditions and explains why the best of his work has the musicality of songs and the casualness of conversation. His steps seem to have beat out a steady rhythm for the poetry, like the metronome of a composer.

One of his best-known poems—"Lines Composed a Few Miles Above Tintern Abbey, on Revisiting the Banks of the Wye During a Tour," to give it its full title—was composed on foot during a walking tour in Wales with Dorothy in 1798. Upon arriving back in Bristol, he jotted the whole thing down and tacked it unrevised onto the *Lyrical Ballads*, where it appears as the last and one of the best poems in his book, his work, and perhaps the English language. Very much a walking poem, "Tintern Abbey" captures that state of musing, of shifting about in time from recollection to experience to hope while

HAYDON *Our walk before breakfast, and after the bath, is the best meal of the day. Anna and*

exploring a place. And like much of his blank verse, it is written in language so close to actual speech that it reads with conversational ease, but speaking it aloud revives the strong rhythms of those walks two hundred years ago.

In 1804 Dorothy wrote to a friend, "At present he is walking and has been out of doors these two hours though it has rained heavily all the morning. In wet weather he takes out an umbrella; chuses the most sheltered spot, and there walks backwards and forwards and though the length of his walk be sometimes a quarter or half a mile, he is as fast bound within the chosen limits as if by prison walls. He generally composes his verses out of doors and while he is so engaged he seldom knows how the time slips by or hardly whether it is rain or fair." There is a path at the top of the small garden at Dove Cottage, where he could see over the house to the lake and most of the ranges rising around it, and it was there he most often paced, composing. Many thousand of the "175 to 180,000 English miles" De Quincey estimated he had walked were walked here, on this terrace about twelve paces long, and on the similar terrace of the larger home he moved to in 1813. Seamus Heaney, writing about the "almost physi-ological relation of a poet composing and the music of the poem," says of Wordsworth's pacing back and forth that it "does not forward a journey but habituates the body to a kind of dreamy rhythm." It also makes composing poetry into physical labor, pacing back and forth like a ploughman turning his furrows up or wandering across the heights like a shepherd in search of a sheep. Perhaps because he was producing beauty out of arduous physical toil, he shamelessly iden-tified himself with the working and walking poor. Though he was basically a rugged and athletic man, the stress of composing gave him headaches and a recurrent pain in the side, so fiercely did he drive himself in this act of poetry as bodily labor. Heaney concludes, "Wordsworth at his best, no less than his worst, is a pedestrian poet."

Had Wordsworth been a perfect Romantic poet, he would have died in his late thirties, still pacing back and forth at humble Dove Cottage, leaving us the first, best version of *The Prelude*, all his early ballads and narratives about the poor, his odes and lyrics of childhood, and his image as a radical intact. Unfortunately for his reputation, though happily enough for self and family, he lingered in Grasmere

----

Elizabeth are beginning to taste and find out how good it is for mind and body. It was meant that the

----

and then in the large house in neighboring Rydal to the age of eighty, becoming increasingly conservative and decreasingly inspired. One might say that he went from being a great Romantic to a great Victorian, and the transition required much renouncement. Though he did not keep faith with his early politics, he kept faith with his walking. And oddly, it is his legacy not as a writer but as a walker that carries on the joyful insurrection of his early years.

One of his own last twinges of democracy came in 1836, when he was sixty-six. He had taken Coleridge's nephew walking on a private estate when, as one biographer recounts it, "the lord who owned the ground came up and told them they were trespassing. Much to his companion's embarrassment, William argued that the public had always walked this way and that it was wrong of the lord to close it off." The nephew recalled that "Wordsworth made his point with somewhat more warmth than I either liked or could well account for. He had evidently a pleasure in vindicating these rights, and seemed to think it a duty." Another version situates the confrontation at Lowther Castle, where Wordsworth, Coleridge's nephew, and the lord in question were dining. The latter declared that his wall had been broken down and he would have horsewhipped the man who did it. "The grave old bard at the end of the table heard the words, the fire flashed into his face and rising to his feet, he answered: 'I broke your wall down, Sir John, it was obstructing an ancient right of way, and I will do it again. I am a Tory, but scratch me on the back deep enough and you will find the Whig in me yet.'"

Of all the other Romantics, only De Quincey seems to have had a lifelong passion for walking comparable to Wordsworth's, and though it is impossible to measure pleasure, it is possible to say something about effects: walking was neither a subject nor a compositional method for the younger writer in the way it had been for the older. His innovations were elsewhere—Morris Marples credits him with being the first to go on a walking tour with a tent, which he slept in during an early sojourn in Wales to save money (the beginnings of the outdoor equipment industry show up here, in the special coats Wordsworth and Robert Jones had a tailor make them for their continental

whole season should be put into us, as it is into a flower.—AMOS BRONSON ALCOTT   *At length, as*

tour, in Coleridge's walking sticks, in De Quincey's tent, in Keats's odd travel outfit). De Quincey's best writing about walking was about prowling the streets of London as a destitute youth, a very different kind of walking—and writing. His fellow essayist William Hazlitt wrote the first essay on walking, but it began another genre of walking literature rather than extending the tradition Wordsworth took up, and it depicts walking as a pastime rather than an avocation. Shelley was too aristocratic an anarchist and Byron too lame an aristocrat to have much to do with walking; they sailed and rode instead.

Coleridge, on the other hand, had a decade of avid walking—1794–1804—which is reflected in his poetry from that time. Even before he met Wordsworth, he set out on a walking tour to Wales with a friend named Joseph Hucks and then another tour in Somerset in southern England with his fellow poet and future brother-in-law Robert Southey. In 1797 Coleridge and Wordsworth began their extraordinary collaborative years, with walks in the same parts of southern England; on one of these tours, when Dorothy joined them, Coleridge composed his most famous poem, the "Rime of the Ancient Mariner" (which is, like his friend's work of the time, a poem about wandering and exile). He and the Wordsworths walked together many more times: there was the epochal walking tour in the Lake District with William Wordsworth and his younger brother John, during which Wordsworth decided to return to this scene of his childhood, and there were many shorter walks after Coleridge and Southey moved to Keswick in the north of that district, as well as one final, blighted tour of Scotland with William, Dorothy, and a donkey cart. The two men got on each other's nerves, split up, and never quite resumed their great friendship. During the course of a solitary and athletic tour of the Lakes, Coleridge also became the first recorded person to reach the summit of Scafell Peak, though he lost some of the glory he might have achieved with this difficult climb by getting stuck on his descent and then tumbling down the mountain. After 1804, Coleridge went on no more long walks. Although the links between walking and writing are neither so explicit nor so profuse in his work as in his friend's, the critic Robin Jarvis does point out that Coleridge ceased to write blank verse when he ceased to walk.

These walking tours on the part of poets who would not walk

----

we plodded along the dusty roads, our thoughts became as dusty as they, all thought indeed stopped,

----

much later suggest that there was indeed an emerging fashion for traveling on foot. Certainly the very unpoetic literature of the guidebook began to address itself to walkers at this point, and the very notion of a walking tour suggests that the parameters of how to walk and what it meant were beginning to be established. Like the garden stroll, the long walk was acquiring conventions of both meaning and doing. This is easily seen in John Keats's one great experiment with walking. In 1818 the young Keats set out on a walking tour for the sake of poetry, suggesting that such an excursion was a familiar rite of passage as well as a refinement of sensibility. "I purpose within a month to put my knapsack at my back and make a pedestrian tour through the north of England, and part of Scotland—to make a sort of prologue to the life I intend to pursue—that is to write, to study and to see all Europe at the lowest expense. I will clamber through the clouds and exist," he wrote, and soon afterward wrote to another friend, "I should not have consented to myself these four Months tramping in the Highlands but that I thought it would give me more experience, rub off more Prejudice, use [me] to more hardship, identifying finer scenes, load me with grander Mountains, and strengthen more my reach in Poetry, than would stopping at home among Books even though I should read Homer." In other words, roughing it and growing acquainted with mountains was poetic training. Yet like the walkers who came after him, he wanted only so much hardship and experience. He turned back from Ireland appalled by the harsh poverty on that oppressed island, and in reading of this rejection of experience one thinks of a key moment in *The Prelude* and evidently in Wordsworth's life. He was walking in France with the revolutionary soldier Michel Beaupuy; they encountered a "hunger-bitten girl . . . who crept along" a lane, and Beaupuy explained that she was the reason they were fighting. Wordsworth had connected walking to both pleasure and suffering, to politics and scenery. He had taken the walk out of the garden, with its refined and restricted possibilities, but most of his successors wanted the world in which they walked to be nothing but a larger garden.

---

thinking broke down, or proceeded only passively in a sort of rhythmical cadence of the confused material

A THOUSAND MILES OF CONVENTIONAL SENTIMENT: The Literature of Walking

I. THE PURE

Other kinds of walk survived, and early in Thomas Hardy's novel *Tess of the D'Urbervilles*, one of them collides with the traditions drawn from romanticism. Tess and her fellow peasant girls are celebrating May Day by going "club-walking," a pre-Christian spring ceremony in which they walk in procession across the countryside. Dressed all in white, the young women and a few older ones go in "a processional march of two and two round the parish" and in the designated meadow begin to dance. Looking on "were three young men of a superior class, carrying small knapsacks strapped to their shoulders and stout sticks in their hands. The three brethren told casual acquaintances that they were spending their Whitsun holidays in a walking tour through the Vale. . . ." Two of these three sons of a devout clergyman are themselves ministers; the third, who is less sure of the order of the world and his own place in it, chooses to leave the road and dance with the celebrants. The peasant women on their procession and the young gentlemen on their walking tour are both engaged in nature rituals, in very different ways. The men, with their

---

of thought, and we found ourselves mechanically repeating some verse of the Robin Hood

---

costumes of knapsack and staff, are being artificially natural, for their version of how to connect to nature involves leisure, informality, and travel. The women, with their highly structured rite handed down from an unremembered past, are being naturally artificial. Their acts speak of the two things specifically excluded from the walking tour, work and sex, since it is a kind of crop fertility rite they are engaged in and since the local young men will come to dance with them when their day's work is done. Nature, after all, is not where they take their vacations but where they lead their lives, and work, sex, and the fertility of the land are part of that life. But pagan survivals and peasant rites are not the dominant cult of nature.

Nature, which had been an aesthetic cult in the eighteenth century and become a radical cult at the end of that century, was by the middle of the next century an established religion for the middle classes and, in England far more than the United States, for much of the working classes as well. Sadly, it had become as pious, sexless, and moral a religion as the Christianity it propped up or supplanted. Going out into "nature" was a devout act for those English, American, and Central European heirs of romanticism and transcendentalism. In a cheerfully malicious essay entitled "Wordsworth in the Tropics," Aldous Huxley asserted, "In the neighborhood of latitude fifty north, and for the last hundred years or thereabouts, it has been an axiom that Nature is divine and morally uplifting. For good Wordsworthians—and most serious-minded people are now Wordsworthians—either by direct inspiration or at second hand—a walk in the country is the equivalent of going to church, a tour through Westmoreland is as good as a pilgrimage to Jerusalem."

The first essay specifically on walking is William Hazlitt's 1821 "On Going a Journey," and it establishes the parameters for walking "in nature" and for the literature of walking that would follow. "One of the pleasantest things in the world is going a journey; but I like to go by myself," it opens. Hazlitt declares that solitude is better on a walk because "you cannot read the book of nature, without being perpetually put to the trouble of translating it for the benefit of others" and because "I want to see my vague notions float like the down of the thistle and not to have them entangled in the briars and thorns of controversy." Much of his essay is about the relationship

---

ballads.—THOREAU, "A WALK TO WACHUSETT"   *On another occasion, when I was walking alone I*

between walking and thinking. But his solitude with the book of nature is very questionable, since in the course of the short piece he manages to quote from other books by Virgil, Shakespeare, Milton, Dryden, Gray, Cowper, Sterne, Coleridge, and Wordsworth, along with the Book of Revelation. He describes a day of walking through Wales launched by reading Rousseau's *Nouvelle Heloise* the night before and quoting Coleridge's landscape poetry as he goes. Clearly, the books set forth the kind of experience of walking in nature he should have—pleasant, mingling thoughts, quotations, and scenery—and Hazlitt manages to have it. If nature is a religion and walking its principal rite, then these are its scriptures being organized into a canon.

Hazlitt's essay became the foundation of a genre. It appears in each of the three anthologies of walking essays—an English one from 1920 and two American ones from 1934 and 1967—that I own, and many of the later essayists cite it. The walking essay and the kind of walking described in it have much in common: however much they meander, they must come home at the end essentially unchanged. Both walk and essay are meant to be pleasant, even charming, and so no one ever gets lost and lives on grubs and rainwater in a trackless forest, has sex in a graveyard with a stranger, stumbles into a battle, or sees visions of another world. The walking tour was much associated with parsons and other Protestant clergymen, and the walking essay has something of their primness. Most of the classic essays cannot resist telling us how to walk. Individually, some of these are very fine pieces of writing. Leslie Stephen, who in his "In Praise of Walking" takes up Hazlitt's theme of the musings of the mind, writes, "The walks are the unobtrusive connecting thread of other memories, and yet each walk is a little drama itself, with a definite plot with episodes and catastrophes, according to the requirements of Aristotle; and it is naturally interwoven with all the thoughts, the friendships, and the interests that form the staple of ordinary life." Which is very interesting in its way, and Stephen, who distinguished himself as a scholar, an early Alpine climber, and an athletic walker, is himself interesting until he goes on to tell us that Shakespeare walked and so did Ben Johnson and many others, on up to, inevitably, Wordsworth. And then moralizing sneaks in; he says of Byron that his "lameness was too severe to admit of walking, and therefore all the unwholesome

---

arrived at the Lizard alone and asked if they could give me a bed. "Is your name Mr. Trevelyan?" they

---

humours which would have been walked off in a good cross-country march accumulated in his brain and caused the defects, the morbid affectation and perverse misanthropy, which half ruined the achievement of the most masculine intellect of his time." Stephen goes on to announce, after throwing in a few dozen more English authors, "Walking is the best of panaceas for the morbid tendencies of authors." And then come the instructional shoulds, the shoulds that none of these essayists seems able to resist. He writes that monuments and landmarks "should not be the avowed goal but the accidental addition to the interest of a walk."

It doesn't take Robert Louis Stevenson nearly as long to get to that fateful word. Two or three pages after he has begun his celebrated 1876 essay "Walking Tours," he declares, "A walking tour should be gone on alone, because freedom is of the essence; because you should be able to stop and go on, and follow this way or that, as the freak takes you; and because you must have your own pace, and neither trot alongside a champion walker, nor mince in time with a girl." He goes on to praise and criticize Hazlitt: "Notice how learned he is in the theory of walking tours. . . . Yet there is one thing I object to in these words of his, one thing in the great master's practice that seems to me not wholly wise. I do not approve of that leaping and running." On his own long walking tour in France's Cévennes mountain range, described in *Travels with a Donkey*, Stevenson carried a pistol but described only picturesque and lightly comic situations. Few of the canonical essayists can resist telling us that we should walk because it is good for us, nor from providing directions on how to walk. In 1913 the historian G. M. Trevelyan begins his "Walking" with "I have two doctors, my left leg and my right. When body and mind are out of gear (and those twin parts of me live at such close quarters that the one always catches the melancholy from the other) I know that I shall have only to call in my doctors and I shall be well again. . . . My thoughts start out with me like blood-stained mutineers debauching themselves on board the ship they have captured, but I bring them home at nightfall, larking and tumbling over each other like happy little boy scouts at play."

The possibility that some of us would prefer that happy little boy scouts keep their distance doesn't occur to him, but it must have to

---

answered. "No," I said, "are you expecting him?" "Yes," they said, "and his wife is here already." This

---

one writer, who blasphemed against the cult in 1918. In his "Going Out on a Walk," the Anglo-German satirist Max Beerbohm exploded, "Whenever I was with friends in the country, I knew that at any moment, unless rain were actually falling, some man might suddenly say, 'Come out for a walk!' in that sharp imperative tone which he would not dream of using in any other connection. People seem to think there is something inherently noble and virtuous in the desire to go for a walk." Beerbohm's heresy goes further; he claims that walking is not at all conducive to thinking because though "the body is going out because the mere fact of its doing so is a sure indication of nobility, probity, and rugged grandeur," the mind refuses to accompany it. He was, however, a voice crying out in a densely populated but otherwise convinced wilderness.

On the other side of the Atlantic, one essay on walking had lurched toward greatness, but even Henry David Thoreau could not resist preaching. "I wish to speak a word for Nature, for absolute freedom and wildness," he famously begins his 1851 essay "Walking," for like all the other essayists he connects walking in the organic world with freedom—but like all the others, he instructs us on how to be free. "I have met with but one or two persons in the course of my life who understood the art of Walking, that is, of taking walks—who had a genius, so to speak, for sauntering." A page later, "We should go forth on the shortest walk, perchance, in the spirit of undying adventure, never to return. . . . If you are ready to leave father and mother, and brother and sister, and wife and child and friends, and never see them again—if you have paid your debts, and made your will, and settled all your affairs, and are a free man, then you are ready for a walk." His are the most daring, wildest instructions, but they are still instructions. Soon afterward comes the other word, *must*: "You must be born into the family of Walkers." And then, "You must walk like a camel, which is said to be the only beast which ruminates when walking. When a traveler asked Wordsworth's servant to show him her master's study, she answered, 'Here is his library, but his study is out of doors.'"

Although the walking essay was officially a celebration of bodily and mental freedom, it was not actually opening up the world for that celebration—that revolution had already taken place. It was instead

---

surprised me, as I knew it was his wedding day. I found her languishing alone, as he had left her at Truro,

---

domesticating the revolution by describing the allowable scope of that freedom. And the sermonizing never let up. In 1970, a century and a half after Hazlitt, Bruce Chatwin wrote an essay that set out to be about nomads but detoured to include Stevenson's *Travels with a Donkey*. Chatwin wrote divinely, but he always declined to distinguish nomadism—a persistent travel by any means, seldom primarily by foot—from walking, which may or may not be travel. Blurring those distinctions by conflating nomadism and his own British walking-tour heritage made nomads Romantics, or at least romantic, and allowed him to fancy himself something of a nomad. Soon after Chatwin cites Stevenson, he falls into step with the tradition: "The best thing is to walk. We should follow the Chinese poet Li Po in 'the hardships of travel and the many branchings of the way.' For life is a journey through a wilderness. This concept, universal to the point of banality, could not have survived unless it were biologically true. None of our revolutionary heroes is worth a thing until he has been on a good walk. Che Guevera spoke of the 'nomadic phase' of the Cuban Revolution. Look what the Long March did for Mao Tse-Tung, or Exodus for Moses. Movement is the best cure for melancholy, as Robert Burton (the author of *The Anatomy of Melancholy*) understood."

A hundred and fifty years of moralizing! A century and a half of gentlemanly exhortation! Doctors have asserted many times over the centuries that walking is very good for you, but medical advice has never been one of the chief attractions of literature. Besides, only a walk that is guaranteed to exclude certain things—assailants, avalanches—is truly wholesome, and only such walking is advocated by these sermonizing gentlemen who seem not to see the boundaries they have put around the act (one of the delights of urban walking is how unwholesome it is). Gentlemen, I say, because all the writers on walking seem to be members of the same club—not one of the real walking clubs, but a kind of implicit club of shared background. They are generally privileged—most of the English ones write as though everyone else also went to Oxford or Cambridge, and even Thoreau went to Harvard—and of a vaguely clerical bent, and they are always male—neither dancing peasant lasses nor mincing girls, with wives rather than husbands to leave, as the above passages make clear. Thoreau considerately adds, "How womankind, who are confined to

----

saying that he could not face the whole day without a little walk. He arrived about ten o'clock at night,

----

the house still more than men, stand it I do not know." Many women after Dorothy Wordsworth went on long solitary walks, and Sarah Hazlitt, Hazlitt's estranged wife, even went on a walking tour alone and kept a journal of it, which like most of these documents of female walking went unpublished in its time. Flora Thompson's account of her journeys on foot across rural Oxfordshire to deliver mail in all seasons and weather is one of the most enchanting descriptions of country walks, but it is not part of the canon because it is by a poor woman, about work (and sex, in that a gamekeeper whose grounds she regularly crosses courts her, unsuccessfully), and buried in a book about many other things. Like the great women travelers of the nineteenth century—Alexandra David-Neel in Tibet, Isabelle Eberhardt in North Africa, Isabella Bird in the Rockies—they are anomalies, these walking women (the reasons why will be dealt with at length later, in chapter 14).

By the late nineteenth century the word *tramp* as both noun and verb was popular among the walking writers, as was *vagabond* and *gypsy* and, far down the road in a different world, *nomad*, but to play at tramp or gypsy is one way of demonstrating that you are not really one. You must be complex to want simplicity, settled to desire this kind of mobility. Bruce Chatwin to the contrary, Bedouins do not go on walking tours. Stephen Graham, an Englishman who early in the twentieth century took remarkable long walks through eastern Europe, Asia, and the Rocky Mountains, wrote, along with his books on specific travels, a hybrid volume called *The Gentle Art of Tramping*. It gives cheery anecdotal instructions on the art for 271 pages, in chapters on boots, "marching songs," "drying after rain," and "trespassers' walk." Thoreau alone seems to get lost in his own thinking and to find himself in surprising places, advocating abandonment, manifest destiny, amnesia, and, for his work, a rare nationalism, but by the time he is advocating the latter, ax-swinging frontiersmen rather than unarmed walkers are his protagonists. Perhaps the limits are implicit in the form of the essay, which is widely regarded as a kind of literary birdcage capable of containing only small chirping subjects, as distinct from the lion's den of the novel and the open range of the poem. Writing and walking were reduced to fit each other, at least in this major tradition in the English-speaking world.

----

completely exhausted, having accomplished the forty miles in record time, but it seemed to me a somewhat

----

II. THE SIMPLE

This belief in walking—rural walking, anyway—as virtuous persists. Examples are everywhere. Recently, I found one particularly annoying essay in a Buddhist magazine asserting that all the world's problems would be solved if only the world's leaders would walk. "Perhaps walking can be the way to peace in the world. Let world leaders walk to the conference site instead of riding in a power-contagious limousine. Take away the conference tables—whatever their size and shape—and have a meeting of minds along the shores of Lake Geneva." Another example by a world leader suggests how dubious this idea is. Ronald Reagan wanted to start his memoirs with the most important moment of his presidency, writes the editor Michael Korda, who tried to make them work as a book. The moment was during his first meeting with Mikhail Gorbachev, near Geneva. Geneva is Rousseau's birthplace, and it was a Rousseauian scene that Reagan described. "Reagan had realized . . . that the summit meeting was going nowhere. The two leaders were surrounded by advisers and specialists as they discussed disarmament, and were unable to make any human contact, so Reagan had tapped Gorbachev on the shoulder and invited him to go for a walk. The two went outside, and Reagan took Gorbachev down toward the shore of Lake Geneva." Reagan went on to say that during "a long, heartfelt discussion" on that occasion they agreed toward mutual inspection and verification as well as the first steps toward nuclear disarmament. Korda objected to an aide that the anecdote as Reagan had told it, though deeply moving, was problematic. Gorbachev and Reagan didn't speak each other's language. If any such walk took place, a retinue of translators and security people must have gone with them to make the event more resemble a state procession than a friendly ramble.

To propose that the world's problems would be solved by two old men walking by a Swiss lakeside (in Rousseau's hometown, what's more) was to propose that the simple, the good, and the natural were still aligned, and that these world leaders who held the power to destroy the earth were themselves simple men (and to suggest that they were simple is to imply that they were good and thus that their regimes were just and their achievements honorable, a series of dom-

---

curious beginning for a honeymoon.—BERTRAND RUSSELL   *"And that's to be life!" said Helen, with a*

inoes lined up behind the first Romantic assumption). The aesthetic of the simple virtues had continued to triumph over the aesthetic of the royal procession, with its signs of complexity, sophistication, its many people signifying society. Jimmy Carter actually walked down Pennsylvania Avenue for his inauguration as president, but Reagan brought a new level of pomp and ceremony to the White House and came as close as American presidents ever have to being a Sun King. He did so by telling us simple stories about our lost innocence, our corruption by education and the arts, our ability to fall back on log-cabin virtues and thereby to dispense with the complex interdependencies of society, economic and otherwise. Portraying himself as a Rousseauian walker was one of those stories. The history of rural walking is full of people who wish to portray themselves as wholesome, natural, a brother to all man and nature, and who in that wish often reveal themselves to be powerful and complicated—though other walkers are true radicals out to undermine the laws and authorities that stifle others as well as themselves.

III. The Far

Just as the walking essay seems to have been the dominant form for writing about walking in the nineteenth century, so the lengthy tale of the very long walk is for the twentieth century. Perhaps the twenty-first will bring us something altogether new. In the eighteenth century, travel literature was commonplace, but the long-distance walkers left little written record of their feats. Wordsworth's walking tour across the Alps, as described in *The Prelude*, was not published until 1850, and *The Prelude* is not exactly travel writing. Thoreau wrote accounts of walks in which his own experience is charted with the same scientific acuity as is the natural world around him, but these are more nature essays than walking literature. The first significant account of a long-distance walk for the sake of walking I know is John Muir's *Thousand Mile Walk to the Gulf*, describing a journey from Indianapolis to the Florida Keys in 1867 (published after his death in 1914). The South he walked through was an open wound still festering from the Civil War, and Civil War historians must be frustrated by Muir's neglect of social

---

catch in her throat. *"How can you, with all the beautiful things to see and do—with music—with*

observation for the sake of botanizing, though it is still the most populated of his many books. The wilderness writings make him a kind of
John the Baptist come back from a suddenly appealing wilderness to
preach its wonders to the rest of us (wilderness because its indigenous
inhabitants had been forcibly removed and decimated before Muir
arrived, but that's another story). For Muir is the United States's evangelist of nature, adapting the language of religion to describe the
plants, mountains, light, and processes that he so loved. As close an
observer as Thoreau, he is far more apt to read religion into what he
sees. He was also one of the great mountaineers of the nineteenth
century, achieving in his woolens and hobnailed boots feats that most
with modern gear would be hard-pressed to follow. Lacking Wordsworth's poetic gifts and Thoreau's radical critique, Muir nevertheless
walked as they only imagined walking, for weeks alone in the wilderness, coming to know a whole mountain range as a friend and
turning his passion for the place into political engagement. But that
came decades after his walk in the South.

A *Thousand Mile Walk* is episodic, as are most such walking books.
In such travel literature there is no overarching plot, except for the
obvious one of getting from point A to point B (and for the more in-
trospective, the self-transformation along the way). In a sense these
books on walks for their own sakes are the literature of paradise, the
story of what can happen when nothing profound is wrong, and so
the protagonist—healthy, solvent, uncommitted—can set out seeking
minor adventure. In paradise, the only things of interest are our own
thoughts, the character of our companions, and the incidents and appearance of the surroundings. Alas, many of these long-distance
writers are not fascinating thinkers, and it's a dubious premise that
someone who would be dull to walk round the corner with must be
fascinating for a six-month trek. To hear about walking from people
whose only claim on our attention is to have walked far is like getting
one's advice on food from people whose only credentials come from
winning pie-eating contests. Quantity is not everything. But Muir has
far more to offer than quantity. An acute and often ecstatic observer
of the natural world around him, he says nothing at all about why he
is walking in A *Thousand Mile Walk to the Gulf*, though it seems clear
enough that it is because he is hardy, poor, and possessed of botanical

----

walking at night—" "Walking is well enough when a man's in work," he answered. "Oh I did talk a lot

----

passions best fulfilled on foot. But though he is one of history's great walkers, walking itself is seldom his subject. There is no well-defined border between the literature of walking and nature writing, but nature writers tend to make the walking implicit at best, a means for the encounters with nature which they describe, but seldom a subject. Body and soul seem to disappear into the surrounding environment, but Muir's body reappears when his paradisiacal luck runs out and he starves waiting for money to arrive and later becomes mortally ill.

Seventeen years after Muir, another young man in his twenties set out to walk more than a thousand miles, from Cincinnati to Los Angeles. Charles F. Lummis says at the outset of his *Tramp Across the Continent*, "But why tramp? Are there not railroads and Pullmans enough, that you must walk? That is what a great many of my friends said when they learned of my determination to travel from Ohio to California on foot; and very likely it is the question that will first come to your mind in reading of the longest walk for pure pleasure that is on record." Which is to say, he starts out thinking of his friends, readers, and the record, as well as pleasure. But he goes on to say, "I was after neither time nor money, but life—not life in the pathetic meaning of the poor health-seeker, for I was perfectly well and a trained athlete; but life in the truer, broader, sweeter sense, the ex-hilarant joy of living outside the sorry fences of society, living with a perfect body and a wakened mind. . . . I am an American and felt ashamed to know so little of my country as I did, and as most Americans do." Seventy-nine pages later, he says of a brief companion, "He was the only live, real walker I met on the whole long journey, and there was a keen zest in reeling off the frosty miles with such a companion." Lummis is vain; there are anecdotes where he outshoots and outtoughs westerners, rattlesnakes, and snowstorms, and his solemn attempts at jokes, in the vein of Twain, often fall flat. But he redeems himself by his great (and for the time, unusual) affection for the people and land of the Southwest and occasional anecdotes at his own expense. It is a remarkable story, of toughness and navigatory ability and adaptability. Long-distance walking in North America never had the gentility of the walking tour. In England, you can walk from pub to pub or inn to inn (or, nowadays, hostel to hostel); in America a long-distance walk is usually a plunge into the wilderness or at least

of nonsense once, but there's nothing like a bailiff in the house to drive it out of you. When I saw him

un-English scale and uninviting spaces such as highways and hostile towns.

There seem to be three motives for these long-distance trips: to comprehend a place's natural or social makeup; to comprehend oneself; and to set a record; and most are a combination of the three. An extremely long walk is often taken up as a sort of pilgrimage, a proof of some kind of faith or will, as well as a means of spiritual and practical discovery. Too, as travel became more common, travel writers often sought out more extreme experience and remote places. One of the implicit premises of the latter kind of writing is that the journey, rather than the traveler, must be exceptional to be worth reading about (though Virginia Woolf wrote a brilliant essay about going out into a London evening to buy a pencil, and James Joyce managed to write the greatest novel of the twentieth century about a pudgy ad salesman trudging Dublin's streets). For writers, the long-distance walk is an easy way to find narrative continuity. If a path is like a story, as I was proposing a few chapters ago, then a continuous walk must make a coherent story, and a very long walk makes a full-length book. Or so goes the logic of these recent books, and to some extent it is true; a walker does not skip over much, sees things close up, and makes herself vulnerable and accessible to local people and places. On the other hand, a walker may be so consumed by athletic endeavor as to be unable to participate in his surroundings, particularly when driven by a schedule or competition. Some of them are happy with these limits, as is Colin Fletcher, one of those inevitable Englishmen whose first long walk is a journey up the eastern side of California in 1958. The resultant book, entitled *The Thousand-Mile Summer*, is a sort of trail mix made up of bite-size epiphanies, moral lessons, blisters, social encounters, and recounted practical details. He took other walks later, and like Graham wrote a guidebook, *The Complete Walker*, still used by backpackers. Another Englishman, John Hillaby, walked the length of Britain—a thousand miles—in 1968 and wrote a best-seller about it, as well as several other books about other walks.

By the time Peter Jenkins set out to walk more than three thousand miles across the United States in 1973 (with *National Geographic* sponsorship), the cross-country expedition had become a kind of rite of

fingering my Ruskins and Stevensons, I seemed to see life straight real, and it isn't a pretty sight."—E. M.

passage of American manhood, though by that time the means were more often vehicular. Crossing the continent seemed to embrace or encompass it at least symbolically, the route wrapped around it like a ribbon around a package. The movie *Easy Rider*, which had recently been released, seemed to draw some of its sensibility from Jack Kerouac's road stories, which themselves often sprawled more like travel books than novels (Kerouac's *Dharma Bums* recounts how the poet and ecologist Gary Snyder got Kerouac out of the car and into the mountains). Jenkins set out to have social encounters; the America he was looking for was, unlike Muir's, made up of people rather than places. Like Wordsworth in his incessant encounters with characters eager to tell their tale, he takes the time to listen to everyone he meets and tells about them in his naively earnest *Walk Across America* and *Walk Across America II*. In part a reaction against the anti-Americanism of the young radicals of the time, Jenkins's journey brings him into close contact and, often, friendship with the white southerners so reviled by northern civil rights activists. In the course of his travels, he stays with an Appalachian living off the land, lives with a poor black family for several weeks, and in Louisiana falls in love with a Southern Baptist seminarian, undergoes a religious conversion, marries the woman, and after several months resumes his walk with her, arriving on the Oregon coast a far different person from the one who set out. This is truly a journey as life, for Jenkins goes as slowly as experience demands.

The literature of the long-distance walk is a sort of downhill slope. Toward the bottom come books by people who are athletic walkers but not necessarily writers, for the necessary combination of silver tongue and iron thighs seems to be a rare one. The most impressive of the contemporary long-distance walkers I have read— there are many now—is Robyn Davidson, who didn't exactly set out to write about walking at all, but did so brilliantly in the course of her *Tracks*, a book recounting her 1,700-mile trek across the Australian outback to the sea with three camels (sponsored, like Jenkins's odyssey, by the National Geographic Society). Midway in her journey, she explains its effect on her mind: "But strange things do happen when you trudge twenty miles a day, day after day, month after month. Things you only become totally conscious of in retrospect.

FORSTER, *HOWARDS END*   *Rhythm is originally the rhythm of the feet. Every human being walks, and*

For one thing I had remembered in minute and Technicolor detail everything that had ever happened in my past and all the people who belonged there. I had remembered every word of conversation I had had or overheard way, way back in my childhood and in this way I had been able to review these events with a kind of emotional detachment as if they had happened to somebody else. I was rediscovering and getting to know people who were long since dead and forgotten. . . . And I was happy, there is simply no other word for it." She brings us back to the territory of the philosophers and the walking essayists, to the relationship between walking and the mind, and she does it from a kind of extreme experience few have had.

The 1970s seem to have been a golden age of long-distance walks; Jenkins, Davidson, and Alan Booth all set out in the mid-1970s. Booth's delightful *Roads to Sata: A Two-Thousand-Mile Walk Through Japan* is a milestone in how far the literature of walking had come. An Englishman who had lived in Japan for seven years and come to know the language and culture well, he is unfailingly humorous and modest, a great evoker of place and recounter of comic conversations, respectful but not reverent about the culture. He describes his trip— dirty socks, hot springs, sake and more sake, comic and tragic figures, sultry weather, lechers of both sexes—with élan. He comments wryly, "In properly developed countries, the inhabitants regard walkers with grave suspicion and have taught their dogs to do the same," but enjoys himself all the same. Yet like most of these travel books, his is not really a book about walking. That is to say, it is not about the acts but about the encounters, just as *A Thousand Mile Walk to the Gulf* is really about botany and natural epiphanies—and *On the Road* and *Easy Rider* are only implicitly about the internal combustion engine and its implications. Walking is only a means to maximize those encounters and perhaps test body and soul.

The test is central to Ffyona Campbell's prolific walks, as recounted in her book *The Whole Story: A Walk Around the World*. The daughter of a harsh military man, she seems to be on a quest to prove herself to him and to herself, with her walking an obsessive activity not unlike her sister's anorexia (which crops up in her book). In 1983, at the age of sixteen, Campbell successfully walked the length of Britain—a thousand miles—sponsored by London's *Evening Standard*

---

since he walks on two legs with which he strikes the ground in turn and since he only moves if he continues

and seeking to raise money for a hospital. She then set out to walk around the world—not literally, for the continuous line that links up many walkers' narratives has nothing to do with her: "The Guinness Book of Records defines a walk around the world as beginning and finishing in the same place, crossing four continents, and covering a total of at least 16,000 miles," says the preface of her book. She set off across the United States two years later, Australia five years later, and the length of Africa eight years later, finishing up eleven years after the English walk with a trek northward from Spain to the English Channel. It is as discontinuous as could be—she flies back to Africa and the States to complete segments she left out earlier—and only a kind of accounting holds it all together as a single act.

Perhaps it is a mistake to include Campbell in the literature of walking, even though she has produced books, but she is certainly part of the culture of walking. There is another ancestry for her in the pedestrian athletes of the late eighteenth and early nineteenth century, who seemed indifferent whether they went their thousand miles around a track or down a road, and who were the subjects of heavy betting. After all, almost no landscape appears in her narratives of several continents; we cannot in it trace an inheritance from Wordsworth. Yet the notion that walking is somehow redemptive and walking farther is more so seems to have taken on a fearful life of its own, and surely this is something of a Victorian heritage, and those Victorians were themselves heirs of Wordsworth. Such is the winding road down which history comes, now with one set of desires in view, now with another. Like Davidson, Campbell seems driven, but Davidson represents a more intellectual, insightful version of the wounded self seeking redemption through an ordeal, and comes equipped with vastly more literary and landscape sensibility. The fierce alienation is much the same, the sense of a young woman clinging to her stubbornness and her arduous goal because that's all she has. Jenkins is softer, less locked up, maybe because it's easier for a man, maybe because he's more openly a seeker: he knows what kind of pilgrimage he is on.

To some extent Campbell resembles the Walkathon walkers, in that she is often walking to raise money for a cause (or more often looking for a cause to represent so she can also raise money for her expeditions, which with support staff, publicity, and so forth were

---

to do this, whether intentionally or not, a rhythmic sound ensues. . . . Animals too have their familiar

---

often expensive). Still, to walk fifty miles in a day is remarkable, to get up and do it again the next day is stunning, and to do it day after day across the Australian outback alongside a road in ugly weather is brutal. Campbell did it, walking 3,200 miles across that continent in ninety-five days, a world record. Her legs are indefatigably strong and relentless in their pursuit, but nothing is left in her walks but accomplishment—no scenery, no pleasure, few encounters. For 20,000 miles she is struggling to understand herself well enough to outwalk her suffering, but she is alarmingly unclear about her values, seeking corporate sponsorship and media attention at some points and condemning journalists and capitalists at others, insulting people who drive cars on her second walk in the United States, after having been trailed across the country by a motor home driven by her support staff the first time. Her book ends with an anecdote that undermines all her effort, one of many passages of fuzzy reverence for indigenous peoples. It is a tale of the military men who challenge some aboriginal Australians to a footrace across the desert, which the latter abandon to track down honeycomb. Telling it, she suggests she is on the side of the aborigines in disdaining rigid goals, quantifiable experience, competition, even record-keeping or -making, as deeply flawed ways of being in the world. The tragedy is that all along she has been on the side of the military men.

Perhaps Campbell shows us pure walking. It is impurity that makes it worthwhile, the views, the thoughts, the encounters—all those things that connect mind and world through the medium of the roving body, that leaven the self-absorption of the mind. These books suggest how slippery a subject is walking, how hard it is to keep one's mind on it. Walking is usually about something else—about the walker's character or encounters, about nature or about achievement, sometimes so much so it ceases to be about walking. Yet together all these things—the canons of walking essays and travel literature—constitute a coherent, if meandering, two-hundred-year history of reasons to walk across the land.

---

gait, their rhythms are often richer and more audible than those of men, hoofed animals flee in herds, like

---

Chapter 9

MOUNT OBSCURITY AND
MOUNT ARRIVAL

Ffyona Campbell's tale of the military men racing across the Australian outback toward the finish line and their aboriginal rivals straying from it to gather honeycomb suggests some of the various ways and reasons to walk and to live, or at least some of the questions. Can one weigh public glory against private pleasure, and are they mutually exclusive? What portions of an act can be measured and compared? What does it mean to arrive, and what to wander without destination? Is competition an ignoble motive? Can the soldiers be imagined as students of discipline and the aboriginal men as students of detachment? After all, there are pilgrims for whom arrival at their journey's end is spiritual consummation, but there are other pilgrims and mystics who wander without cease or destination, from the Chinese sages of antiquity to the anonymous nineteenth-century Russian peasant who wrote *The Way of a Pilgrim*. These questions about how one travels and why become most pressing, or at least most evident, with mountaineering.

Mountaineering is the art of getting up mountains by foot and occasionally by hand, and though the climbing is usually emphasized, most ascents are mostly a matter of walking (and since good climbers climb with their legs as much as possible, climbing could be called the art of taking a vertical walk). In the steepest places the steady semiconscious rhythm of walking slows down, every step can become a

---

regiments of drummers. The knowledge of the animals by which he was surrounded, which threatened him

separate decision about direction and about safety, and the simple act of walking is transformed into a specialized skill that often calls for elaborate equipment. Here I want to address mountaineering that includes climbing but leave aside the separate discipline of climbing without mountaineering, a somewhat artificial division, but one with reasons. The latter is a recently explored side canyon in the history of mountaineering in which technique has been vastly refined to ascend ever-more-challenging surfaces. A supremely hard climb can be less than a hundred feet long, and a single move can become a famous "problem" to be worked out by intense application and training. And while mountaineering is traditionally motivated by a taste for mountain scenery, technical climbing seems to involve other pleasures. Since the eighteenth century, nature has been imagined as scenery, and scenery is what is seen at a certain distance, but climbing puts one face-to-face with the rock, with a wholly different kind of engagement. Perhaps tactile encounters, sensations of gravity (and, sometimes, mortality), and the kinesthetic pleasures of one's body moving at its limit of ability are an equally valid if less culturally hallowed experience of nature. With climbing, sometimes scenery disappears altogether, at lest in the rapidly proliferating indoor climbing gyms. Too, walking fosters one kind of awareness in which the mind can stray away from and return to the immediate experience of traversing a particular place; rock climbing, on the other hand, is demanding enough that one guide told me, "Climbing is the only time my mind doesn't wander." Climbing is about climbing. Mountaineering, on the other hand, is still about mountains.

Most standard histories of mountaineering and of landscape aesthetics start with the poet Petrarch, "the first man to climb a mountain for its own sake, and to enjoy the view from the top," as the art historian Kenneth Clark put it. Long before Petrarch climbed Italy's Mount Ventoux in 1335, there were others ascending mountains in other parts of the world. Petrarch prefigures the Romantic-generated practice of traveling among mountains for aesthetic pleasure and getting to their summits for secular reasons. This history of mountaineering really begins in Europe in the late eighteenth century, when curiosity and changed sensibilities spurred a few bold individuals not just to travel through the Alps but to try to get to their

---

and which he hunted, was man's oldest knowledge. He learnt to know animals by the rhythm of their

---

summits. The practice was gradually consolidated into mountain-eering, a set of skills and assumptions—for example, the assumption that getting to the top of a mountain is a uniquely meaningful act, distinct from walking among the passes or foothills. In Europe moun-taineering developed largely as a gentleman's pastime and a guide's profession, since the former so often relied upon the latter; in North America the first recorded ascents were made by explorers and sur-veyors in far remoter places (some ascents in the Alps could and can be watched through telescopes from the villages below; some in North America took weeks of wilderness trekking to reach). Of course, as the great surveyor and mountaineer Clarence King re-counts, when in 1871 he got to the top of Mount Whitney, the highest point in the contiguous forty-eight states, he found that "a small mound of rock was piled on the peak, and solidly built into it an Indian arrow-shaft, pointing due west." Mountains attracted attention and walkers long before romanticism spawned mountaineering.

A lone peak or high point is a natural focal point in the landscape, something by which both travelers and locals orient themselves. In the continuum of landscape, mountains are discontinuity—culminating high points, natural barriers, unearthly earth. On mountains, lati-tude's imperceptible changes can become altitude's striking transfor-mations. Ecology and climate change rapidly from balmy foothills to glacial heights: there's the timberline and, farther up, what could be called the lifeline, beyond which nothing lives or grows, and, above about 18,000 feet, what mountaineers call the death zone, the icy low-oxygen realm where the body starts to die, judgment is impaired, and even the most acclimated alpinists lose brain cells. Up high, bi-ology vanishes to reveal a world shaped by the starker forces of ge-ology and meteorology, the bare bones of the earth wrapped in sky. Mountains have been seen around the world as thresholds between this world and the next, as places where the spirit world comes close. In most parts of the world, sacred meanings are ascribed to moun-tains, and though the spirit world may be terrifying, it is seldom evil. Christian Europe seems to be alone in having seen mountains as ugly and almost hellish realms. In Switzerland, dragons, the souls of the unhappy dead, and the Wandering Jew were supposed to haunt the heights (sentenced in the legend to wander the earth until the Second

---

movement. The earliest writing he learnt to read was that of their tracks; it was a kind of rhythmic

---

Coming because he slighted Jesus, the Wandering Jew suggests that
European Christians often took a dim view of wandering as well as of
Jews). Many seventeenth-century English writers express their detes-
tation of mountains as "high and hideous," "rubbish of the earth," "deformities," and even damage caused to a formerly smooth earth by the
Deluge. So though Europeans led the world in the development of
modern mountaineering, that mountaineering came out of romanticism's recovery of an appreciation for natural places that much of the
rest of the world had never lost.

One of the first individuals whose ascent of a mountain is recorded is China's "First Emperor," who in the third century B.C. drove
his chariot up T'ai Shan against the advice of his sages, who thought
he should walk. Better known for starting the Great Wall and for
burning all the books so that Chinese history would start with him,
the First Emperor may have eradicated the record of those who ascended before him. Most people since have walked to T'ai Shan's
summit—for many centuries on the 7,000-step staircase leading from
the City of Peace at the foot of the mountain through three Heavenly
Gates to the Temple of the Jade Emperor on top. American writer and
Buddhist Gretel Ehrlich walked up T'ai Shan and other mountain pilgrimage sites in China and wrote, "The Chinese phrase for 'going on
a pilgrimage,' ch'ao-shan chin-hsiang, actually means 'paying one's
respects to the mountain,' as if the mountain was an empress or an
ancestor before whom one must kneel." In the fourth century A.D. a
very different kind of pilgrim climbed mountains on the other side of
Eurasia: the Christian pilgrim Egeria. Almost no trace of her but her
pilgrimage diary survives, though that manuscript suggests she was
an abbess or other religious figure of some stature and that Mount
Sinai deep in the Egyptian desert was among the sites of Christian
pilgrimage then. She was guided by resident holy men through "the
vast and very flat valley where the children of Israel tarried during
those days when the holy man Moses climbed the mountain of God"
on their flight from slavery in Egypt. She and her unnamed companions scaled the nearly 9,000-foot peak of Mount Sinai on foot—
"straight up, as if scaling a wall." Egeria noted that "this seems to be a
single mountain all around; however, once you enter the area you see
there are many, but the whole range is called the Mountain of God."

---

notation imprinted on the soft ground. . . . the large numbers of the herd which they hunted blended into

For Egeria, Sinai was the mountain on which God had descended and Moses had ascended to receive the Tablets of the Law: climbing it was a profession of faith in Scripture and a return to the site of its greatest moments. Since her time stairs have been built up Sinai too, and one fourteenth-century mystic ascended them every day as his religious expression.

Mountains, like labyrinths and other built structures, function as metaphorical and symbolic space. There is no more clear geographical equivalent to the idea of arrival and triumph than the topmost peak beyond which there is no farther to go (though in the Himalayas many pilgrims circumambulate mountains, believing it would be sacrilegious to stand on the summit). The athletically gifted and enormously ambitious Victorian mountaineer Edward Whymper said of reaching the top of the Matterhorn, "There is nothing to look up to; all is below," in a telling mix of literal and figurative language. "The man who is there is somewhat in the position of one who has attained all that he desires—he has nothing to aspire to." The appeal of climbing to the top of mountains may also be drawn from language metaphors. English and many other languages associate altitude, ascent, and height with power, virtue, and status. Thus we speak of being on top of the world or at the top of one's field, at the height of one's ability, on the way up; of peak experiences and the peak of a career; of rising and moving up in the world; to say nothing of social climbers, upward mobility, high-minded saints and lowly rascals, and of course the upper and the lower classes. In Christian cosmology, heaven is above us and hell below, and Dante portrays Purgatory as a conical mountain he arduously ascends, conflating spiritual and geographical travel (starting with what modern climbers would call a chimney: "We climbed up through the narrow cleft, / rock pressed in on us from either side, / and that ground needed both feet and hands"). A walk uphill traverses these metaphysical territories; a goalless ramble across the same mountain moves through very different metaphysics.

In Japan mountains have been imagined as the centers of vast mandalas spreading across the landscape like, in one scholar's words, "overlapping flowers," and approaching the center of the mandala means approaching the source of spiritual power—but the approach may be indirect. In a labyrinth one can be farthest from the destination when one is closest; on a mountain, as Egeria found, the mountain itself changes shape again and again as one ascends. The famous Zen parable about the master for whom, before his studies, mountains were only mountains, but during his studies mountains were no longer mountains, and afterward mountains were again mountains could be interpreted as an allegory about this perceptual paradox. Thoreau noticed it and wrote, "To the traveller, a mountain outline varies with every step, and it has an infinite number of profiles, though absolutely but one form," and that form is best apprehended from a distance. In every print but one of the Japanese artist Hokusai's famous *Thirty-Six Views of Mount Fuji*, the perfect cone of Mount Fuji looms largely nearby or small far away, giving orientation and continuity to city, road, field, and sea. Only in the print of pilgrims actually ascending the mountain does the familiar shape that unites the other prints vanish. When we are attracted, we draw near; when we draw near, the sight that attracted us dissolves: the face of the beloved blurs or fractures as one draws near for a kiss, the smooth cone of Mount Fuji becomes rough rock rising from underfoot to blot out the sky in Hokusai's print of the mountain pilgrims. The objective form of the mountain seems to dissolve into subjective experience, and the meaning of walking up a mountain fragments.

A walk, I have claimed, is like a life in miniature, and a mountain ascent is a more dramatic walk: there is more danger and more awareness of death, more uncertainty about the outcome, more triumph at what is more unequivocally arrival. "To climb up rocks is like all the rest of your life, only simpler and safer," wrote the British mountaineer Charles Montague in 1924. "Each time that you get up a hard pitch, you have succeeded in life." What fascinates me about mountaineering is how one activity can mean so many disparate things. Though the idea of pilgrimage almost always seems to be present, many ascents derive their meaning from sports and military action as well. Pilgrimage draws meaning from following hallowed routes to established destinations, while the most revered mountaineers are often those who are first on a route or summit, who like athletes make a record. Mountaineering has often been seen as a pure form of the imperial mission, calling into play all its skills and heroic virtues, with none of its material gains or oppositional violence (which is why the superb French alpinist Lionel Terray

---

state of communal excitement which I shall call the rhythmic or throbbing crowd. The means of achieving

called his memoir *Conquistadors of the Useless*). On March 17, 1923, while on a speaking tour to raise money for an Everest expedition, the great mountaineer George Mallory apparently got exasperated with the continual questions about why he wanted to climb it, and uttered the most famous line in mountaineering history, the one sometimes cited as a Zen koan: "Because it's there." His usual reply was, "We hope to show that the spirit that built the British Empire is not yet dead." Mallory and his companion Andrew Irvine themselves died on that expedition, and mountaineering historians still debate whether they got to the summit before vanishing. (Mallory's battered, frozen body was discovered seventy-five years later, on May 1, 1999.)

The measurable part of an experience translates most easily, so the highest peaks and worst disasters are the best known aspects of mountaineering, along with all the records—first ascent, first ascent by the north face, first American, first Japanese, first woman, fastest, first without this or that piece of gear. Mount Everest has always been about these calculables for Westerners, to whose attention it first came through trigonometry. In 1852 a clerk in the office of the British Trigonometrical Survey in India calculated that what they called "Peak XV" and the Tibetans "Chomalungma" was taller than all the Himalayan peaks clustered around it. The man who had just measured it named it after a man who had never noticed it, former surveyor general of India Sir George Everest (thereby giving it a kind of sex change, since *chomalungma* means "goddess of the place"). The locals consider Chomalungma one of the less significant sacred mountains, but mountaineering writers sometimes call Everest (which is at the same latitude as southern Florida) the top or the roof of the world, as though our spherical planet were instead some kind of pyramid. The widely traveled mountaineer and religious scholar Edwin Bernbaum writes, wryly, "Whatever Western society regards as number one tends to take on an aura of ultimacy that makes it seem more real and worthwhile than anything else—in a word, sacred." And number one is generally determined by measurement. Triumph in mountaineering, as in sports, is measured in firsts, fastests, and mosts.

Like sports, mountaineering is exertion with only symbolic results, but the nature of that symbolism dictates everything—why, for example, French mountaineer Maurice Herzog could consider his

---

this state was first of all the rhythm of their feet, repeating and multiplied.—ELIAS CANETTI, *CROWDS*

1950 expedition to Annapurna, the world's seventh-highest mountain, a great victory because they made it to the top and not a failure because he got so severely frostbitten he lost all his fingers and toes and had to be carried down by sherpas. Perhaps it was that Herzog trod the terrain of history as selflessly as Egeria did Scripture. In the mid-1960s, David Roberts led the second-ever ascent to the summit of Alaska's Mount Huntington. As he recounted it in his book _The Mountain of My Fear_, the expedition seems to have begun in Massachusetts with his study of photographs of the mountain, his surmise of a new route up it, and his desire to do something that hadn't been done before. That is, the expedition began with visual representation and desire to situate himself in the historical record, with months of planning, fund-raising, recruiting, collecting gear, and writing lists; it only became a bodily engagement with the mountain long afterward. This tension between history and experience, between aspiration, memory, and the moment, fascinates me, and though it exists throughout human activity, it seems to become, so to speak, more transparent at high altitudes. History, let me clarify, means an act imagined as being situated in the context of other such acts and as it will be perceived by others; it arises from a social imagination of how one's private acts fit into public life. History is carried in the mind to the remotest places to determine what one's acts mean even there, and who can say how much it weighs for those who carry it?

Because mountain heights are usually so remote from inhabited earth, because mystics and outlaws have so often gone there to vanish from sight, because climbing is "the only time my mind doesn't wander," making history in the mountains seems a particularly paradoxical idea and mountaineering a particularly paradoxical sport—when it is regarded as a sport. Being first up a mountain means entering the unknown, but for the sake of putting the place into human history, of making it known. There are those who decline to record their ascents or name their climbs, who see their mountaineering as a retreat from history. Gwen Moffat, who became Britain's first certified woman climbing guide in 1953, wrote about the immediate satisfactions: "And before I started to move I felt the familiar feeling that came when I was about to do something hard. Mental and physical relaxation, a loosening of the muscles so complete that even the face relaxes and the eyes widen; one's body becomes light and supple—a pliable and co-ordinated entity to be shown a climb as a horse is shown a jump. In that exquisite moment before the hard move, when one looks and understands, may lie an answer to the question why one climbs. You are doing something hard, so hard that failure could mean death, but because of knowledge and experience you are doing it safely." She and a partner once decided to set the record for the slowest traverse ever of a ridge on the Isle of Skye and, with the help of a surprise blizzard, probably succeeded.

European mountaineering history had its beginnings in a competition of sorts. Decades before Mont Blanc was climbed, the glacier coming down from it to the Chamonix Valley and the valley itself had become tourist destinations (as they have remained ever since). The locals were, like those of north Wales and the Lake District, beneficiaries of travelers' growing taste for wild and rugged scenery. One outcome of this growing tourist economy was that a twenty-year-old gentleman scientist from Geneva, Horace Benedict de Saussure, arrived in 1760, became so fascinated by glaciers that he dedicated the rest of his life to studying them, and posted a handsome prize for the first person to reach the 15,782-foot summit of Mont Blanc. Mont Blanc, the highest point in Europe, was a magnet in the early years of the cult of mountains and a cultural icon for landscape romantics, the subject of a major poem by Shelley, the first measure of ambition of mountaineers. In 1786 a local doctor reached the summit with the aid of a local hunter. An attempt a few years before had so frightened the four guides who tried it that they declared it unclimbable, and no one in Europe was then certain whether human beings could survive at such high altitudes. Chamonix resident Dr. Michel Gabriel Paccard, writes a notable later mountaineer, Eric Shipton, "turned his keen intelligence and his already keen experience as a climber to the problems of mountain survival. . . . He did not seek notoriety, and he spoke little of his exploits, many of which showed great resolution and physical stamina. His wish to climb Mont Blanc was apparently inspired more by his desire to be first for France—and in the interests of science—than by any desire to win fame for himself. Among other things he was anxious to make barometric observations at the top. . . ."

After four unsuccessful attempts, the doctor hired Jacques Balmat,

---

admired, beauty of, enjoyment of, "know thyself" philosophy, and men, mysticism, romantic feeling

---

a strong climber who made a living as a hunter and collector of crystals. They set out on a night of the full moon in August, without the ropes and ice axes of modern mountaineering, crossing the deep ice crevasses with nothing more than a pair of long poles. When they reached the dread Valley of Snow, the deep recess surrounded by icy walls where the four guides had given up earlier, Balmat begged to turn back, but Paccard convinced him to continue, and they climbed up a snow ridge in a high wind. They reached the summit early in the evening, fourteen hours after they had set out; Paccard made his measurements, and they descended to spend the night under a boulder. In the morning both were badly windburned and frostbitten, and Paccard was snow-blind as well and had to be led downward. "Judged by sheer physical effort alone, the first ascent of Mont Blanc was a remarkable performance," Shipton concludes. But the story didn't end there. The scheming Balmat began to spread stories that it was he who had explored the route and led the expedition, and that Paccard was little more than baggage he dragged along. His stories grew until he was claiming that Paccard had collapsed several hundred feet below the summit and Balmat alone had completed the ascent. It wasn't until the twentieth century that the truth was uncovered and the brave doctor was restored to his place among the heroes of mountaineering. One of the mountaineers had betrayed his companion and the truth for the sake of history, publicity, and the reward (and a century later the explorer Frederick Cook lied and faked photographs to claim he had made the first ascent of Alaska's Mount Denali; for him history counted for everything, experience for nothing).

As soon as Mont Blanc was proved climbable, many others began to climb it. By the middle of the nineteenth century, forty-six parties, many of them English, had reached the summit, and the focus turned to other Alpine peaks and routes. Though there are many greater mountaineers, I can't get over my affection for Henriette d'Angeville. It may be the effusiveness of her *My Ascent of Mont Blanc* that charms me, since it proves that great physical stamina need not be coupled with stoicism, or it may be that real mountaineering literature is for real mountaineers, who love passages full of hand-jams, mantelshelf moves, crampon or belaying technique, and so forth. D'Angeville was

towards, scientific attitude, sublimity, sunrise seen from, walking, and women.—INDEX ENTRY IN

forty-four when she went up the mountain in 1838, though she had grown up among the Alps and walked in them before. She cleared up the inevitable question about why she climbed early in her book, writing, "The soul has needs, as does the body, peculiar to each individual. . . . I am among those who prefer the grandeur of natural landscapes to the sweetest or most charming views imaginable . . . and that is why I chose Mont Blanc." Later she earned popular scorn by quipping that she climbed it to become as famous as the novelist George Sand, but she continued to climb mountains into her sixties without receiving further attention and wrote, "It was not the puny fame of being the first woman to venture on such a journey that filled me with the exhilaration such projects always called forth; rather it was the awareness of the spiritual well-being that would follow." Her climb is a tender drama of arduous ascent bracketed by extravagant packing lists beforehand and a victory dinner with her ten guides afterward. Guiding was already becoming a profession, and technique and tools had evolved much since Paccard's time.

Golden ages usually end with a fall, and the golden age of mountaineering was no exception. Usually described as the period between 1854 and 1865 when many of the Alps were climbed for the first time, it was a largely British golden age in which climbing mountains became a recognized, but by no means popular, sport (far more people continued to walk in the Alps without striving for summits). About half the major first ascents of that age were made by well-heeled British amateurs with local guides. The Alpine Club, founded in 1857 as a sort of cross between a gentleman's club and a scientific society, has so long been an accepted part of the mountaineering world that its oddness— a British club focused on Continental mountains—has seldom been remarked. But in those years the Alps were almost the exclusive focus of this new sport, or pastime, or passion; mountains farther afield and smaller, more technically demanding climbs in places like the Peak District or the Lakes had yet to receive much attention, and climbing in North America took place in a radically different context. The British audience for this activity was far larger than its practitioners, and in Europe mountaineers and climbers still sometimes become celebrities. Albert Smith's popular entertainment *Mont Blanc*, based on

MORRIS MARPLES'S *SHANK'S PONY*   *Since the days of Leslie Stephen, the intelligent justifications for a*

his 1851 ascent, ran in a London theater for years, and books like Alfred Wills's *Wanderings Among the Alps* and the Alpine Club's *Peaks, Passes and Glaciers* series were well received.

Lured by this literature, the twenty-year-old engraver Edward Whymper managed to get an assignment to make images of the Alps. He spent his spare time exploring the mountains and turned out to have a talent for getting up them. Though he made a number of first ascents elsewhere, it was the Matterhorn that captured his imagination. Between 1861 and 1865 he made seven unsuccessful attempts on the spectacular peak, racing against other climbers to be the first. He finally succeeded, and his success is said to have ended the golden age. Whether it ended because Whymper had brought a different or at least more overtly ambitious spirit to the enterprise, because the Matterhorn was the last major Alpine peak to be summitted, or because of the ensuing disaster is unclear. His eighth ascent had been made in collaboration with the greatest amateur climber of the time, the Reverend Charles Hudson, two other young Englishmen, and three local guides. On the descent Hudson, the two young men, and the outstanding guide Michel Croz, who were all roped together, fell to their deaths when one of them slipped. The Victorian equivalent of a media circus ensued, with much condemnation of mountaineering itself as unjustifiably dangerous and much muttering about whether Whymper and the guides had behaved professionally and ethically. Whymper's *Scrambles in the Alps* became a classic anyway, and perhaps it's why the Matterhorn has become a ride at Disneyland.

The history of mountaineering is about the firsts, mosts, and disasters, but behind the dozens of famous faces are countless mountaineers whose rewards have been entirely private and personal. What is recorded as history seldom represents the typical, and what is typical seldom becomes visible as history—though it often becomes visible as literature. Something of this dichotomy is present in the two major genres of mountaineering book, the epics that the general public generally reads and the memoirs that seem to have a far smaller audience. The epics are heroic accounts of an attempt on a major summit; they are books about History and, almost always, Tragedy (high-altitude mountaineering literature, with its emphasis on bodily suffering, survival through sheer will, and the grisly details

---

life of climbing have been few, far between, and (vide Mallory) cryptic, at best. . . . Love of nature, by

---

of frostbite, hypothermia, high-altitude dementia, and fatal falls, often reminds me of books about concentration camps and forced marches, except that mountaineering is voluntary and, for some, deeply satisfying). In contrast, the cheerful memoirs by even some of the greatest climbers—Joe Brown, Don Whillens, Gwen Moffat, Lionel Terray—often read as humorous idylls that deemphasize difficulty. The satisfactions in these narratives come from minor and major excursions, from friendships, freedoms, love of mountains, refinement of skill, low ambition, and high spirits, with only an occasional tragedy on the rocks. The best books' merit comes from the vividness rather than the historic importance of the events they recount.

If we look for private experience rather than public history, even getting to the top becomes an optional narrative rather than the main point, and those who only wander in high places become part of the story. That is to say, we can leave behind sports and records, and when we do, the discipline of the destination is once again balanced with the discipline of detachment. Smoke Blanchard, the guide who came of age climbing the Oregon peaks during the Great Depression, wrote in my favorite of all mountain memoirs, *Walking Up and Down in the World*, "For half a century I have tried to promote the idea that mountaineering is best approached as a combination of picnic and pilgrimage. Mountain picnic-pilgrimage is short on aggression and long on satisfaction. I hope that I can show that mild mountaineering can be happily pursued through a long lifetime without posting records. Can a love affair be catalogued?" The convivial, humorous Blanchard was as much a walker as a mountaineer, and among the pleasures he recounts are long hikes along the Oregon coast and across the width of California, from the White Mountains east of the Sierra Nevada to the sea, as well as many ascents in the Sierra Nevada and Pacific Northwest. Like a lot of other Pacific Coast mountaineers from John Muir to Gary Snyder, he approached the mountains in a way that reconciles wandering and arriving and recalls the older mountain traditions on the other side of the ocean, in China and Japan.

It wasn't ascending so much as being in the mountains that those poets, sages, and hermits celebrated, and the mountains so frequently portrayed in Chinese poetry and paintings were a contemplative

---

the way, seems to have little to do with it. Superclimbers are, on the whole, uncheerful about hiking,

retreat from politics and society. In China, wandering was celebrated—"To 'wander' is the Taoist code word for becoming ecstatic," writes a scholar—but arriving was sometimes regarded with ambiguity. One of the eighth-century poet Li Po's compositions is titled "On Visiting a Taoist Master in the Tai-T'ien Mountains and Not Finding Him," a common theme in Chinese poetry then. Mountains had both physical and symbolic geography, so that literal walking has metaphorical overtones:

> People ask the way to Cold Mountain
> Cold Mountain? There is no road that goes through. . . .
> How can you hope to get there by aping me?
> Your heart and mine are not alike.

writes Li Po's contemporary, the ragged, humorous Buddhist hermit Han-Shan.

In Japan mountains have had religious significance since prehistoric times, though Bernbaum writes, "Before the sixth century A.D. the Japanese did not climb their sacred mountains, which were regarded as a realm apart from the ordinary world, too holy for human presence. The people built shrines at their feet and worshipped them from a respectful distance. With the introduction of Buddhism from China in the sixth century came the practice of climbing the sacred peaks all the way to their summits, there to commune directly with the gods." Afterward, though monks and ascetics wandered, the indeterminate geography of wandering was overshadowed by a determinate one of pilgrimage to the mountains. Climbing mountains became a central part of religious practice, notably in Shugendō, which is more or less a Buddhist mountaineering sect. "Every aspect of Shugendō is conceptually or physically related to the power of sacred mountains and the benefit of reverential behavior within sacred mountains," writes the foremost Western scholar of this sect, H. Byron Earhart. Though festivals, temple ceremonies, and extended periods of mountain asceticism were also part of Shugendō, ascending mountains was central to it both for priests and for lay people, and a kind of priestly guide service emerged. Mountains themselves were perceived as Buddhist mandalas, and ascent paralleled the six stages of spiritual progress

---

impatient with the weather, insensitive to the subtleties of landscape.—DAVID ROBERTS   *. . . the lure of*

---

toward enlightenment (one stage involved dangling initiates over an abyss while they confessed their sins). The seventeenth-century Zen poet Bashō ascended some of Shugendō's most sacred mountains in the course of his meanders, as he recounts in his haiku-and-travel-narrative masterpiece *The Narrow Road to the Deep North:* "I . . . set off with my guide on a long march of eight miles to the top of the mountain. I walked through mists and clouds, breathing the thin air of high altitudes and stepping on slippery ice and snow, till at last through a gateway of clouds, as it seemed, to the very paths of the sun and moon, I reached the summit, completely out of breath and nearly frozen to death." After Shugendō was banned late in the nineteenth century, it ceased to be a major religion in Japan, but it still has shrines and practitioners, Mount Fuji remains a major pilgrimage site, and the Japanese remain among the most avid mountaineers in the world.

Gary Snyder, pretty good mountaineer and great poet, seems to unite the spiritual and secular traditions. After all, he studied Buddhism in Asia but learned how to climb mountains far earlier, with Oregon's Mazamas (a mountaineering club founded atop Mount Hood in Oregon in the 1890s). In an afterword to his book-length poem named after a Chinese scroll, *Mountains and Rivers Without End,* begun in 1956 and finished nearly forty years later, he writes, "I had been introduced to the high snow peaks of the Pacific Northwest when I was thirteen and had climbed a number of summits even before I was twenty. East Asian landscape paintings, seen at the Seattle Art Museum from the age of ten on, also presented such a space." During his years in Japan, he practiced walking meditation and made contact with surviving Shugendō practitioners, "and I was given a chance to see how walking the landscape can become both ritual and meditation. I did the five-day pilgrimage on the Omine ridge and established a tentative relationship with the archaic Buddhist mountain deity Fudo. This ancient exercise has one visualizing the hike from peak to valley floor as an inner linking of the womb and diamond mandala realms of Vajrayana Buddhism."

In 1956, just before he left for Japan, Snyder led Jack Kerouac on an overnight hike to the sea and back across Mount Tamalpais, a 2,571-foot peak on the other side of the Golden Gate from San Francisco. On that walk, Snyder told his footsore companion, "The closer

challenge and testing, the delight of achievement, the contact with our ancestral simplicity, the escape from

you get to real matter, rock air fire wood, boy, the more spiritual the world is." The scholar David Robertson comments, "This sentence states what is perhaps not only the central idea of Gary Snyder's poetry and prose, but the fixed point around which rotate the thought and practice of many who take to trails. If one habit beats in the ritual heart of the lives and their literature, surely this is it: the practice of 'mattering,' of repeatedly accessing the thing that is at one and the same time both spirit and matter. . . . Hiking for Snyder is a way of furthering a political, social, and spiritual revolution. . . . The essential nature of things is not an Aristotelian plot nor a Hegelian dialectic, and does not lead to a goal. Therefore, it cannot be the object of a quest, as for the grail. Instead, it goes round and round and on and on, rather like the hike that Kerouac and Snyder took and even more like the poem that Snyder projected writing and told Kerouac about as they walked."

That poem is _Mountains and Rivers Without End,_ and one of the pieces in it, "The Circumambulation of Mount Tamalpais," describes the sites and recounts the chants of his daylong excursion there with Philip Whalen (now a Zen _roshi_) and Allen Ginsberg in 1965 "to show respect and to clarify the mind." The Himalayan-style circumambulation has been taken up by local Buddhists to become a several-times-a-year walk of some fifteen miles and ten stations, from near the foot to Tam's east peak and down again (which, when I went on it, preserved Snyder's humor by reading from his "Smoky the Bear Sutra" in the penultimate station, a roadside parking lot appropriately sprinkled with cigarette butts). The peak isn't a culmination, just one of the ten stations on this circuit, which wraps the mountain in a spiral of interpretations drawn from mostly Asian religious sources. Mountains have been a recurrent subject of Snyder's poetry. He once reset Muir's description of his ascent of Mount Ritter to make a poem of it, wrote his own _Cold Mountain Poems_ after Han-Shan, and described not only climbing and walking in mountains but living in them and working in them as a fire lookout and a trail builder. In _Mountains and Rivers Without End,_ says Snyder, "I translate space from its physical sense to the spiritual sense of space as emptiness—spiritual transparency— in Mahayana Buddhist philosophy." The book opens with what at first seems to be a long description of landscape but is in fact a description

_____

a normal petty existence, the finding of values, of beauty, of vision. These are worth living for—
_____

of a Chinese painting, and Snyder travels through all kinds of space—paintings, cities, wildernesses—in the same spirit. In "Walking the New York Bedrock / Alive in the Sea of Information," Snyder traverses Manhattan, thinking of the Indians' encounters early on with European settlers, seeing the skyscrapers as corporate deities—"Equitable god" and "Old Union Carbide god"—seeing trees, peregrines nesting "at the thirty-fifth floor," homeless people wandering amid the street-canyons whose buildings become "arêtes and buttresses rising above them." But real mountains delight him in ways Manhattan does not, as a short poem with a long title, "On Climbing the Sierra Matterhorn Again After Thirty-One Years," suggests:

> Range after range of mountains
> Year after year after year.
> I am still in love.

Chapter 10

OF WALKING CLUBS AND
LAND WARS

I. THE SIERRA NEVADA

"Another perfect Sierra Nevada day," said Michael Cohen, hovering over his coffee and managing to sound caustic about it, to me, hovering over my tea, as we looked at the morning light glittering on the lake. Valerie Cohen wasn't up yet, and I wasn't all that awake myself early that morning in the Cohens' June Lake cabin, on the eastern side of the Sierra Nevada, a little southeast of the highlands of Yosemite National Park. I don't remember what made him say all of a sudden, "The Sierra Club likes to say that John Muir founded the Sierra Club. But California culture founded it." We were products of that culture ourselves. The Cohens had both grown up in the Los Angeles area and spent a lot of time in the Sierra from an early age. I wasn't nearly so dedicated a wilderness explorer as they were, nor so athletic, even if my father's parents had met in immigrant hiking clubs in L.A. The high Sierra was definitely the Cohens' territory, in which they had skied, climbed, hiked, worked, and even gotten married thirty years before, and I let them pick the destination for our day's hike.

It was a gorgeous cloudless day in mid-August, and winter had been so late and so wet that the meadows were still green and wildflowers were everywhere. So were other hikers. Valerie led off at a good clip down the trail that began southwest of Tuolumne Meadows, and for the first mile or so through the pines she reminisced for my benefit about when she was a law enforcement ranger, this trail was her beat, and she was responsible for dealing with the demented and drugged of the high country. In the meadow where the trail had been trodden into a narrow trench several inches deep she told me about the time the campers complained that there was a crazy guy in their campground who stayed up all night walking in circles muttering to himself, a guy who turned out to be an eminent but deranged mathematician. Somewhere in the course of these stories—maybe during the one about the babysitter and the amanita mushrooms— Michael commented that the consequence of the theory that nature is supposed to make you happy is that those most desperately in search of happiness tend to show up there. Certainly they do, along with a few million others every year, in Yosemite National Park, one of the most famous and heavily visited natural places in the world.

Yosemite is also a major historical site, not least for the history of walking, mountaineering, and the environmental movement. It was my good fortune that Michael had written the history of the Sierra Club and an intellectual biography of John Muir, so that in walking along the Mono Pass trail we were traversing the terrain of his scholarship. Dorothy and William Wordsworth walking together through the Pennines just before the nineteenth century began seem lonely figures, choosing an unpopular activity in an unpopulated countryside, and John Muir tramping across Yosemite and the Sierra Nevada in the decades after his arrival in California in 1868 seems part of that tradition of solitary wandering, pursuing the aesthetic while those all around pursued the utilitarian. But Muir, as Michael had been saying, was a founder—if not the founder—of the Sierra Club, and the club would further transform the social landscape in its efforts to keep the natural landscape untransformed (except by trails; trail building was an important activity of the early club). A little more than a hundred years after the Wordsworths set off on their lonely winter walk, almost a century before the Cohens and I set out from the roadside bustle, ninety-six Sierra Club members—including their president, Muir—spent two weeks walking, mountaineering, and camping in Tuolumne Meadows. That first Sierra Club High Trip in July 1901 is a milestone in the history of the taste for walking in the landscape. Not the only such milestone, for club secretary William Colby wrote, "An excursion of this sort, if properly conducted, will do

heading out across the desert in a straight line to a sinuous weaving through undergrowth. Descending

an infinite amount of good toward awakening the proper kind of interest in the forests and other natural features of our mountains, and
will also tend to create a spirit of good-fellowship among our members.
The Mazamas and Appalachian Clubs have for many years shown
how successful and interesting such trips may be made." Walking had
become so entrenched a part of the culture that it could be, by means
of walking clubs, a foundation for further change.

Since English mountaineers founded the Alpine Club in 1857,
outdoor organizations had been proliferating across Europe and
North America, many, like the Alpine and Appalachian Clubs, com-
bining the pleasures of a social club with the publications and explora-
tions of a scientific society. But the Sierra Club was different. "The
proper kind of interest in the forests and other natural features" was,
in the ideology of the club, a political interest. Mountaineering and
hiking were ends in themselves for most of the clubs, but the Sierra
Club had been founded as a dual-purpose organization. In 1890 Muir
and such friends as the painter William Keith and the lawyer Warren
Olney had started meeting to discuss defending Yosemite National
Park from the developers who sought to raid its timber and mineral
resources. They had merged with professors at the University of
California, Berkeley, who were considering founding a mountaineering club, and the new organization's name came from the range
they would explore, just as the Appalachian Club's name came from
its members' local mountains. On June 4, 1892, the Sierra Club was
formed.

To pretend that the world is a garden is an essentially apolitical
act, a turning away from the woes that keep it from being one. But to
try to make the world a garden is often a political endeavor, and it is
this taste that the more activist walking clubs around the world have
taken up. Walking in the landscape had long been considered a
vaguely virtuous act, but Muir and the club had at last defined that
virtue as defense of the land. This made it a self-perpetuating virtue,
securing the grounds of its existence, and made the club an ideological organization. Walking—or hiking and mountaineering, as the
club tended to call it—became its ideal way of being in the world: out
of doors, relying on one's own feet, neither producing nor destroying.
The club's mission statement said its purpose was "to explore, enjoy,

---

rocky ridges and talus slopes is a specialty in itself. It is an irregular dancing—always shifting—step

and render accessible the mountain regions of the Pacific Coast; To publish authentic information concerning them; To enlist the support and cooperation of the people and the government in preserving the forests and other natural features of the Sierra Nevada Mountains."

From the beginning, the Sierra Club had a lot of built-in contradictions. It was founded as a combination mountaineering and preservation society, because Muir and some of the other founders believed that those who spent time in the mountains would come to love them, and that that love would be an active love, a love willing to go into political battle to save them. Though the premise proved to be good enough, there are plenty of mountaineers whose love has no political dimension and plenty of environmentalists who, for various reasons, don't travel in remote places. The other contradiction had to do with the fact that environmental devastation is usually done in the name of economic growth. The middle-class club found itself fighting innumerable battles in a war that dare not speak its name, a war against economic exploitation of the environment in the name of progress and free enterprise. John Muir took a stand against anthropocentrism, against the idea that trees, animals, minerals, soil, water, are there for humans to use, let alone to destroy, but by positioning wilderness as a place apart from society and the economy he avoided addressing the wider politics of land and money. For most of its history, the club would tender the milder, more anthropocentric argument that the exploitation of beautiful places destroyed them as recreational sites. Eventually it became clear that recreation was destroying Yosemite Valley almost as much as resource extraction had destroyed the neighboring Hetch-Hetchy Valley by damming it as a reservoir for San Francisco during World War I. The club would have to rescind the "render accessible" clause in its mission statement and begin advocating for survival of species, then ecosystems, then the planet, as nature began to be imagined as a necessity rather than a pleasure.

Most of the Sierra Club's troubles and transformations were far ahead, however, when the first High Trip set out in July 1901. The world was far larger and less paved then than now, and they spent three days walking from relatively accessible Yosemite Valley to Tuolumne Meadows, accompanied by a vast caravan of mules and horses

---

of walk on slabs and scree. The breath and eye are always following this uneven rhythm. It is never paced

---

carrying stoves, blankets, camp beds, and quantities of food (nowadays, the Meadows are a few hours by car from the valley). Once arrived, they settled into their large camp, from which smaller parties made forays into the surrounding mountains and canyons. It was a strange halcyon era between the violent settlement of California by avaricious Yankees and the overdevelopment of the state later. Wrote Ella M. Sexton of that first High Trip, "There were solemn hours, too, when the mountaineers looked disdainfully at us feeble 'tenderfeet' as we set off with trusty alpenstocks, a light lunch, and much courage to conquer the jagged peaks, loose talus, and snow-fields of Mt. Dana. . . . The climbers were so delayed by ten long miles to the foot of the mountain, the hard ascent and a weary ten miles back to camp, that relief parties had to go out to kindle fires at stream-crossings, and it was nine o'clock before the last straggler was ferried over on the shaky raft." With unbridged rivers and unmapped routes, it was a far wilder place than it is today. Few besides club members, fishermen, and the surviving Indians ventured into these regions then, and during those early years the club included many leading mountaineers among its members and sponsored many first ascents.

But it is the ordinary experiences that give something of the flavor of these mass expeditions into the mountains. Nelson Hackett was a high-school student when two of his female teachers recruited him, and the experience did exactly what it was supposed to: incorporate him into a community of activist nature lovers. He later became editor of the _Sierra Club Bulletin_ and a member of the board. While on the 1908 High Trip to the Kings Canyon region of the Sierra, he wrote to his parents about the club leaders, "Mr. Colby goes like lightning and Mr. Parsons is very fat and very slow so there is no need of not finding someone to suit your pace. Of course when half a dozen are ahead you can't miss the trail for the tracks they make. I sort of imagined that the 120 people would all walk along in a row but they are so scattered that you are hardly ever in sight of more than a half dozen or so." Several days later: "Next morning we turned out in the cold star-light at 3:30 and started at 4:30 for Mt. Whitney. The climb is easy but tedious and the rocks are hard on the feet. I arrived at the top at nine o'clock. We ate lunch and made some chocolate sherbet, enjoyed the view for a couple of hours and then returned. We

---

or clocklike, but flexing—little jumps—sidesteps—going for the well-seen place to put a foot on a rock,

---

could see the desert, and Owen's Lake, eleven thousand feet below us." And in a second letter written that day, July 18, 1908, "I had a long talk with Mr. Muir this aft—or rather he did all the talking—about a thousand mile walk he took thro the South, the year after the Rebellion. Also, how he first became interested in Botany. Camp-fire ready. Good-bye. . . ."

Light on tradition and heavy on new juxtapositions, California has long been a wellspring of fresh cultural possibilities. At the turn of the century, a regional culture including painters, bad poets, and good architects was responding to the state's distinctive influences and environment, and the early Sierra Club was part of this response. Unlike organizations such as the Alpine Club, which excluded women, the Sierra Club made them welcome and seems to have provided many with opportunities for mountaineering they might have found in few other places. At a time when a woman could hardly go unchaperoned around London, it says something for the freedom of the West Coast or the club that women seem to have gone wherever they liked, with whomever they liked, in the mountains. Populated by professionals of both sexes, the Sierra Club had some intellectual force behind it in its early days, and evenings around the campfire were lively with discussion, music, and performances. Muir was the most influential member then, but the club would later become a home for the men who invented American nature photography—Ansel Adams and Eliot Porter—and for many who redefined American wilderness in both law and imagination, such as George Marshall and David Brower. But California culture didn't come out of nowhere. The campers on those early trips were making their own culture, but much of the material came from points east. It's not hard to trace the lineage. After all the dean of New England transcendentalism, Ralph Waldo Emerson, had visited with both Wordsworth and John Muir, seeming to transmit—and transmute—the legacy from the peripatetic poet who had walked through the French Revolution to the evangelical mountaineer who died at the beginning of World War I. The members of the Sierra Club had imported their taste for nature, but it may have been the nature itself—the colossal wilderness of the West—that transformed that taste into something new.

That first High Trip in 1901 suggested how far the culture of

---

bit flat, move on—zigzagging along and all deliberate. The alert eye looking ahead, picking the

---

walking had come from the aristocratic stroll in the garden and the solitary ramble in the woods when it arrived in the mountains of California. It had become not only mainstream culture but politics, and if landscape could shape its walkers, these California walkers were returning the favor by shaping that landscape through both legislation and cultural representation. In recent decades the Sierra Club has been excoriated by younger environmental organizations for its compromises and missteps and for having been of its time rather than ours on such issues as dams and nuclear power. But environmental awareness and the Sierra Club grew up together. In the postwar years the club began to expand its scope and eventually its membership. It went from being a regional club whose several thousand members were mostly participants in outdoor activities to being a national organization whose half-million members include many who have never participated in a club excursion. It was the first major force for environmental protection in the United States and remains one of the most influential, achieving major victories on forests, air, water, species, parks, toxics. And it still sponsors thousands of local hikes and wilderness outings every year.

Our own walk came out of the woods and proceeded across some beautiful meadows with streams running through them. Michael and Valerie had led some of the last High Trips the club held, in 1968, when "the old man of the mountains," the legendary climber and curmudgeon Norman Clyde, still came by, but the impact of large-scale camps and expeditions was beginning to dismay the club, and soon afterward the tradition came to an end. As we approached Mono Pass, we came across the same wildflowers blooming here that I had seen in the Marin Headlands in March, not three hundred miles away. And then we reached the saddleback that the signpost announced as Mono Pass, 10,600 feet high, and sat down in the gravel and tufts of lupine. The crest of the Sierra Nevada is one of the few real borders in the world, besides the ebbing and flowing borders dividing land from water. These mountains scrape off the stormclouds sailing in from the west, and the clouds' bounty becomes snowmelt running westward again to water some of the greatest temperate forests of the world, the sequoia, ponderosa, and fir groves of the Sierra, and thence to the valleys and salmon-run rivers to the ocean—and farms and

----

footholds to come, while never missing the step of the moment. The body-mind is so at one with this rough

----

cities—below. Though a little of the mountain runoff flows down the east side of the Sierra, everything east of the peaks is desert. At Mono Pass, we were sitting facing a bright green meadow full of tender wildflowers, and a few miles behind us began a thousand miles of aridity. We were also sitting within view of the results of two great battles over land. Yosemite National Park's boundaries had been set in the 1890s, and John Muir had drawn them up. Mono Lake, the blue oval in the dusty east, had been saved in the 1990s when environmentalists, after many years of fighting, finally prevented Los Angeles from diverting some of the lake's tributaries into the vast hydraulic system that waters the city.

We fell to talking about the Sierra Club again. Although I admire the club's staunch work over the decades, I worry that equating the love of nature with certain kinds of leisure activity and visual pleasure leaves out those with other tastes and tasks. Walking in the landscape can be a demonstration of a specific heritage, and when it is mistaken for a universal experience, those who don't participate can be seen as less sensitive to nature, rather than less acculturated to the northern European romantic tradition. Michael told me about a Sierra Club outing he led and Valerie cooked for in which some well-meaning members brought along two inner-city African-American boys who were totally bewildered. The wilderness alarmed them, and the point of exerting oneself in it escaped them. Only the man who took them fishing and the hamburgers Valerie made them every day redeemed the experience. Michael wrote about it in *The Pathless Way*, his book on Muir: "We were shocked to discover firsthand that the taste for wilderness was culturally determined, a privilege enjoyed only by the sons and daughters of a certain comfortable class of Americans. One could cultivate a sense of utopian community on the outings only by beginning with a group of people who already agreed closely about certain basic values." (Since then, the Sierra Club and other organizations have sponsored "inner-city outings" better equipped to mediate the experience.) Afterward we left the trail and went cross-country, straying near a small lake tucked under a dark cliff face that seemed to increase its depths and then venturing through marshy green expanses of wild onion splashed with the scarlet of Indian paintbrush to a windswept slope above Bloody Canyon.

----

world that it makes these moves effortlessly once it has had a bit of practice. The mountain keeps up with

----

II. The Alps

One of the monuments to John Muir—along with the John Muir Trail from Mount Whitney to Yosemite Valley and dozens of California public schools—is the expanse of redwoods called Muir Woods, on the foothills of Mount Tamalpais, a dozen miles north of the Golden Gate Bridge. Mount Tamalpais is the small peak where Gary Snyder and friends instituted the Buddhist practice of ritual circumambulation, but there are other ways to interpret mountains and walking, and this one has had many interpreters. Above Muir Woods, there's an inconspicuous trail that runs for half a mile or so, then comes around a bend to a very different and disorienting monument on the steep slope above the woods. It looks like a perfect Alpine chalet, with its outdoor dance floor, pitched roof, and tiers of balconies made of pine planks cut out in folkloric designs, and it is one of the few surviving American outposts of the Austria-based organization Die Naturfreunde. The Naturfreunde, or Nature Friends, was founded in Vienna in 1895 by teacher Georg Schmiedl, blacksmith Alois Rohrauer, and student Karl Renner, at a time when the Hapsburg monarchy and other elites still controlled access to most of the Austrian mountains. "Berg frei"—free mountains—was their slogan. They were socialists and antimonarchists, and they were immensely successful. Sixty people attended the organization's first meeting, and within a few decades there were 200,000 members, mostly in Austria, Germany, and Switzerland. Each local chapter bought land and built a clubhouse, which was open to all members of the Naturfreunde. They sponsored hikes, environmental consciousness, and folk festivals, and advocated access to the mountains for working people.

The late nineteenth to early twentieth century was a golden age of organizations. Some provided social cohesion for the displaced of a rapidly changing world; others offered resistance to industrialization's inhuman appetite for the time, health, energy, and rights of workers. Many were organized around utopian ideals or pragmatic social change, and all of them created communities—of Zionists, feminists, labor activists, athletes, charities, and intellectuals. Walking clubs were part of this larger movement, and each of the major political walking clubs was founded in some kind of opposition to the

---

the mountain.—GARY SNYDER, "BLUE MOUNTAINS CONSTANTLY WALKING"   *And the meaning of*

---

mainstream of its society. For the Sierra Club, this mainstream was the rampant destruction of a pristine ecosystem by a rapidly developing country. In most of Europe, the remaining open space was in more stable but less accessible condition. For the Austrian Naturfreunde as well as many British groups, the aristocratic monopoly on open space was the problem. Manfred Pils, the current Naturfreunde secretary general, wrote me, "The Friends of Nature were founded because leisure time and tourism was a privilege for upper class people at that time. They wanted to open up such opportunities also for common people . . . it was the Friends of Nature who campaigned against the efforts to exclude people from private meadows and forests in the Alps. The campaign was called 'Der verbotene Weg' (the forbidden path). So the Friends of Nature achieved finally a legalistic regulation which guaranteed access by walking to forests and alpine meadows for everyone." As a result, "the Alps are not a national territory, they stayed in private property but we (and all tourists) have access to all footpaths and generally to forests and alpine meadows."

When German and Austrian radicals arrived in the United States, they brought their organization with them. In San Francisco, immigrants who met at the German workers' hall on Valencia Street went forth in big groups to hike on Mount Tam. After the 1906 earthquake, local Naturfreunde historian Erich Fink told me, many more craftsmen arrived in the region, the number of weekend hikers mushroomed, and they decided to buy property to start their own branch of the Naturfreunde. Five young people bought a whole steep hillside on Mount Tam for two hundred dollars, and the members built themselves a rural outpost. Fink's wife told me that until the 1930s you had to show a union card to join. This Bavarian lodge perched above the redwoods provided a workers' alternative to the Sierra Club, a local place for people who had only the weekend in which to escape the city.

The Naturfreunde paid for its success. Its socialism provoked the Nazi regime to repress it in Austria and Germany, while the Germanness of the organization made it suspect in the United States during that era. After the end of World War II, socialism became an issue in the United States too. McCarthyism in the United States so traumatized the organization that one local leader was still reluctant

Earth completely changes: with the legal model, one is constantly reterritorializing around a point of

to talk to me about the club's history. "They are very political today in Europe," he said in a heavy Teutonic accent, "which we cannot be. We stay away from any politics because they almost took away what we built up through all the years." During the years when being or having been a socialist or Communist was a dangerous offense, all the branches of the Naturfreunde in the eastern United States collapsed, and the clubhouses bought, built, and owned by the members fell into private hands. Only three California outposts survived by being adamantly apolitical, and a fourth one recently opened up in northern Oregon. Of the 600,000 Naturfreunde members in twenty-one countries, less than a thousand remain in the United States, and they are anomalies for their apolitical stance.

The German youth movement, the Wandervogel, did not survive World War II, but its history demonstrates that no ideology had a monopoly on walking. A reaction against the authoritarianism of the German family and government, it began inauspiciously enough in a suburb of Berlin in 1896, where a group of shorthand students began to go on expeditions together to the woods nearby and then farther away. By 1899 they were setting off for weeks at a time to wander in the mountains. The most charismatic member of that circle, Karl Fischer, transformed the organization, formalizing its behavior and spreading its ideas. When the Wandervogel Ausschuss für Schuler-fahrten (Wandervogel Committee for Schoolboys' Rambles) was founded on November 4, 1901, it was a Romantic rambling society. *Wandervogel* means a magical bird; a word taken from a poem, it suggests the free and weightless identity the members would seek. Medieval wandering scholars were the first role models for the thousands of boys who joined up, and rambling on long excursions together was their principal activity. There were other cultural activities—the lasting legacy of the Wandervogel, and, according to historians, its only first-rate cultural contribution, was the revival of folk songs. Most of its members were in the throes of adolescence's heady idealism, and heated philosophical debates as well as music filled their evenings. The movement seemed to be forever splintering over some

---

view, on a domain, according to a set of constant relations; but with the ambulant model, the process of

---

minor point or other. "On the main thing—rambling—we are in complete agreement," concluded a Wandervogel statement.

Theirs was an odd antiauthoritarianism, since the Wandervogel was exclusive, hierarchical, organized into small groups giving unquestioning obedience to a leader, with semiformal uniforms (usually shorts, dark shirts, and neckerchiefs) and initiation rituals of various degrees of difficulty and danger. Though the Wandervogel was detached from practical politics, most members subscribed to an ethnic nationalism, and so the folk culture that meant working-class culture for the Naturfreunde meant ethnic identity for the Wandervogel. The members were almost exclusively middle-class; girls were admitted to some groups after 1911 or encouraged to form their own groups. "The Jewish problem" meant that Jews—and often, Catholics—were generally unwelcome (though certainly one prominent Jew, Walter Benjamin, was involved in a radical splinter group of the youth movement in his own youth). At its height, the Wandervogel had about sixty thousand members. The Wandervogel seems to have started out as a real rebellion against German authoritarianism, and to this extent it was a political club, but it had neither the strength nor the insight to truly oppose its country's slide toward fascism.

There were other organizations for young people to join, church groups and the Protestant Youth Movement and, after 1909, a German version of the Boy Scouts, while working-class youths had Communist and socialist youth clubs. The Boy Scouts, like the Wandervogel, like so many situations in the history of walking, raise the question of when walking becomes marching. Most walking clubs were groups come together to celebrate and protect individual and private experience, but some embraced authoritarianism. Marching subordinates the very rhythms of individual bodies to group and to authority, and any group that marches is marching toward militarism if it is not already there. The scouting movement was adapted by the Boer War veteran Sir Baden-Powell from ideas of his own and ideas plagiarized from the Anglo-Canadian Ernest Thompson Seton. Seton's goal had been to introduce boys to outdoor life with a strong focus on Native American skills and values, and he is sometimes credited with starting the pagan revival among adults instead. Baden-Powell brought

---

deterritorialization constitutes and extends the territory itself.—DELEUZE & GUATTARI, *TREATISE ON*

a more militaristic, conservative sensibility to the idea of living in the woods. Even now, each scouting group seems to have its own style; some teach outdoor skills, some train the boys as little soldiers. After World War I, the Wandervogel collapsed, but the German Boy Scouts—the Pathfinders, they were called—rebelled against their adult leaders and largely replaced the original movement.

Werner Heisenberg, the physicist most famous for his uncertainty principle, became the leader of one of these New Pathfinder troops. Playing at adventure must have been a relief to him after the war, during which he and his brother had undergone real risks smuggling food into besieged Munich. Like many other Germans, he had a tradition of hiking and love of mountains to draw on: his paternal grandfather had gone on the "wander year" that was a rite of passage for young artisans, and his maternal grandfather was an avid hiker who went on long walking tours. But the Pathfinder movement, with its idealism and its camaraderie, had other attractions. The movement instilled in him a love of his country and close ties to his peers that made him deeply ambivalent and deeply troubled during World War II, when he was in charge of the Nazi program to develop an atomic bomb. "After 1919, the militant dictatorships in Russia, Italy, and Germany built up youth organizations of their own," writes one historian of the era. "The Hitler Youth took over many of the symbols and rituals of the original Youth Movement, but it was no more than a caricature."

III. The Peak District and Beyond

Everywhere but Britain, organized walking seems to become hiking, then camping, and eventually something as nebulous as, in contemporary terminology, outdoor recreation or wilderness adventure. The clubs are "walking and" organizations: walking and climbing and environment activism, walking and socialism and folk songs, walking and adolescent dreaming and nationalism. Only in Britain has walking remained the focus all along, even if the word *rambling* is often used to describe it. Walking has a resonance, a cultural weight, there that it does nowhere else. On summer Sundays, more than eighteen million

---

NOMADOLOGY   . . . *malformations of the spine are very frequent among mill-hands, some of them*

Britons head for the country, and ten million say they walk for recreation. In most British bookstores walking guides occupy a lot of shelf space, and the genre is so well established that there are classics and subversive texts—among the former, Alfred Wainwright's handwritten, illustrated guides to the wilder parts of the country, and among the latter the Sheffield land-rights activist Terry Howard's itinerary of walks that are all trespasses. The American magazine *Walking* is nothing but a health and fitness publication aimed at women—walking appears there as just another exercise program—but Britain has half a dozen outdoor magazines in which walking is about the beauty of landscape rather than the body. "Almost a spiritual thing," the outdoor writer Roly Smith told me, "a religion almost. A lot of people walk for the social aspects—there are no barriers on the moors and you say hello to everyone—overcome our damn British reserve. Walking is classless, one of the few sports that is classless."

But accessing the land has been something of a class war. For a thousand years, landowners have been sequestering more and more of the island for themselves, and for the past hundred and fifty, landless people have been fighting back. When the Normans conquered England in 1066, they set aside huge deer parks for hunting, and ever since, the penalties for poaching and interfering with hunting land have been fierce—castration, deportation, and execution were some of the punishments meted out over the centuries (after 1723, for example, taking rabbits or fish, let alone deer, was an offense punishable by death). The commons were usually privately owned land to which locals retained rights to gather wood and graze animals, while the traditional rights-of-way—footpaths across the fields and woods that the public had the right to walk no matter whose property they traversed—were necessary for work and travel. In Scotland, common land was abolished by an act of Parliament in 1695, and in England enclosure acts and unauthorized but fiercely enforced seizures of hitherto common land accelerated in the eighteenth century.

Corollaries of the glorious open gardens of the era, the lucrative enclosures were vast areas fenced off and filled with sheep or farmed by a single large landowner, and they were often created by shutting landworkers out from agricultural and common land. In the nineteenth century, an upper-class mania for hunting inspired many more

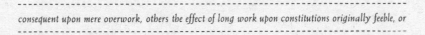

consequent upon mere overwork, others the effect of long work upon constitutions originally feeble, or

landowners to sequester public land that formerly supported many people. The Highland Clearances of 1780–1855 in Scotland were particularly brutal, displacing quantities of people, many of whom emigrated to North America, while some were driven to the coast, where they eked out a bare survival on small farms. Hunting grouse, pheasant, and deer for a few weeks annually has become the excuse for denying access to thousands of miles of Britain's wildest countryside year-round, and while hunting in the United States is sometimes a source of food for poor, rural, and indigenous people, hunting in Britain is an elite sport. Armies of gamekeepers patrolled and patrol such land, and some have used extreme measures to keep people out: spring guns and mantraps, dogs, brandished guns and shots fired overhead, assault with sticks or fists, threats, and usually, the support of local law enforcement.

When Britain was still a rural economy of landworkers, the struggle over access was about economics. But by the middle of the nineteenth century, half the nation's population lived in cities and towns, and nowadays more than 90 percent does. The cities they moved to, particularly the new industrial cities, were often bleak. Densely built, without adequate fresh water, sewers, or garbage collection systems and with a constant pall of soot in the air from the coal-burning mills and homes, the English cities of the nineteenth century were foul places, and the poor lived in the foulest. It's a chicken-and-egg question as to whether the taste for the rural or the awfulness of the cities came first, but the British have always sworn allegiance to footpaths, not boulevards. People wanted to get out of the cities whenever they could, and many of these cities were still compact enough that one could walk out of them into the country. During this period, the conflict over the commons and the rights-of-way stopped being about economic survival and became about psychic survival—about a reprieve from the city.

As more and more people chose to spend their spare time walking, more and more of the traditional rights-of-way were closed to them. In 1815 Parliament passed an act allowing magistrates to close any path they considered unnecessary (and throughout these land wars, the administration of rural Britain has been largely in the hand of landowners and their associates). In 1824 the Association for

----

weakened by bad food. Deformities seem even more frequent than these diseases; the knees were bent inward,

----

the Protection of Ancient Footpaths was formed near York, and in 1826 a Manchester association of the same name was formed. The Scottish Rights of Way Society, formed in 1845, is the oldest surviving such society, but the Commons, Open Spaces, and Footpath Preservation Society, founded in 1865, is still active as the Open Space Society. It fought and won the war of Epping Forest near London. In 1793 the forest was a 9,000-acre expanse used by the public; by 1848 it had been reduced to 7,000 acres, and a decade later it was fenced off. Three laborers who cut wood there were given harsh sentences, and in protest of the sentences and the fences—which had been ordered removed by a court order—five to six thousand people came out to exercise their right to be there. In 1884 the Forest Ramblers' Club was formed by London businessmen to "walk through Epping Forest and report obstructions we have seen." Countless other walking clubs were formed in these years.

The conflict is over two ways of imagining the landscape. Imagine the countryside as a vast body. Ownership pictures it divided into economic units like internal organs, or like a cow divided into cuts of meat, and certainly such division is one way to organize a food-producing landscape, but it doesn't explain why moors, mountains, and forests should be similarly fenced and divided. Walking focuses not on the boundary lines of ownership that break the land into pieces but on the paths that function as a kind of circulatory system connecting the whole organism. Walking is, in this way, the antithesis of owning. It postulates a mobile, empty-handed, shareable experience of the land. Nomads have often been disturbing to nationalism because their roving blurs and perforates the boundaries that define nations; walking does the same thing on the smaller scale of private property.

Certainly one of the pleasures of walking in England is this sense of cohabitation right-of-way paths create—of crossing stiles into sheep fields and skirting the edges of crops on land that is both utilitarian and aesthetic. American land, without such rights-of-way, is rigidly divided into production and pleasure zones, which may be one of the reasons why there is little appreciation for or awareness of the immense agricultural expanses of the country. British rights-of-way are not impressive compared to those of other European

---

the ligaments very often relaxed and enfeebled, and the long bones of the legs bent. The thick ends of these

countries—Denmark, Holland, Sweden, Spain—where citizens retain much wider rights of access to open space. But rights-of-way do preserve an alternate vision of the land in which ownership doesn't necessarily convey absolute rights and paths are as significant a principle as boundaries. Nearly 90 percent of Britain is privately owned, so gaining access to the countryside means gaining access to private land, while in the United States a lot of land remains public—if not always conveniently located for Sunday strolls. Thus the Sierra Club fought for boundaries, while British walking activists fight against them, but the boundaries laid down in America are to keep the land public, wild, and indivisible, to keep private enterprise out, while in Britain they kept the public out.

When I went to see the great garden of Stowe, I ran into a docent who told me that the gardens were built by destroying the village around the church and relocating "the dirty little people" a mile or so away. She added that these people were not allowed back in unless they wore smocks that made them picturesque. Three hours later I ran into this subversive, charming docent near that church, now hidden behind trees and shrubs, and we fell to talking again. About right of access, she said that as a little girl she lived near a farmer whose signs said, "Trespassers Will Be Prosecuted." She believed this meant they would be executed and used to wonder how a man who would chop people's heads off had the temerity to show himself in church. Later in life, she lived in Russia with her diplomat husband, and there and many other places, she said, trespassing was hardly a concept. Most of the British people I spoke with had a sense that the landscape was their heritage and they had a right to be there. Private property is a lot more absolute in the United States, and the existence of vast tracts of public land serves to justify this, as does an ideology in which the rights of the individual are more often upheld than the good of the community.

So I was thrilled when I got to England and discovered a culture in which trespassing is a mass movement and the extent of property rights is open to question. If walking sews together the land that ownership tears apart, then trespassing does so as a political statement. The Liberal member of Parliament James Bryce, who introduced an unsuccessful bill to allow access to privately held moors and moun-

---

long bones were especially apt to be bent and disproportionately developed, and these patients came from

tains in 1884, declared a few years later, "Land is not property for our unlimited and unqualified use. Land is necessary so that we may live upon it and from it, and that people may enjoy it in a variety of ways; and I deny therefore, that there exists or is recognized by our law or in natural justice, such a thing as an unlimited power of exclusion." This position is widely held by British moderates as well as radicals. The author of a pleasant guidebook to Derbyshire remarks of the Peak District, "It is the one thing that is unpleasant, this watchful herding of holiday-makers, where in a space so wide all must keep to a path a few paces across. I have thought: what an incitement to any who believe in the public ownership of all land." Unfortunately even the rights-of-way "a few paces across" are limited, and though it is legal to travel on them, sitting, picnicking, and straying may be illegal. Most footpaths were established for practical purposes and don't go through some of the wildest and most spectacular parts of Britain.

Thus came the great trespasses and walks that changed the face of the English countryside. They took place in the Peak District, where the laborers of the industrial north converged by foot, bicycle, and train during their time off. In the south of England, Leslie Stephen got by with "a little judicious trespassing," and his gentlemanly Sunday Tramps could and did intimidate the gamekeepers they met, while for serious expeditions there were always the Alps. "By the last quarter of the nineteenth century, in all parts of urban Britain, especially the industrial towns, the people's rambling movement was emerging, and gradually they began to take over the leadership of the struggle for access," writes Howard Hill. "The major reason for this was the growing popularity of the Swiss mountains which, being completely free to walk and climb on, drew the gentlemen ramblers and climbers away from Britain." The YMCA was one early sponsor of walking clubs, and in the 1880s members of the Manchester YMCA Rambling Club would walk seventy miles between Saturday afternoon, when work ended, and Sunday evening. In 1888 the Polytechnic Club of London was founded as a walking club; in 1892 the West of Scotland Ramblers' Alliance was organized; in 1894 women teachers formed the Midlands Institute of Ramblers; in 1900 the Sheffield Clarion Ramblers—a socialist organization—was founded by G. B. H. Ward;

the factories in which long work-hours were of frequent occurrence."—FRIEDRICH ENGELS, *THE CONDITION*

in 1905 a London Federation of Rambling Clubs was formed; in 1907
came the Manchester Rambling Club; in 1928 the nationwide British
Workers Sports Federation—the BWSF; and in 1930 the Youth Hostels
Association began to provide, like the Naturfreunde, lodging for
young and poor travelers (the YHA had its beginnings in Germany in
1907; among the rules in Britain in early years was one that no one
could arrive in an automobile). So many went out walking in the first
three or four decades of the twentieth century that some speak of it
as a movement. As historian Raphael Samuels put it, "Hiking was a
major, if unofficial, component of the socialist lifestyle." Laborers had
developed—or retained from their peasant parents and grandparents—
a passion for the land, and a whole culture of working-class botanists
and naturalists emerged, as did a legion of walkers. Walking in groups
was partly a matter of safety—there were the gamekeepers, and one
Sheffield rambler reported "a genuine hatred of ramblers by coun-
tryfolk who sometimes 'beat up' those they found walking alone."

Before the industrial revolution, the Peak District was a major
tourist destination: the Wordsworths went there, and so did Carl
Moritz, and Jane Austen sent the heroine of *Pride and Prejudice* to scenic
spots there. Afterward it became an anomaly, a forty-mile-wide open
space wedged between the great manufacturing cities of Manchester
and Sheffield, much loved by its locals. The Peak District encompasses
every variety of terrain, from the luxurious grounds of Chatsworth,
landscaped by Capability Brown, to gentle Dove Dale, to the rough
moors with their superb gritstone rock-climbing areas (in which two
Manchester plumbers, Joe Brown and Don Whillans, carried out "the
working class revolution in climbing" in the 1950s, taking the art to new
levels of difficulty). In between Chatsworth's gardens and the gritstone
climbs is Kinder Scout, focus of the most famous battle for access. The
highest and wildest point in the Peak District, it was "king's land"—that
is, public land—until 1836, when an enclosure act divided the land
up among the adjacent landowners, giving the lion's share to the duke
of Devonshire, owner of Chatsworth. The fifteen square miles of
Kinder Scout became completely inaccessible to the public, for no
footpath went near its summit. Walkers called it "the forbidden
mountain." An old Roman road across the base had been the main way
of traversing the region, but in 1821 this right-of-way—Doctor's Gate,

OF THE WORKING CLASS IN ENGLAND   *"The fellow, by his agent, or secretary, or somebody, writes to me,*

it was called—was illegally closed by the land's owner, Lord Howard. At the end of the nineteenth century, negotiations to open it began, and the rambling clubs of Manchester and Sheffield began to take direct action. In 1909 the Sheffield Clarion Ramblers walked the length of Doctor's Gate, and the Manchester Ramblers "defiantly" walked it for five years. The lord continued posting No Road signs, wiring up the gate at one end, padlocking the way, but he finally lost. Today Doctor's Gate is, with a few minor changes of route, the public route it had been for nearly two millennia before.

The crest of Kinder Scout posed a bigger problem. Benny Rothman, the secretary of the Manchester branch of the British Workers Sports Federation, writes of the grim cities during the industrial depression of the 1930s, "Town dwellers lived for weekends when they could go camping in the country, while unemployed young people would return home just to 'sign on' at the Labor Exchanges and collect their dole money. Rambling, cycling and camping clubs grew in membership. . . . The feeling of being close to nature receded as the crowds grew, and ramblers looked longingly at the acres of empty peat bogs, moorlands and the tops, which were forbidden territory. They were not just forbidden, they were guarded by gamekeepers armed with sticks, which some were not afraid to use against solitary walkers." In 1932 the BWSF decided to organize a mass trespass to publicize the situation, and Rothman gave interviews to newspapers. Though opposed by other ramblers' clubs, the young radicals drew four hundred ramblers to the nearby town of Hayfield anyway, along with a third of the Derbyshire Police Force. Partway up, Rothman gave a stirring speech about the history of the access-to-mountains movement, to much applause. Farther up the steep approach to Kinder Scout's plateau, about twenty to thirty gamekeepers appeared, shouting, threatening the walkers with their sticks, and getting the worst of the scuffles they initiated. At the crest, the trespassers were joined by members of the Sheffield clubs and latecomers from Manchester.

For this temporary victory and scenic view, Rothman and five others were arrested. One case was dismissed. The others received jail sentences from two to six months for "incitement to riotous assembly." Outrage over the sentences galvanized other ramblers and

----

'Sir Leicester Dedlock, Baronet, presents his compliments to Mr. Lawrence Boythorn, and has to call his
----

members of the public and sent both the curious and the committed to Kinder Scout. Annual rallies protesting lack of access had been held at Winants Pass in the Peak before, but the one that year brought in ten thousand ramblers, and further mass trespasses and demonstrations were held in the wake of this verdict. The politics of walking heated up. In 1935 the national federation of rambling associations became the Ramblers' Association, which stepped up activism on behalf of access, and in 1939 a bill for access was put before Parliament, unsuccessfully. In 1949 a stronger bill succeeded. The National Parks and Access to the Countryside Act changed the rules. The national parks didn't amount to much, but the access did. Every county council in England and Wales was required to map all the rights-of-way in its jurisdiction, and once the paths had been mapped, they were considered definitive. The burden had shifted to the landowner to prove that a right-of-way didn't exist, rather than to the walker to prove that one did. And these rights-of-way have ever since appeared on the Ordnance Survey maps, making the routes accessible to everyone. Local councils were also required to create a "review map" of appropriate open-space areas and then to negotiate for them to become accessible to walkers—not as strong as an absolute right of access, but a great improvement. More recently a number of long-distance trails have been created, making it possible for people to walk or backpack across Britain for days or weeks. In recent years, walkers have grown restless. On the fiftieth anniversary of its creation, the Ramblers' Association began holding "Forbidden Britain" mass trespasses of its own, and in 1997 the Labour party campaigned on a promise to support "right to roam" legislation that would, more than a century since Bryce's 1884 bill, at last open the countryside to the citizens. Recently, more radical new groups such as This Land Is Ours and Reclaim the Streets have taken direct action to enlarge the public sphere, and though walking is less central to their democratic and ecological agenda, the same populist issues of access and preservation prevail.

And this is the great irony—or poetic justice—of the history of rural walking, that a taste that began in aristocratic gardens should end up as an assault on private property as an absolute right and privilege. The gardens and parks in which the culture of walking had

---

attention to the fact that the green pathway by the old parsonage-house, now the property of Mr.

begun were closed spaces, often walled or secured by ditches, accessible to a privileged few, and sometimes created on land seized by enclosure. Yet a democratic principle had been implicit in the development of the English garden, in the way trees, water, and land were allowed their natural contours rather than being pushed into geometric forms, in the dissolution of the walls around the garden, in the increasingly mobile experience of walking through these increasingly informal spaces. The spread of the taste for walking in the landscape obliged some of the descendants of these aristocrats to live up to the principles implicit in their gardens. It may yet open the whole of Britain to walkers.

Walking for pleasure had joined the repertoire of human possibilities, and some of those enjoying the expansion returned the favor and changed the world, making it into a version of the garden—this time a public garden without walls. The terrain shaped by walkers' clubs is spread differently across different countries. In the United States it's a patchwork of wild places and a broad political movement bent on saving the organic world. Radiating from Austria are several hundred lodges scattered across twenty-one countries and more than half a million outdoor types with their own environmentalist bent. In Britain it's 140,000 miles of paths and a truculent attitude about the landed gentry. Walking has become one of the forces that has made the modern world—often by serving as a counterprinciple to economics.

The impulse to organize around walking is at first an odd one. After all, those who value walking often speak of independence, solitude, and the freedom that comes from lack of structure and regimentation. But there are three prerequisites to going out into the world to walk for pleasure. One must have free time, a place to go, and a body unhindered by illness or social restraints. These basic freedoms have been the subjects of countless struggles, and it makes perfect sense that the laborers' organizations campaigning first for eight- or ten-hour workdays and then for five-day work weeks—struggling for free time—should also concern themselves with securing space in which to enjoy this hard-won time. Others too have campaigned for space, and though I have focused on wilderness and

----

Lawrence Boythorn, is Sir Leicester's right of way, being in fact a portion of the park of Chesney Wold,

----

rural space, another rich history concerns the development of urban parks such as Central Park, a democratic and Romantic project to bring the rural virtues to city dwellers without resources to leave the city. The unhindered body is a more subtle subject. The early Sierra Club, with unchaperoned women sleeping on pine-bough beds and climbing mountains in bloomers, suggests that in California, liberation—or some genteel degree of it—was a by-product: for Victorian clothing imprisoned women in the proprieties of shallow breaths, short steps, precarious balance. The nudism of early German and Austrian outdoor clubs suggests that for some, heading for the hills was part of a wider project of embracing the natural, a natural defined to include the erotic, and even for those who remained clothed, the clothes were the informal shorts that displayed the body. As for British workers—one only has to read Friedrich Engels's *Making of the English Working Class*, about living and working conditions so dire they deformed and diseased factory workers' bodies, to understand why striding across open space under clean skies was a liberation many were willing to fight for. Walking in the landscape was a reaction against the transformations that were making the middle-class body an anachronism locked away in homes and offices and laborers' bodies part of the industrial machinery.

The writers at the beginning of this history of walking in the landscape, Rousseau and Wordsworth, linked social liberation with a passion for nature (though, fortunately, neither of them could have envisioned the Boy Scouts, the outdoor equipment industry, and other far-flung effects of the culture of walking). The walking clubs brought many ordinary people closer to their notion of the ideal walker, moving without impediments across the landscape.

---

and that Sir Leicester finds it convenient to close up the same.' "—CHARLES DICKENS, *BLEAK HOUSE*

Part III

LIVES OF THE
$\mathcal{S}$TREETS

THE SOLITARY STROLLER

AND THE CITY

I lived in rural New Mexico long enough that when I came back home to San Francisco, I saw it for the first time as a stranger might. The exuberance of spring was urban for me that year, and I finally understood all those country songs about the lure of the bright lights of town. I walked everywhere in the balmy days and nights of May, amazed at how many possibilities could be crammed within the radius of those walks and thrilled by the idea I could just wander out the front door to find them. Every building, every storefront, seemed to open onto a different world, compressing all the variety of human life into a jumble of possibilities made all the richer by the conjunctions. Just as a bookshelf can jam together Japanese poetry, Mexican history, and Russian novels, so the buildings of my city contained Zen centers, Pentecostal churches, tattoo parlors, produce stores, burrito places, movie palaces, dim sum shops. Even the most ordinary things struck me with wonder, and the people on the street offered a thousand glimpses of lives like and utterly unlike mine.

Cities have always offered anonymity, variety, and conjunction, qualities best basked in by walking: one does not have to go into the bakery or the fortune-teller's, only to know that one might. A city always contains more than any inhabitant can know, and a great city always makes the unknown and the possible spurs to the imagination. San Francisco has long been called the most European of American

----

There are few greater delights than to walk up and down them in the evening alone with thousands of

----

cities, a comment more often made than explained. What I think its speakers mean is that San Francisco, in its scale and its street life, keeps alive the idea of a city as a place of unmediated encounters, while most American cities are becoming more and more like enlarged suburbs, scrupulously controlled and segregated, designed for the noninteractions of motorists shuttling between private places rather than the interactions of pedestrians in public ones. San Francisco has water on three sides and a ridge on the fourth to keep it from sprawling, and several neighborhoods of lively streets. Truly urban density, beautiful buildings, views of the bay and the ocean from the crests of its hills, cafés and bars everywhere, suggest different priorities for space and time than in most American cities, as does the (gentrification-threatened) tradition of artists, poets, and social and political radicals making lives about other things than getting and spending.

My first Saturday back, I sauntered over to nearby Golden Gate Park, which lacks the splendor of a wilderness but has given me many compensatory pleasures: musicians practicing in the reverberant pedestrian underpasses, old Chinese women doing martial arts in formation, strolling Russian émigrés murmuring to each other in the velvet slurp of their mother tongue, dog walkers being yanked into the primeval world of canine joys, and access by foot to the shores of the Pacific. That morning, at the park's bandshell, the local radio variety show had joined forces with the "Watershed Poetry Festival," and I watched for a while. Former poet laureate of the United States Robert Hass was coaching children to read their poetry into the microphone onstage, and some poets I knew were standing in the wings. I went up to say hello to them, and they showed me their brand-new wedding rings and introduced me to more poets, and then I ran into the great California historian Malcolm Margolin, who told me stories that made me laugh. This was the daytime marvel of cities for me: coincidences, the mingling of many kinds of people, poetry given away to strangers under the open sky.

Margolin's publishing house, Heydey Press, was displaying its wares along with those of some other small presses and literary projects, and he handed me a book off his table titled *920 O'Farrell Street*. A memoir by Harriet Lane Levy, it recounted her own marvelous

---

other people, up and down, relishing the lights coming through the trees or shining from the facades,

---

experiences growing up in San Francisco in the 1870s and 1880s. In her day, walking the streets of the city was as organized an entertainment as a modern excursion to the movies. "On Saturday night," she wrote, "the city joined in the promenade on Market Street, the broad thoroughfare that begins at the waterfront and cuts its straight path of miles to Twin Peaks. The sidewalks were wide and the crowd walking toward the bay met the crowd walking toward the ocean. The outpouring of the population was spontaneous as if in response to an urge for instant celebration. Every quarter of the city discharged its residents into the broad procession. Ladies and gentlemen of imposing social repute; their German and Irish servant girls, arms held fast in the arms of their sweethearts; French, Spaniards, gaunt, hardworking Portuguese; Mexicans, the Indian showing in reddened skin and high cheekbone—everybody, anybody, left home and shop, hotel, restaurant, and beer garden to empty into Market Street in a river of color. Sailors of every nation deserted their ships at the water front and, hurrying up Market Street in groups, joined the vibrating mass excited by the lights and stir and the gaiety of the throng. 'This is San Francisco,' their faces said. It was carnival; no confetti, but the air a criss-cross of a thousand messages; no masks, but eyes frankly charged with challenge. Down Market from Powell to Kearny, three long blocks, up Kearny to Bush, three short ones, then back again, over and over for hours, until a glance of curiosity deepened to one of interest; interest expanded into a smile, and a smile into anything. Father and I went downtown every Saturday night. We walked through avenues of light in a world hardly solid. Something was happening everywhere, every minute, something to be happy about. . . . We walked and walked and still something kept happening afresh." Market Street, which was once a great promenade, is still the city's central traffic artery, but decades of tearing it up and redeveloping it have deprived it of its social glory. Jack Kerouac managed to have two visions on it late in the 1940s or early in the 1950s, and he would probably embrace its freeway-shadowed midtown population of panhandlers and people running sidewalk sales out of shopping carts. Levy's downtown stretch is now trod by office workers and shoppers and by tourists swarming around the Powell Street cable car turnaround; more than a mile farther uptown, Market Street finally bursts

---

listening to the sounds of music and foreign voices and traffic, enjoying the smell of flowers and good

---

into vigorous pedestrian life again for a few blocks before it crosses Castro Street and begins its steep ascent of Twin Peaks.

The history of both urban and rural walking is a history of freedom and of the definition of pleasure. But rural walking has found a moral imperative in the love of nature that has allowed it to defend and open up the countryside. Urban walking has always been a shadier business, easily turning into soliciting, cruising, promenading, shopping, rioting, protesting, skulking, loitering, and other activities that, however enjoyable, hardly have the high moral tone of nature appreciation. Thus no similar defense has been mounted for the preservation of urban space, save by a few civil libertarians and urban theorists (who seldom note that public space is used and inhabited largely by walking it). Yet urban walking seems in many ways more like primordial hunting and gathering than walking in the country. For most of us the country or the wilderness is a place we walk through and look at, but seldom make things in or take things from (remember the famous Sierra Club dictum, "Take only photographs, leave only footprints"). In the city, the biological spectrum has been nearly reduced to the human and a few scavenger species, but the range of activities remains wide. Just as a gatherer may pause to note a tree whose acorns will be bountiful in six months or inspect a potential supply of basket canes, so an urban walker may note a grocery open late or a place to get shoes resoled, or detour by the post office. Too, the average rural walker looks at the general—the view, the beauty—and the landscape moves by as a gently modulated continuity: a crest long in view is reached, a forest thins out to become a meadow. The urbanite is on the lookout for particulars, for opportunities, individuals, and supplies, and the changes are abrupt. Of course the city resembles primordial life more than the country in a less charming way too; while nonhuman predators have been radically reduced in North America and eliminated in Europe, the possibility of human predators keeps city dwellers in a state of heightened alertness, at least in some times and places.

Those first months at home were so enchanting that I kept a walking journal and later that glorious summer wrote, "I suddenly

---

food and the air from the nearby sea. The sidewalks are lined with small shops, bars, stalls, dance halls,

realized I'd spent seven hours at the desk without a real interruption and was getting nervous and hunchbacked, walked to the Clay Theater on upper Fillmore via a passage on Broderick I'd never seen before—handsome squat old Victorians near the housing projects—and was pleased as ever when the familiar yielded up the unknown. The film was *When the Cat's Away*, about a solitary young Parisienne forced to meet her Place de Bastille neighbors when her cat vanishes, full of uneventful events and people with seesaw strides and rooftops and mumbling slang, and when it got out I was exhilarated and the night was dark with a pearly mist of fog on it. I walked back fast, first along California, past a couple—her unexceptional, him in a well-tailored brown suit with the knock knees of someone who'd spent time in leg braces—and ignored the bus, and did the same on Divisadero with that bus. Slowed down at an antique store window to look at a big creamy vase with blue Chinese sages painted on it, then a few doors down saw a balding Chinese man holding a toddler boy up to the glass of a store, where a woman on the inside was playing with him through the glass. To their confusion, I beamed. There's a way the artificial lights and natural darkness of nightwalks turn the day's continuum into a theater of tableaux, vignettes, set pieces, and there's always the unsettling pleasure of your shadow growing and shrinking as you move from streetlight to streetlight. Dodging a car as a traffic light changed, I broke into a canter and it felt so good I loped along a few more blocks without getting winded, though I got warm.

"All along Divisadero keeping an eye on the other people and on the open venues—liquor stores and smoke shops—and then turned up my own street. At a cross street a young black guy in a watch cap and dark clothes was running downhill at me at a great clip, and I looked around to suss up my options just in case—I mean if Queen Victoria was moving toward you that fast you'd take note. He saw my hesitation and assured me in the sweetest young man's voice, 'I'm not after you, I'm just *late*' and dashed past me, so I said, 'Good luck' and then, when he was into the street and I had time to collect my thoughts, 'Sorry to look suspicious, but you were kind of speedy.' He laughed, and then I did, and in a minute I recalled all the other encounters I'd had around the 'hood lately that might have had the

movies, booths lighted by acetylene lamps, and everywhere are strange faces, strange costumes, strange

earmarks of trouble but unfolded as pure civility and was pleased that I'd been prepared without being alarmed. At that moment, I looked up and saw in a top-floor window the same poster of Man Ray's *A l'heure de l'observatoire*—his painting of the sunset sky with the long red lips floating across it—that I'd seen in another window somewhere else in town a night or two before. This poster was bigger, and this night was more exuberant; seeing *A l'heure* twice seemed magic. Home in about twenty minutes at most."

Streets are the space left over between buildings. A house alone is an island surrounded by a sea of open space, and the villages that preceded cities were no more than archipelagos in that same sea. But as more and more buildings arose, they became a continent, the remaining open space no longer like the sea but like rivers, canals, and streams running between the land masses. People no longer moved anyhow in the open sea of rural space but traveled up and down the streets, and just as narrowing a waterway increases flow and speed, so turning open space into the spillways of streets directs and intensifies the flood of walkers. In great cities, spaces as well as places are designed and built: walking, witnessing, being in public, are as much part of the design and purpose as is being inside to eat, sleep, make shoes or love or music. The word *citizen* has to do with cities, and the ideal city is organized around citizenship—around participation in public life.

Most American cities and towns, however, are organized around consumption and production, as were the dire industrial cities of England, and public space is merely the void between workplaces, shops, and dwellings. Walking is only the beginning of citizenship, but through it the citizen knows his or her city and fellow citizens and truly inhabits the city rather than a small privatized part thereof. Walking the streets is what links up reading the map with living one's life, the personal microcosm with the public macrocosm; it makes sense of the maze all around. In her celebrated *Death and Life of Great American Cities*, Jane Jacobs describes how a popular, well-used street is kept safe from crime merely by the many people going by. Walking maintains the publicness and viability of public space. "What

---

and delightful impressions. To walk up such a street into the quieter, more formal part of town, is to be

distinguishes the city," writes Franco Moretti, "is that its spatial structure (basically its concentration) is functional to the intensification of mobility: spatial mobility, naturally enough, but mainly social mobility."

The very word *street* has a rough, dirty magic to it, summoning up the low, the common, the erotic, the dangerous, the revolutionary. A man of the streets is only a populist, but a woman of the streets is, like a streetwalker, a seller of her sexuality. Street kids are urchins, beggars, and runaways, and the new term *street person* describes those who have no other home. *Street-smart* means someone wise in the ways of the city and well able to survive in it, while "to the streets" is the classic cry of urban revolution, for the streets are where people become the public and where their power resides. *The street* means life in the heady currents of the urban river in which everyone and everything can mingle. It is exactly this social mobility, this lack of compartments and distinctions, that gives the street its danger and its magic, the danger and magic of water in which everything runs together.

In feudal Europe only city dwellers were free of the hierarchical bonds that structured the rest of society—in England, for example, a serf could become free by living for a year and a day in a free town. The quality of freedom within cities then was limited, however, for their streets were usually dirty, dangerous, and dark. Cities often imposed a curfew and closed their gates at sunset. Only in the Renaissance did the cities of Europe begin to improve their paving, their sanitation, and their safety. In eighteenth-century London and Paris, going out anywhere at night was as dangerous as the worst slums are supposed to be nowadays, and if you wanted to see where you were going, you hired a torchbearer (and the young London torch carriers—link boys, they were called—often doubled as procurers). Even in daylight, carriages terrorized pedestrians. Before the eighteenth century, few seem to have walked these streets for pleasure, and only in the nineteenth century did places as clean, safe, and illuminated as modern cities begin to emerge. All the furniture and codes that give modern streets their orderliness—raised sidewalks, streetlights, street names, building numbers, drains, traffic rules, and traffic signals—are relatively recent innovations.

---

part of a procession, part of a ceaseless ceremony of being initiated into the city and rededicating the

---

Idyllic spaces had been created for the urban rich—tree-lined promenades, semipublic gardens and parks. But these places that preceded the public park were anti-streets, segregated by class and disconnected from everyday life (unlike the pedestrian *corsos* and paseos of the plazas and squares of Mediterranean and Latin countries and Levy's Market Street promenade—or London's anomalous Hyde Park, which accommodated both carriage promenades for the rich and open-air oratory for the radical). Though politics, flirtations, and commerce might be conducted in them, they were little more than outdoor salons and ballrooms. And from the mile-long Cours de la Reine built in Paris in 1616 to Mexico City's Alameda to New York's Central Park built during the 1850s, such places tended to attract people whose desire to display their wealth was better served by promenading in carriages than walking. On the Cours de la Reine, the carriages would gather so thickly a traffic jam would result, which may be why in 1700 a fashion for getting out and dancing by torchlight on the central round developed.

Though Central Park was shaped by more-or-less democratic impulses, English landscape garden aesthetics, and the example of Liverpool's public park, poor New Yorkers often paid to go to private parks akin to Vauxhall Gardens instead, where they might drink beer, dance the polka, or otherwise engage in plebeian versions of pleasure. Even those who wished only to have an uplifting stroll, as the park's codesigner Frederick Law Olmsted had intended them to, found obstacles. Central Park became a great promenade for the rich, and once again carriages segregated the society. In their history of the park and its city, Ray Rosenzweig and Elizabeth Blackmar write, "Earlier in the [nineteenth] century the late afternoon, early evening, and Sunday promenades of affluent New Yorkers had evolved into parades of high fashion; the wide thoroughfares of Broadway, the Battery, and Fifth Avenue had become a public setting in which to see and be seen. By midcentury, however, the fashionable Broadway and Battery promenades had declined as 'respectable' citizens lost control over these public spaces. . . . Both men and women wanted grander public space for a new form of public promenading—by carriage. In the mid-nineteenth century, carriage ownership was becoming a defining feature of urban upper-class status." The rich went to Central Park, and a populist jour-

city itself.—J. B. JACKSON, "THE STRANGER'S PATH"   . . . after scrambling under bellies of horses,

nalist said, "I hear that pedestrians have acquired a bad habit of being accidentally run over in that neighborhood."

Just as poorer people continued to promenade in New York's Battery, so their Parisian counterparts strolled along the peripheries of the city, often under avenues of trees planted to shade just such excursions. After the Revolution, Paris's Tuileries could be entered by anyone the guards deemed properly dressed. Private pleasure gardens modeled after London's famous Vauxhall Gardens, including Ranelagh and Cremorne Gardens in London itself; Vienna's Augarten; New York's Elysian Fields, Castle Gardens, and Harlem Gardens; and Copenhagen's Tivoli Gardens (sole survivor of them all) sorted out people by the simpler criterion of ability to pay. Elsewhere in these cities, markets, fairs, and processions brought festivity to the sites of everyday life, and the stroll was not so segregated. To me, the magic of the street is the mingling of the errand and the epiphany, and no such gardens seem to have flourished in Italy, perhaps because they were unneeded.

Italian cities have long been held up as ideals, not least by New Yorkers and Londoners enthralled by the ways their architecture gives beauty and meaning to everyday acts. Since at least the seventeenth century, foreigners have been moving there to bask in the light and the life. Bernard Rudofsky, nominally a New Yorker, spent a good deal of time in Italy and sang its praises in his 1969 *Streets for People: A Primer for Americans*. For those who consider New York the exemplary American pedestrian city, Rudofsky's conviction that it is abysmal is startling. His book uses primarily Italian examples to demonstrate the ways plazas and streets can function to tie a city together socially and architecturally. "It simply never occurs to us to make streets into oases rather than deserts," he says at the beginning. "In countries where their function has not yet deteriorated into highways and parking lots, a number of arrangements make streets fit for humans. . . . The most refined street coverings, a tangible expression of civic solidarity— or, should one say, of philanthropy, are arcades. Apart from lending unity to the streetscape, they often take the place of the ancient forums." Descendants of the Greek stoa and *peripatos*, arcaded streets blur the boundaries between inside and out and pay architectural tribute to the pedestrian life that takes place beneath them. Rudofsky

---

through wheels, and over posts and rails, we reached the gardens, where were already many thousand

---

singles out Bologna's famous *portici*, a four-mile-long covered walkway running from the central square to the countryside; Milan's Galleria, less strictly commercial in its functions than the upscale shopping malls modeled and named after it; the winding streets of Perugia; the car-free streets of Siena; and Brisinghella's second-story public arcades. He writes with passionate enthusiasm about the Italian predinner stroll—the *passaggiata*—for which many towns close down their main streets to wheeled traffic, contrasting it with the American cocktail hour. For Italians, he says, the street is the pivotal social space, for meeting, debating, courting, buying, and selling.

The New York dance critic Edwin Denby wrote, about the same time as Rudofsky, of his own appreciation of Italian walkers. "In ancient Italian towns the narrow main street at dusk becomes a kind of theatre. The community strolls affably and looks itself over. The girls and the young men, from fifteen to twenty-two, display their charm to one another with a lively sociability. The more grace they show the better the community likes them. In Florence or in Naples, in the ancient city slums the young people are virtuoso performers, and they do a bit of promenading any time they are not busy." Of young Romans, he wrote, "Their stroll is as responsive as if it were a physical conversation." Elsewhere, he instructs dance students to watch the walk of various types: "Americans occupy a much larger space than their actual bodies do. This annoys many Europeans; it annoys their instinct of modesty. But it has a beauty of its own, that a few of them appreciate. . . . For myself I think the walk of New Yorkers is amazingly beautiful, so large and clear." In Italy walking in the city is a universal cultural activity rather than the subject of individual forays and accounts. From Dante pacing out his exile in Verona and Ravenna to Primo Levi walking home from Auschwitz, Italy has not lacked great walkers—but urban walking itself seems to be more part of a universal culture than the focus of particular experience (save that by foreigners, copiously recorded, and the cinematic strolls of such characters as the streetwalker in Federico Fellini's *Nights of Cabiria* and the protagonists in Vittorio De Sica's *Bicycle Thief* and in many of Michelangelo Antonioni's films). However, the cities that are neither so accommodating as Naples nor so forbidding as Los Angeles—London, New York—have produced their own fugitive culture of walking. In

persons. . . .We walked twice round and were rejoiced to come away, though with the same difficulties as at

London, from the eighteenth century on, the great accounts of walking have to do not with the cheerful and open display of ordinary life and desires but with nocturnal scenes, crimes, sufferings, outcasts, and the darker side of the imagination, and it is this tradition that New York assumes.

In 1711 the essayist Joseph Addison wrote, "When I am in a serious Humour, I very often walk by my self in Westminster Abbey; where the Gloominess of the Place, and the Use to which it is applied . . . are apt to fill the Mind with a kind of Melancholy, or rather Thought-fulness, that is not disagreeable." At the time he wrote, walking the city streets was perilous, as John Gay pointed out in his 1716 poem *Trivia; or, The Art of Walking the Streets of London.* Travel through the city was as dangerous as cross-country travel: the streets were full of sewage and garbage, many of the trades were filthy, the air was already bad, cheap gin had ravaged the city's poor the way crack did American inner cities in the 1980s, and an underclass of criminals and desperate souls thronged the streets. Carriages jostled and mangled pedestrians without fear of reprisal, beggars solicited passersby, and street sellers called out their wares. The accounts of the time are full of the fears of the wealthy to go out at all and of young women lured or forced into sexual labor: prostitutes were everywhere. This is why Gay focuses on urban walking as an *art*—an art of protecting oneself from splashes, assaults, and indignities:

> Though you through cleanlier allies wind by day,
> To shun the hurries of the publick way,
> Yet ne'er to those dark paths by night retire;
> Mind only safety, and contemn the mire.

Like Dr. Johnson's 1738 poem "London," Gay's *Trivia* uses a classical model to mock the present. Divided into three books—the first on the implements and techniques of walking the streets, the second on walking by day, the third on walking by night—the poem makes it clear that the minutia of everyday life can only be observed scorn-fully. The high-flown style cannot but contrast abrasively with such small subjects, with something of the same mockery he brought to his *Beggars' Opera.* Gay tries—

> Here I remark each walker's diff'rent face,
> And in their look their various bus'ness trace.

—but he ends by despising everyone, assuming he can read their tawdry lives in their faces. At the end of Gay's century Wordsworth "goes forward with the crowd," seeing a mystery in the face of each stranger; while William Blake wanders "each charter'd street / And mark in every face I meet / Marks of weakness, marks of woe:"—the cry of a chimney sweep, the curse of a young harlot. Earlier eighteenth-century literary language was not supple enough or personal enough to connect the life of the imagination to that of the street. Johnson had been one of those desperate London walkers in his early years there—in the late 1730s, when he and his friend, the poet and rogue Richard Savage, were too poor to pay for lodgings, they used to walk the streets and squares all night talking insurrection and glory—but he didn't write about it. Boswell did in his *Life of Johnson,* but for Boswell, the darkness of night and anonymity of the streets were a less reflective opportunity, as his London diary records: "I should have been at Lady Northumberland's rout tonight, but my barber fell sick [meaning his hair was not properly powdered]; so I sallied to the streets, and just at the bottom of our own, I picked up a fresh, agreeable young girl called Alice Gibbs. We went down a lane to a snug place. . . ." Of Alice Gibbs's impression of the streets and the night, we have no record.

That few women other than prostitutes were free to wander the streets and that wandering the street was often enough to cause a woman to be considered a prostitute are matters troubling enough to be taken up elsewhere. Here I merely want to comment on their presence in the street and in the night, habitats in which they more than almost any other kind of walker became natives. Until the twentieth century women seldom walked the city for their own pleasure, and prostitutes have left us almost no records of their experience. The eighteenth century was immodest enough to have a few famous novels about prostitutes, but Fanny Hill's courtesan life was all indoors,

---

VAUXHALL GARDENS, 1769   *Walking the streets so constantly as he did . . . gave him an opportunity*

Moll Flanders's was entirely practical, and both of them were creations of male authors whose work was at least partly speculative. Then as now, however, a complex culture of working the streets must have existed, each city mapped according to safety and the economics of male desire. There have been many attempts to confine such activity; Byzantine-era Constantinople had its "street of harlots," Tokyo from the seventeenth to the twentieth century had a gated pleasure district, nineteenth-century San Francisco had its notorious Barbary Coast, and many turn-of-the-century American cities had red-light districts, the most famous of which was New Orleans's Storyville, where jazz is reputed to have been born. But prostitution wandered outside these bounds, and the population of such women was enormous: 50,000 in 1793, when London had a total population of one million, estimated one expert. By the mid-nineteenth century they were to be found in the most fashionable parts of London too: social reformer Henry Mayhew's report refers to "the circulating harlotry of the Haymarket and Regent Street," as well as to the women working in the city's parks and promenades.

Twenty-odd years ago a researcher on prostitution reported, "Prostitution streetscapes are composed of *strolls*, loosely defined areas where the women solicit. . . . On the stroll the prostitute moves around to entice or enjoin customers, reduce boredom, keep warm and reduce visibility [to the police]. Part of most streetscapes resemble common greens, areas to which all have unimpeded access. Here women assemble in groups of two to four, laughing, talking and joking among themselves. . . . Working the same stroll infuses much needed predictability into an illegal, sometimes dangerous environment." And Dolores French, an advocate for prostitutes' rights, worked the streets herself and reports that her fellow streetwalkers "think that women who work in whorehouses have too many restrictions and rules" while the street "welcomed everyone democratically. . . . They felt they were like cowboys out on the range, or spies on a dangerous mission. They bragged about how free they were. . . . They had no one to answer to but themselves." The same refrains—freedom, democracy, danger—come up in this as in the other ways of occupying the streets.

In the eighteenth-century city, a new image of what it means to be human had arisen, an image of one possessed of the freedom and

----

of examining into the condition of every poor person that he met. Which he did, with so well practiced a

----

isolation of the traveler, and travelers, however wide or narrow their scope, became emblematic figures. Richard Savage proposed this early with a 1729 poem called *The Wanderer,* and the aptly named George Walker inaugurated the new century with his novel *The Vagabond,* followed in 1814 by Fanny Burney's *Wanderer.* Wordsworth had his *Excursion* (whose first two sections were titled "The Wanderer" and "The Solitary"); Coleridge's Ancient Mariner was condemned like the Wandering Jew to roam; and the Wandering Jew himself was a popular subject for Romantics in Britain and on the continent.

The literary historian Raymond Williams remarks, "Perception of the new qualities of the modern city had been associated, from the beginning, with a man walking, as if alone, in its streets." He cites Blake and Wordsworth as founders of this tradition, but it was De Quincey who wrote of it most poignantly. In the beginning of *Confessions of an English Opium Eater,* De Quincey tells of how at the age of seventeen he had run away from a dull school and his unsympathetic guardians and landed in London. There he was afraid to contact the few people he knew and unable to seek work without connections. So for sixteen weeks in the summer and fall of 1802 he starved, having found no other support in London but a home in an all-but-abandoned mansion whose other resident was a forlorn female child. He fell into a spectral existence shared with a few other children, and he wandered the streets restlessly. Streets were already a place for those who had no place, a site to measure sorrow and loneliness in the length of walks. "Being myself at that time, of necessity, a peripatetic, or walker of the streets, I naturally fell in more frequently with those female peripatetics who are technically called street-walkers. Many of these women had occasionally taken my part against watchmen who wished to drive me off the steps of houses where I was sitting." He was befriended by one, a girl named Ann—"timid and dejected to a degree which showed how deeply sorrow had taken hold of her young heart"—who was younger than he and who had turned to the streets after being cheated of a minor inheritance. Once when they were "pacing slowly along Oxford Street, and after a day when I had felt unusually ill and faint, I requested her to turn off with me into Soho Square," and he fainted. She spent what little she had on hot spiced

---

sagacity, as could seldom be imposed upon. And every man that follows his example, will soon find, that

wine to revive him. That he was never able to find her again after his fortune changed was, he declares, one of the great tragedies of his life. For De Quincey, his sojourn in London was one of the most deeply felt passages in his long life, though it had no sequel: the rest of his book is given over to its putative subject, the effects of opium, and the rest of his life to rural places.

Charles Dickens was different, in that he chose such urban walking and his writing explored it thoroughly over the years. He is the great poet of London life, and some of his novels seem as much dramas of place as of people. Think of *Our Mutual Friend*, where the great euphemistic piles of dust, the dim taxidermy and skeleton shop, the expensively icy interiors of the wealthy, are portraits of those associated with them. People and places become one another—a character may only be identified as an atmosphere or a principle, a place may take on a full-fledged personality. "And this kind of realism can only be gained by walking dreamily in a place; it cannot be gained by walking observantly," wrote one of his best interpreters, G. K. Chesterton. He attributed Dickens's acute sense of place to the well-known episode in his boyhood when his father was locked up in a debtor's prison and Dickens himself was put to work in a blacking factory and lodged in a nearby roominghouse, a desolate child abandoned to the city and its strangers. "Few of us understand the street," Chesterton writes. "Even when we step into it, we step into it doubtfully, as into a house or room of strangers. Few of us see through the shining riddle of the street, the strange folk that belong to the street only—the street-walker or the street arab, the nomads who, generation after generation have kept their ancient secrets in the full blaze of the sun. Of the street at night many of us know less. The street at night is a great house locked up. But Dickens had, if ever man had, the key of the street. . . . He could open the inmost door of his house—the door that leads onto the secret passage which is lined with houses and roofed with stars." Dickens is among the first to indicate all the other things urban walking can be: his novels are full of detectives and police inspectors, of criminals who stalk, lovers who seek and damned souls who flee. The city becomes a tangle through which all the characters wander in a colossal game of hide and seek, and only a vast city

---

this practice will lead him into the exercise of more charity, than is possible to be practiced in carriages of

could allow his intricate plots so full of crossed paths and overlapping lives. But when he wrote about his own experiences of London, it was often an abandoned city.

"If I couldn't walk fast and far, I should explode and perish," he once told a friend, and he walked so fast and far that few ever managed to accompany him. He was a solitary walker, and his walks served innumerable purposes. "I am both a town traveller and a country traveller, and am always on the road," he introduces himself in his essay collection _The Uncommercial Traveller._ "Figuratively speaking, I travel for the great house of Human Interest Brothers, and have rather a large connection in the fancy goods way. Literally speaking, I am always wandering here and there from my rooms in Covent-garden, London." This metaphysical version of the commercial traveler is an inadequate description of his role, and he tried on many others. He was an athlete: "So much of my travelling is done on foot, that if I cherished better propensities, I should probably be found registered in sporting newspapers under some such title as the Elastic Novice, challenging all eleven stone mankind to competition in walking. My last special feat was turning out of bed at two, after a hard day, pedestrian and otherwise, and walking thirty miles into the country to breakfast. The road was so lonely in the night that I fell asleep to the monotonous sound of my own feet, doing their regular four miles an hour." And a few essays later, he was a tramp, or a tramp's son: "My walking is of two kinds: one straight on end to a definite goal at a round pace; one, objectless, loitering, and purely vagabond. In the latter state, no gypsy on earth is a greater vagabond than myself; it is so natural to me, and strong with me, that I think I must be the descendant, at no great distance, of some irreclaimable tramp." And he was a cop on the beat, too ethereal to arrest anyone but in his mind: "It is one of my fancies, that even my idlest walk must always have its appointed destination. . . . On such an occasion, it is my habit to regard my walks as my beat, and myself as a higher sort of police-constable doing duty on the same."

And yet despite all these utilitarian occupations and the throngs who populate his books, his own London was often a deserted city, and his walking in it a melancholy pleasure. In an essay on visiting abandoned cemeteries, he wrote, "Whenever I think I deserve par-

----

any kind.—PATRICK DELANY, OBSERVATIONS UPON LORD ORRERY'S "REMARKS ON THE LIFE AND

----

ticularly well of myself, and have earned the right to enjoy a little treat, I stroll from Covent-garden into the City of London, after business-hours there, on a Saturday, or—better yet—on a Sunday, and roam about its deserted nooks and corners." But the most memorable of them all is "Night Walks," the essay that begins, "Some years ago, a temporary inability to sleep, referable to a distressing impression, caused me to walk about the streets all night, for a series of several nights." He described these walks from midnight till dawn as curative of his distress, and during them "I finished my education in a fair amateur experience of houselessness"—or what is now called homelessness. The city was no longer as dangerous as it had been in Gay's and Johnson's time, but it was lonelier. Eighteenth-century London was crowded, lively, full of predators, spectacles, and badinage between strangers. By the time Dickens was writing about houselessness in 1860, London was many times as large, but the mob so feared in the eighteenth century had in the nineteenth been largely domesticated as the crowd, a quiet, drab mass going about its private business in public: "Walking the streets under the pattering rain, Houselessness would walk and walk and walk, seeing nothing but the interminable tangle of streets, save at a corner, here and there, two policemen in conversation, or the sergeant or inspector looking after his men. Now and then in the night—but rarely—Houselessness would become aware of a furtive head peering out of a doorway a few yards before him, and, coming up with the head, would find a man standing bolt upright to keep within the doorway's shadow, and evidently intent upon no particular service to society. . . . The wild moon and clouds were as restless as an evil conscience in a tumbled bed, and the very shadow of the immensity of London seemed to lie oppressively upon the river." And yet he relishes the lonely nocturnal streets, as he does the graveyards and "shy neighborhoods" and what he quixotically called "Arcadian London"—London out of season, when society had gone en masse to the country, leaving the city in sepulchral peace.

There is a subtle state most dedicated urban walkers know, a sort of basking in solitude—a dark solitude punctuated with encounters as the night sky is punctuated with stars. In the country one's solitude is geographical—one is altogether outside society, so solitude has a

WRITINGS OF JONATHAN SWIFT," 1754   *They don't want anything to do with a roof and four walls.*

sensible geographical explanation, and then there is a kind of communion with the nonhuman. In the city, one is alone because the world is made up of strangers, and to be a stranger surrounded by strangers, to walk along silently bearing one's secrets and imagining those of the people one passes, is among the starkest of luxuries. This uncharted identity with its illimitable possibilities is one of the distinctive qualities of urban living, a liberatory state for those who come to emancipate themselves from family and community expectation, to experiment with subculture and identity. It is an observer's state, cool, withdrawn, with senses sharpened, a good state for anybody who needs to reflect or create. In small doses melancholy, alienation, and introspection are among life's most refined pleasures.

Not long ago I heard the singer and poet Patti Smith answer a radio interviewer's question about what she did to prepare for her performances onstage with "I would roam the streets for a few hours." With that brief comment she summoned up her own outlaw romanticism and the way such walking might toughen and sharpen the sensibility, wrap one in an isolation out of which might come songs fierce enough, words sharp enough, to break that musing silence. Probably her roaming the streets didn't work so well in a lot of American cities, where the hotel was moated by a parking lot surrounded by six-lane roads without sidewalks, but she spoke as a New Yorker. Speaking as a Londoner, Virginia Woolf described anonymity as a fine and desirable thing, in her 1930 essay "Street Haunting." Daughter of the great alpinist Leslie Stephen, she had once declared to a friend, "How could I think mountains and climbing romantic? Wasn't I brought up with alpenstocks in my nursery, and a raised map of the Alps, showing every peak my father had climbed? Of course, London and the marshes are the places I like best." London had more than doubled in size since Dickens's night walks, and the streets had changed again to become a refuge. Woolf wrote of the confining oppression of one's own identity, of the way the objects in one's home "enforce the memories of our own experience." And so she set out to buy a pencil in a city where safety and propriety were no longer considerations for a no-longer-young woman on a winter evening, and in recounting—or inventing—her journey, wrote one of the great essays on urban walking.

How can that be compared to the street? The vice of the children is the street itself. The street is an

"As we step out of the house on a fine evening between four and six," she wrote, "we shed the self our friends know us by and become part of that vast republican army of anonymous trampers, whose society is so agreeable after the solitude of one's room." Of the people she observes she says, "Into each of these lives one could penetrate a little way, far enough to give one the illusion that one is not tethered to a single mind, but can put on briefly for a few minutes the bodies and minds of others. One could become a washerwoman, a publican, a street singer." In this anonymous state, "the shell-like covering which our souls have excreted for themselves, to make for themselves a shape distinct from others, is broken, and there is left of all these wrinkles and roughnesses a central oyster of perceptiveness, an enormous eye. How beautiful a street is in winter! It is at once revealed and obscured." She walked down the same Oxford Street De Quincey and Ann had, now lined with windows full of luxuries with which she furnished an imaginary house and life and then banished both to return to her walk. The language of introspection that Wordsworth helped develop and De Quincey and Dickens refined was her language, and the smallest incidents—birds rustling in the shrubbery, a dwarf woman trying on shoes—let her imagination roam farther than her feet, into digressions from which she reluctantly returns to the actualities of her excursion. Walking the streets had come into its own, and the solitude and introspection that had been harrowing for her predecessors was a joy for her. That it was a joy because her identity had become a burden makes it modern.

Like London, New York has seldom prompted unalloyed praise. It is too big, too harsh. As one who knows only smaller cities intimately, I continually underestimate its expanse and wear myself out on distances, just as I do by car in Los Angeles. But I admire Manhattan: the synchronized beehive dance of Grand Central Station, the fast pace people set on the long grids of streets, the jaywalkers, the slower strollers in the squares, the dark-skinned nannies pushing pallid babies before them through the gracious paths of Central Park. Wandering without a clear purpose or sense of direction, I have often disrupted the fast flow of passersby intent on some clear errand or

---

addiction even stronger than the thinner they buy in the hardware store. . . . Only the street is theirs.

---

commute, as though I were a butterfly strayed into the beehive, a snag in the stream. Two-thirds of all journeys around downtown and midtown Manhattan are still made on foot, and New York, like London, remains a city of people walking for practical purposes, pouring up and down subway stairs, across intersections—but musers and the nocturnal strollers move to a different tempo. Cities make walking into true travel: danger, exile, discovery, transformation, wrap all around one's home and come right up to the doorstep.

The Italophile Rudofsky uses London to scorn New York: "On the whole North America's Anglo-Saxomania has had a withering effect on its formative years. Surely, the English are not a desirable model for an urban society. No other nation developed such a fierce devotion to country life as they did. And with good reason, their cities have been traditionally among Europe's least wholesome. Englishmen may be intensely loyal to their towns, but the street—the very gauge of urbanity—does not figure large in their affections." New York's streets do figure large in the work of some of its writers. "Paris, c'est une blonde," goes the French song, and Parisian poets have often made their city a woman. New York, with its gridded layout, its dark buildings and looming skyscrapers, its famous toughness, is a masculine city, and if cities are muses, it is no wonder this one's praises have been sung best by its gay poets—Walt Whitman, Frank O'Hara, Allen Ginsberg, and the prose-poet David Wojnarowicz (though everyone from Edith Wharton to Patti Smith has paid homage to this city and its streets).

In Whitman's poems, though he often speaks of himself as happy in the arms of a lover, the passages in which he appears as a solitary walking the streets in quest of that lover—a precursor of the gay cruiser—ring more true. In "Recorders Ages Hence," the immodest Whitman states for the record that he was one "Who often walk'd lonesome walks thinking of his dear friends, his lovers." A few poems later in the final version of *Leaves of Grass*, he begins another poem with the oratorical address "City of orgies, walks and joys." After listing all the possible criteria for a city's illustriousness—houses, ships, parades— he chooses "not these, but as I pass O Manhattan, your frequent and swift flash of eyes offering me love": the walks rather than the orgies, the promises rather than the delivery, are the joys. Whitman was a

It consoles them for their loneliness and lack of love. It has a dizzying appeal. It gives them the money

great maker of inventories and lists to describe variety and quantity and one of the first to love the crowd. It promised new liaisons; it expressed his democratic ideals and oceanic enthusiasms. A few poems past "City of Orgies" comes "To a Stranger": "Passing stranger! You do not know how longingly I look upon you. . . ." For Whitman the momentary glimpse and the intimacy of love were complementary, as were his own emphatic ego and the anonymous mass of crowds. Thus he sang the praises of the swelling metropolis of Manhattan and the new possibilities of urban scale.

Whitman died in 1892, just as everyone else was beginning to celebrate the city. For the first half of the new century, the city seemed emblematic—the capital of the twentieth century, as Paris had been of the nineteenth century. Destiny and hope were urban for both radicals and plutocrats in those days, and New York with its luxury steamers docking and immigrants pouring off Ellis Island, with its skyscrapers even Georgia O'Keeffe couldn't resist painting during her time as a New Yorker, was the definitive modern city. In the 1920s a magazine was devoted to it, the *New Yorker*, whose Talk of the Town section compiled minor street incidents made incandescent by its writers in the tradition of eighteenth-century London's *Spectator* and *Rambler* essays, and it had jazz and the Harlem Renaissance uptown and radical Bohemia down in the Village (and in Central Park was the Ramble, an area so well known for gay cruising it was nicknamed "the fruited plain"). Before World War II, Berenice Abbott roamed New York's streets photographing buildings, and after it, Helen Levitt photographed children playing in the streets while Weegee photographed the underworld of fresh corpses on sidewalks and prostitutes in paddy wagons. One imagines them wandering purposefully like hunter-gatherers with the camera a sort of basket laden with the day's spectacles, the photographers leaving us not their walks, as poets do, but the fruits of those walks. Whitman, however, had no successor until after the war, when Allen Ginsberg stepped into his shoes, or at least his loose long lines of celebratory ranting.

Ginsberg is sometimes claimed as a San Franciscan, and he found his poetic voice during his time there and in Berkeley in the 1950s, but he is a New York poet, and the cities of his poems are big, harsh cities. He and his peers were passionate urbanists at a time when the

they never got at home. It gives them rhythm, a tempo, and immediate compensation.—ELENA

white middle class was abandoning city life for the suburbs (and though many of the so-called Beats gathered in San Francisco, most wrote poetry about things more personal or more general than the streets they thronged, or used the city as a gateway to Asia and the western landscape). He did write about suburbs, notably in his "Supermarket in California," in which he summoned up a supermarket where the abundance of produce and shopping families makes wry comedy of the dead gay poets—Whitman and Federico García Lorca (a New Yorker from 1929 to 1930)—cruising the aisles. But otherwise his early poems burst with snow, tenements, and the Brooklyn Bridge. Ginsberg walked considerably in San Francisco and in New York, but in his poems walking is always turning into something else, since the sidewalk is always turning into a bed or a Buddhist paradise or some other apparition. The best minds of his generation were "dragging themselves through the negro streets at dawn looking for an angry fix," but they immediately commenced to see angels staggering on tenement roofs, eat fire, hallucinate Arkansas and Blake-light tragedy, and so on, even if they did afterward stumble to unemployment offices and walk "all night with their shoes full of blood on the snowbank docks waiting for a door in the East River to open. . . ."

For the Beats, motion or travel was enormously important, but its exact nature was not (save for Snyder, the true peripatetic of the bunch). They caught the tail end of the 1930s romance of freighthoppers, hobos, and railroad yards, they led the way to the new car culture in which restlessness was assuaged by hundreds of miles at 70 m.p.h. rather than dozens at 3 or 4 on foot, and they blended such physical travel with chemically induced ramblings of the imagination and a whole new kind of rampaging language. San Francisco and New York seem pedestrian anchors on either side of the long rope of the open road they traveled. In the same mode, one can see the shift in country ballads: sometime in the 1950s disappointed lovers stopped walking away or catching the midnight train and began driving, and by the 1970s the apotheosis of eighteen-wheeler songs had arrived. Had he lived that long, Kerouac would've loved them. Only in the first section of _Kaddish_, when Ginsberg gives over singing of his generation and his pals to mourn his mother, do the act and the place remain particular. The streets are repositories of history, walking a

----

PONIATOWSKA, "IN THE STREET" *In a charming sally, Mme de Girardin one day said that for the*

----

way to read that history. "Strange now to think of you, gone without corsets & eyes, while I walk on the sunny pavement of Greenwich Village," it opens, and as he walks Seventh Avenue he thinks of Naomi Ginsberg in the Lower East Side, "where you walked 50 years ago, little girl—from Russia / . . . then struggling in the crowds of Orchard Street toward what? /—toward Newark—" in an antiphony of her city and his, joined in later sections by their shared experiences during his childhood.

Handsome as a marble statue, Frank O'Hara was as unlike Ginsberg as a gay poet born the same year could be, and he wrote about far more delicate diurnal adventures. Ginsberg's poetry was oratorical— jeremiads and hymns to be shouted from the rooftops; O'Hara's poetry is as casual as conversation and sequenced by strolls in the street (among his book titles are *Lunch Poems*—not about eating but about lunchtime excursions from his job at the Museum of Modern Art—*Second Avenue*, and the essay collection *Standing Still and Walking in New York*). While Ginsberg tended to speak to America, O'Hara's remarks often addressed a "you" who seemed to be an absent lover in a silent soliloquy or a companion on a stroll. The painter Larry Rivers recalls, "It was the most extraordinary thing, a simple walk" with O'Hara, and O'Hara wrote a poem titled "Walking with Larry Rivers." Walking seems to have been a major part of his daily repertoire, as well as a kind of syntax organizing thought, emotion, and encounter, and the city was the only conceivable site for his tender, street-smart, and sometimes campy voice celebrating the incidental and the inconsequential. In the prose-poem "Meditations in an Emergency" he affirmed, "I can't even enjoy a blade of grass unless I know there's a subway handy, or a record store or some other sign that people do not totally *regret* life. It is more important to affirm the least sincere; the clouds get enough attention as it is. . . ." The poem "Walking to Work" ends

> I'm becoming
> the street.
>   Who are you in love with?
> me?
> Straight against the light I cross.

Parisian, walking is not taking exercise—it is searching. . . . The Parisian truly seems an explorer,
---

Yet another walking poem begins:

> I'm getting tired of not wearing underwear
> And then again I like it
>   strolling along
>   feeling the wind blow softly on my genitals

and goes on to speculate on "who dropped that empty carton / of cracker jacks," before turning to the clouds, the bus, his destination, the "you" to whom he speaks, Central Park. The texture is that of everyday life and of a connoisseur's eye settling on small things, small epiphanies, but the same kind of inventory that studs Whitman's and Ginsberg's poems recurs in O'Hara's. Cities are forever spawning lists.

David Wojnarowicz's *Close to the Knives: A Memoir of Disintegration* reads like a summary of all the urban experience that came before him. Like De Quincey he was a runaway, but like De Quincey's friend Ann he supported himself as a child prostitute, and like Dickens and Ginsberg he brought an incandescant, hallucinatory clarity to the moods and scenes of his city. Most who took up the Beat subject of the urban underworld of the erotic, the intoxicated, and the illegal took it up in William Burroughs's amoral vein, more interested in its coolness than its consequences or its politics, but Wojnarowicz raged at the system that created such suffering, that created his suffering as a runaway child, a gay man, a person with AIDS (of which he died in 1991). He writes in a collage of memories, encounters, dreams, fantasies, and outbursts studded with startling metaphors and painful images, and in his writings walking appears like a refrain, a beat: he always returns to the image of himself walking alone down a New York street or a corridor. "Some nights we'd walk seven or eight hundred blocks, practically the whole island of Manhattan," he wrote of his hustling years, for walking remained the recourse for those with nowhere to sleep, as it had been for Johnson and Savage.

Wojnarowicz's 1980s New York had come full circle to resemble Gay's early eighteenth-century London. It had the scourges of AIDS, of the vast new population of homeless people, and of the drug-damaged staggering around like something out of William Hogarth's Gin Lane, and it was notoriously violent, so that the well-to-do feared

always ready to set off again, or, better like some marvelous alchemist of life.—F. BLOCH, *TYPES DU*

its streets as they once had London's. Wojnarowicz writes of seeing "long legs and spiky boots and elegant high heels and three prostitutes suddenly surround a business man from the waldorf and they're saying: 'Come on honey' and rubbing his dick . . . and his wallet appears behind his back in the hands of one of them and they all drop away as he continues to giggle" and we're back to Moll Flanders stripping a passed-out trick of his silver gloves, snuffbox, and even his periwig. He writes of the years when he was suffering from malnutrition and exposure, living on the streets until he was eighteen, "I had almost died three times at the hands of people I'd sold my body to in those days and after coming off the street. . . . I could barely speak when in the company of other people. . . . That weight of image and sensation wouldn't come out until I picked up a pencil and started putting it down on paper." "Coming off the street": the phrase describes all streets as one street and that street as a whole world, with its own citizenry, laws, language. "The street" is a world where people in flight from the traumas that happen inside houses become natives of the outside.

One of the book's sections, "Being Queer in America: A Journal of Disintegration," is as tidy a chronicle of the uses of walking for a queer man of the streets in 1980s urban America as *Pride and Prejudice* is of the uses of walking for a country lady almost two centuries before. "I'm walking through these hallways where the windows break apart a slow dying sky and a quiet wind follows the heels of the kid as he suddenly steps through a door frame ten rooms down," it opens. He follows the kid into the room, which resembles the long wharves and warehouses he used to cruise, sucks him off, and a few sections later his walking becomes mourning for his friend, the photographer Peter Hujar, dead of AIDS. "I walked for hours through the streets after he died, through the gathering darkness and traffic, down into the dying section of town where bodies litter the curbsides and dogs tear apart the stinking garbage by the doorways. There was a green swell to the clouds above the buildings. . . . I turned and left, walking back into the gray haze of traffic and exhaust, past a skinny prostitute doing the junkie walk bent over at the waist with knuckles dragging the sidewalk." He meets a friend—"man on second avenue at 2:00 am"—who tells him about a third man being jumped on West Street

BOULEVARD   *To rove about, musing, that is to say loitering, is, for a philosopher, a good way of*

by a carload of kids from Jersey and brutally beaten for being gay. And then comes his refrain, "I walk this hallway twenty-seven times and all I can see are the cool white walls. A hand rubbing slowly across a face, but my hands are empty. Walking back and forth from room to room trailing bluish shadows I feel weak. . . ." His city is not hell but limbo, the place in which restless souls swirl forever, and only passion, friendship, and visionary capacity redeem it for him.

I began walking my own city's streets as a teenager and walked them so long that both they and I changed, the desperate pacing of adolescence when the present seemed an eternal ordeal giving way to the musing walks and innumerable errands of someone no longer wound up so tight, so isolated, so poor, and my walks have now often become reviews of my own and the city's history together. Vacant lots become new buildings, old geezer bars are taken over by young hipsters, the Castro's discos become vitamin stores, whole streets and neighborhoods change their complexion. Even my own neighborhood has changed so much it sometimes seems as though I have moved two or three times from the raucous corner I started out on just before I turned twenty. The urban walkers I have surveyed suggest a kind of scale of walking, and on it, I have moved from near the Ginsberg-Wojnarowicz end of the spectrum to that of a low-rent Virginia Woolf.

Two days before the end of the year, I went to one of the local liquor stores for milk early one Sunday morning. Around the corner a guy was sitting in a doorway drinking and singing falsetto, with that knack some local drunks have for sounding like fallen angels. The word *Alooooone* trilled out of nowhere, echoing beautifully in the stairwell. On my way back I saw him weaving so intently down the street he didn't notice me pass a few feet away. Merely walking seemed to take all the singer's concentration, as though he were forcing himself through an atmosphere that had become thick around him. When I started watering the tree in front of my building, he was still winding around the corner. The old lady who always wears a dress and always speaks so politely in word-salad non sequiturs was walking in the other direction. I said hello to her as she passed me, but she didn't notice me

---

spending time, especially in that kind of mock rurality, ugly but odd, and partaking of two natures,

---

any more than he did. All of a sudden, when she had reached the same point on her side of the street that he had on his, she broke into a sort of soft-shoe shuffle that carried on until she turned out of sight down the facing corner. The two of them seemed to be listening to some inaudible music that carried them along and made them joyous as well as haunted.

Later on the churchgoers would appear. When I first moved here, there were no cafés, and all the churchgoers walked—on Sunday mornings the streets were busy and sociable with black women in resplendent hats, walking in all directions to their churches, not with the dogged steps of pilgrims but with the festive stride of celebrants. That was long ago; gentrification has dispersed the Baptist congregations to other neighborhoods, from which many now drive to church. Young African-American men still saunter by, their legs nonchalant while their arms and shoulders jump around as though staking a bodily territory, but most of the churchgoers have been replaced on the sidewalks these weekend mornings by joggers and dog walkers pumping towards that great secular temple of the middle class, the garden as represented by Golden Gate Park, while the hung-over drift towards the cafés. But this early the street belonged to us three walkers, or to the two of them, for they made me feel like a ghost drifting through their private lives out in public on that cold, sunny Sunday morning, in the communal solitude of urban walkers.

Chapter 12

PARIS, OR BOTANIZING

ON THE ASPHALT

Parisians inhabit their public gardens and streets as though they were salons and corridors, and their cafés face the street and overflow into it as though the theater of passersby were too interesting to neglect even for the duration of a drink. Nude bronze and marble women are everywhere out of doors, standing on pedestals and springing from walls as though the city were both museum and boudoir, while victory arches and pillars punctuate the avenues like the yonis and lingams of a militant sexuality. Streets turn into courtyards, the largest buildings wrap around other courtyards that are actually parks, the national buildings are as long as avenues, and avenues are lined with trees and chairs just like the parks. Everything—houses, churches, bridges, walls—is the same sandy gray so that the city seems like a single construction of inconceivable complexity, a sort of coral reef of high culture. All this makes Paris seem porous, as though private thought and public acts were not so separate here as elsewhere, with walkers flowing in and out of reveries and revolutions. More than any other city, it has entered the paintings and novels of those under its sway, so that representation and reality reflect each other like a pair of facing mirrors, and walking Paris is often described as reading, as though the city itself were a huge anthology of tales. It exerts a magnetic attraction over its citizens and its visitors, for it has always been the capital of refugees and exiles as well as of France.

---

stroll aimlessly about and look at things; and by way of curing our state we buy two old Saint-Cloud

---

"Now a landscape, now a room," Walter Benjamin wrote of the walker's experience of Paris. Benjamin is one of the great scholars of cities and the art of walking them, and Paris drew him into its recesses as it had drawn so many before, coming to overshadow all the other subjects of his writing during the last decade before his death in 1940. He first visited Paris in 1913 and returned for longer and longer visits until he finally settled there at the end of the 1920s. Even in writing of his birthplace, Berlin, Benjamin's words wandered toward Paris. "Not to find one's way in a city may well be uninteresting and banal. It requires ignorance—nothing more. But to lose oneself in a city—as one loses oneself in a forest—that calls for quite a different schooling. Then signboards and street names, passers-by, roofs, kiosks, or bars must speak to the wanderer like a crackling twig under his feet, like the startling call of a bittern in the distance, like the sudden stillness of a clearing with a lily standing erect at its center. Paris taught me this art of straying," he said in his essay on his Berlin childhood. "It fulfilled a dream that had shown its first traces in the labyrinths on the blotting pages of my school exercise books." He had been brought up as a good turn-of-the-century German to revere mountains and forests—a photograph of him as a child shows him holding an alpenstock before some painted Alps, and his wealthy family often took long vacations in the Black Forest and Switzerland—but his enthusiasm for cities was both a rejection of that musty romanticism and an immersion in modernism's urbanism.

Cities fascinated him as a kind of organization that could only be perceived by wandering or by browsing, a spatial order in contrast to the tidily linear temporal order of narratives and chronologies. In that Berlin essay, he speaks of a revelation he had in a Paris café—"it had to be in Paris, where the walls and quays, the places to pause, the collections and the rubbish, the railings and the squares, the arcades and the kiosks, teach a language so singular"—that his whole life could be diagrammed as a map or a labyrinth, as though space rather than time were its primary organizing structure. His *Moscow Diary* mixed his own life into an account of that city, and he wrote a book whose form seems to mimic a city, *One-Way Street*, a subversive confection of short passages titled as though they were city sites and signs—Gas Station, Construction Site, Mexican Embassy, Manorially Furnished Ten-Room

---

teapots, embossed with silver gilt in a box with a fleur-de-lys lock.—THE GONCOURT JOURNALS, 1856

Apartment, Chinese Curios. If a narrative is like a single continuous path, this book's many short narratives are like a warren of streets and alleys.

He was himself a great wanderer of streets. I picture Benjamin walking the streets of Paris—"I don't think I ever saw him walk with his head erect. His gait had something unmistakable about it, something pensive and tentative, which was probably due to his shortsightedness," said one friend—passing without noticing another exile with worse eyesight, James Joyce, who lived there from 1920 to 1940. There is a sort of symmetry between the exiled Catholic who had written a novel studded and layered with obscure information about a Jew wandering the streets of Dublin and the exiled Berlin Jew strolling the Paris streets while writing lyrical histories about a Catholic—Charles Baudelaire—walking and writing the streets of Paris. The kind of homage Joyce received in his lifetime has come far later to Benjamin, with the rediscovery of his works first in Germany in the 1960s and 1970s and later in English. He has become the patron saint of cultural studies, and his writing has spawned hundreds more essays and books. It may be the hybrid nature of this writing—more or less scholarly in subject, but full of beautiful aphorisms and leaps of imagination, a scholarship of evocation rather than definition—that has made him so rich a source for further interpreters. His Parisian studies have been of particular interest—he left a huge collection of quotes and notes for an unwritten book, the *Arcades Project*, which would have expanded further on the linked subjects of Baudelaire, Paris, the Parisian arcades, and the figure of the flâneur. It was he who named Paris "the capital of the nineteenth century" and he who made the flâneur a topic for academics at the end of the twentieth.

What exactly a flâneur is has never been satisfactorily defined, but among all the versions of the flâneur as everything from a primeval slacker to a silent poet, one thing remains constant: the image of an observant and solitary man strolling about Paris. It says something about the fascination public life exerted over Parisians that they developed a term to describe one of its types, and something about French culture that it theorized even strolling. The word only became common usage in the early nineteenth century, and its origins are shrouded. Priscilla Parkhurst Ferguson says it comes from "old

---

The end of her life was terrible. She who had been the most beautiful and most desired woman of her time

---

Scandinavian (*flana, courir etourdiment ca et la* [to run giddily here and there])," while Elizabeth Wilson writes, "The nineteenth-century *Encylopedie Larousse* suggests that the term may be derived from an Irish word for 'libertine.' The writers of this edition of Larousse devoted a long article to the flâneur, whom they defined as a loiterer, a fritterer away of time. They associated him with the new urban pastimes of shopping and crowd watching. The flâneur, Larousse pointed out, could exist only in the great city, the metropolis, since provincial towns would afford too restricted a stage for his strolling."

Benjamin himself never clearly defined the flâneur, only associated him with certain things: with leisure, with crowds, with alienation or detachment, with observation, with walking, particularly with strolling in the arcades—from which it can be concluded that the flâneur was male, of some means, of a refined sensibility, with little or no domestic life. The flâneur arose, Benjamin argues, at a period early in the nineteenth century when the city had become so large and complex that it was for the first time strange to its inhabitants. Flâneurs were a recurrent subject of the feuilletons—the serialized novels in the newly popularized newspapers—and the *physiologies*, those popular publications that purported to make strangers familiar but instead underscored their strangeness by classifying them as species one could identify on sight, like birds or flowers. In the nineteenth century, the idea of the city so intrigued and overwhelmed its inhabitants that they eagerly devoured guidebooks to their own cities as modern tourists peruse those of other cities.

The crowd itself seemed to be something new in human experience— a mass of strangers who would remain strange—and the flâneur represented a new type, one who was, so to speak, at home in this alienation: "The crowd is his domain, just as the air is the bird's, and water that of the fish," wrote Baudelaire in a famous passage often used to define flâneurs. "His passion and his profession is to merge with the crowd. For the perfect idler, for the passionate observer it becomes an immense source of enjoyment to establish his dwelling in the throng, in the ebb and flow, the bustle, the fleeting and infinite. To be away from home and yet to feel at home anywhere . . ." The flâneur, Benjamin wrote in his most famous passage on the subject, "goes botanizing on the asphalt. But even in those days it was not possible to stroll about

---

became a half-mad woman driven by a mania for walking. She was living at 26 bis, Place Vendôme, and

everywhere in the city. Before Haussmann [remodeled the city] wide pavements were rare, and the narrow ones afforded little protection from vehicles. Strolling could hardly have assumed the importance it did without the arcades." "Arcades," he wrote elsewhere, "where the flâneur would not be exposed to the sight of carriages that did not recognize pedestrians as rivals were enjoying undiminished popularity. There was the pedestrian who wedged himself into the crowd, but there was also the flâneur who demanded elbow room and was unwilling to forego the life of a gentleman of leisure." One demonstration of this leisureliness, Benjamin goes on to say, was the fashion, around 1840, for taking turtles for walks in the arcades. "The flâneurs liked to have the turtles set the pace for them. If they had their way, progress would have been obliged to accommodate itself to this pace."

His final, unfinished work, the _Arcades Project_, was devoted to teasing out the meanings of these shopping arcades that had arisen during the first decades of that century. The arcades intensified the blurring of interior and exterior: they were pedestrian streets paved with marble and mosaic and flanked by shops, they had roofs made of the new building materials of steel and glass, and they were the first places in Paris to be lit by the new gaslight. Precursors of Paris's great department stores (and later America's shopping malls), they were elegant environments for selling luxury goods and accommodating idle strollers. The arcades allowed Benjamin to link his fascination with the stroller to other, more Marxist themes. The flâneur, visually consuming goods and women while resisting the speed of industrialization and the pressure to produce, is an ambiguous figure, both resistant to and seduced by the new commercial culture. The solitary walker in New York or London experiences cities as atmosphere, architecture, and stray encounters; the promenader in Italy or El Salvador encounters friends or flirts; the flâneur, the descriptions suggest, hovers on the fringes, neither solitary nor social, experiencing Paris as an intoxicating abundance of crowds and goods.

The only problem with the flâneur is that he did not exist, except as a type, an ideal, and a character in literature. The flâneur is often described as detectivelike in his aloof observation of others, and feminist scholars have debated whether there were or could be female

---

every evening, dressed in black, her face hidden by veils, dragging along two miserable, fat, asthmatic

---

flâneurs—but no literary detective has found and named an actual individual who qualifies or was known as a flâneur (Kierkegaard, were he less prolific and less Danish, might be the best candidate). No one has named an individual who took a tortoise on a walk, and all who refer to this practice use Benjamin as their source (though during the flâneurs' supposed heyday, the writer Gérard de Nerval famously took a lobster on walks, with a silk ribbon for a leash, but he did so in parks rather than arcades, and for metaphysical rather than foppish reasons). No one quite fulfilled the idea of the flâneur, but everyone engaged in some version of flâneury. Benjamin to the contrary, it was not only "possible to stroll about everywhere in the city" but widely done. The solitary walker in other cities has often been a marginal figure, shut out of the private life that takes place between intimates and inside buildings—but in nineteenth-century Paris, real life was in public, on the street and among society.

Paris before Baron Georges-Eugène Haussmann's massive remodeling of the city from 1853 to 1870 was still a medieval city. "Narrow crevices," Victor Hugo calls "those obscure, contracted, angular lanes, bordered by ruins eight stories high. . . . The street was narrow and the gutter wide, the passerby walked along a pavement which was always wet, beside shops that were like cellars, great stone blocks encircled with iron, immense garbage heaps. . . ." It was a remarkably unsegregated city: the very courtyard of the Louvre had a kind of slum built into it, and the outdoor arcade-courtyard of the Palais Royale offered sex, luxury goods, books, and drinks for sale, while social spectacles and political discourse were free. In 1835 the writer Frances Trollope went out to shop in a fashionable boutique "which I reached without any other adventure than being splashed twice and nearly run over thrice"; on her way back with her parcels she stopped to look at "the monuments raised over some half-dozen or half-score of revolutionary heroes, who fell and were buried on a spot at no great distance from the fountain" and eavesdropped, along with a gathering crowd, on an artisan telling his daughter why he and those heroes fought in 1830. On other days she reported on an uprising and on the fashionable promenaders of the boulevard des Italiens. Shopping and revolution, ladies and artisans, mingled on these dirty, enchanted streets.

---

dogs, she would leave her house, taking care not to be recognized, go to the arcades of the street whose

A Moroccan who had visited Paris in 1845–46 was impressed by the pedestrian life there: "In Paris there are places where people take walks, which is one of their forms of entertainment. A fellow takes the arm of his friend, man or woman, and together they go to one of the spots known for it. They stroll along, chatting and taking in the sights. Their idea of an outing is not eating or drinking, and certainly not sitting. One of their favorite promenades is a place called the Champs-Élysées." The popular places for strolling were the Champs-Élysées, the Tuileries gardens, the avenue de la Reine, the Palais Royale, and the boulevard des Italiens, all on the right bank, and the Jardin des Plantes and Luxembourg Gardens on the left, where Baudelaire had grown up. Writing to his mother in 1861, Baudelaire recollected their "long walks and constant affection! I recall the quais, so sad in the evenings," and a friend remembered that when the poet was young they had "strolled about all evening on the boulevards and in the Tuileries" together.

People went to the boulevards for society. They went to the streets and alleys for adventure, proud that they could navigate the vast network that had yet to be adequately mapped. Even before the French Revolution, some writers and walkers had cherished this idea of the city as a kind of wilderness, mysterious, dark, dangerous, and endlessly interesting. Restif de la Bretonne's *Les Nuits de Paris, ou le Spectateur Nocturne* is the classic book of prerevolutionary walks (and after the revolution it was expanded to include *Les Nuits de la Revolution,* the book was first published in 1788 and expanded in the 1790s). A peasant who became a printer, then a Parisian, then a writer, Bretonne is one of French literature's great eccentrics, little remembered now. He wrote book after book, emulating Rousseau's *Confessions* with a sixteen-volume autobiography, imitating while claiming to revile the marquis de Sade's *Justine* in his own *Anti-Justine* (sold, like Sade's works, in a pornographic bookshop in the Palais Royale's arcades), and producing dozens of novels, as well as some journalism about Paris that foreshadows the *physiologies* of the nineteenth century. The *Nuits* is unique, a collection of hundreds of anecdotes about his adventures on hundreds of nights on the streets of Paris. Each brief chapter covers a night, and the pretext of the book is that he was working as a missionary to rescue distressed maidens and bring them to his patroness,

---

name she was once so proud to bear, in the direction of the Rue de Rivoli. For hours and hours she

the marquise de M————, though he has many other kinds of adventure as well. The episodic quality of the book makes it recall the innumerable adventures of the Native American trickster Coyote or of the comic-strip Spiderman.

In his late-night wanderings, Bretonne meets with shop girls, blacksmiths, drunkards, servants, and of course prostitutes, spies on politicians in debate and aristocrats in adultery (notably in the Tuileries), sees crimes, fires, mobs, cross-dressers, a freshly murdered corpse. He writes of Paris in the way many others would later: as a book, a wilderness, and a sort of erogenous zone, or bedroom. The Île Saint-Louis was his favorite haunt, and from 1779 to 1789 he chiseled onto its stone walls dates of great personal significance, along with a few evocative words. Thus Paris became both the source of his adventures and a book recording them, a tale to be both written and read by walking. As Proust's famous madeleine served to recall his past, so did these inscriptions for Bretonne: "Whenever I had stopped along the parapet [of the Île Saint-Louis] to ponder some sorrowful thought, my hand would trace the date and the thought that had just stirred me. I would walk on then, wrapped in the darkness of the night whose silence and loneliness were touched with a horror I found pleasing." He reads the first date he had carved: "I cannot describe the emotion I felt as I thought back to the year before. . . . A rush of memories came to me; I stood motionless, preoccupied with linking the present moment to the preceding year's, to make them one." He relived love affairs, nights of desperation, and ruptured friendships. His Paris is a bedroom full of liaisons in the gardens and lechery on the streets (fittingly enough, Bretonne was a foot fetishist and sometimes followed women with small feet and high heels). In his Paris, the privacy of erotic life is constantly spilling forth in public, and the city is a wilderness, because its public and private spaces and experiences are so intermingled and because it is lawless, dark, and full of dangers.

In the nineteenth century, the theme of the city as wilderness would come up again and again in novels, poems, and popular literature. The city was called a "virgin forest," its explorers were sometimes, in Benjamin's famous phrase, naturalists "botanizing on the asphalt," but its indigenous inhabitants were often "savages." "What are the dangers of the forest and the prairie compared with the daily

---

would walk, coming back to her house only when the dawn was beginning to disperse the darkness

---

shocks and conflicts of civilization?" wrote Baudelaire in a passage
Benjamin cites. "Whether a man grasps his victim on a boulevard or
stabs his quarry in unknown woods—does he not remain both here
and there the most perfect of all beasts of prey?" In admiration of
James Fenimore Cooper's novels of the American wilderness, Alexandre Dumas titled a novel _Mobicans du Paris;_ it features the adventures
of a flâneur-detective who lets a blowing scrap of paper lead him to
adventures that always involve crimes; a minor novelist, Paul Feval,
installed an unlikely Native American character in Paris, where he
scalps four enemies in a cab; Balzac, says Benjamin, refers to "Mohicans in spencer jackets" and "Hurons in frock coats"; later in the
century, loiterers and petty criminals were nicknamed "Apaches."
These terms invested the city with the allure of the exotic, turning its
types into tribes, its individuals into explorers, and its streets into a
wilderness. One of its explorers was George Sand, who found that
"on the Paris pavement I was like a boat on ice. My delicate shoes
cracked open in two days, my pattens sent me spilling, and I always
forgot to lift my dress. I was muddy, tired and runny-nosed, and I
watched my shoes and my clothes . . . go to rack and ruin with
alarming rapidity." She put on men's clothes, and though that act is
frequently described as a subversive social one, she described it as a
practical one. Her new costume gave her a freedom of movement she
reveled in: "I can't convey how much my boots delighted me. . . . With
those steel-tipped heels I was solid on the sidewalk at last. I dashed
back and forth across Paris and felt I was going around the world. My
clothes were weatherproof too. I was out and about in all weathers,
came home at all hours, was in the pits of all the theaters."

But it was not the same medieval wilderness Bretonne had ventured into. In Baudelaire some of the same figures recur—the prostitute, the beggar, the criminal, the beautiful stranger—but he does
not speak to them, and the content of their lives remains speculative
to him. Window-shopping and people-watching have become indistinguishable activities; one may attempt to buy but not to know them.
"Multitude, solitude: identical terms, and interchangeable by the
active and fertile poet," wrote Baudelaire. "The man who is unable to
people his solitude is equally unable to be alone in a bustling crowd.
The poet enjoys the incomparable privilege of being able to be

himself or someone else. . . . Like those wandering souls who go looking for a body, he enters as he likes into each man's personality. For him alone everything is vacant." Baudelaire's city is like a wilderness in another way: it is lonely.

The old Paris was cut down like a forest by Baron Haussmann, who carried out Napoleon III's vision of a splendid—and manageable—modern city. Since the 1860s it has been popular to say that Haussmann's destruction of the medieval warrens of streets and his creation of the grand boulevards was a counterrevolutionary tactic, an attempt to make the city penetrable by armies, indefensible by citizens. After all, citizens had revolted in 1789, 1830, and 1848, in part by building barricades across the narrow streets. But this does not explain the rest of Haussmann's project. The wide new avenues accommodated the flow of a vastly increased population, commerce, and on occasion troops, but below them were new sewers and waterways, eliminating some of the stench and disease of the old city—and the Bois de Boulogne was landscaped as a great public park in the English style. As a political project, it seems an attempt not to subdue but to seduce Parisians; as a development project, it displaced the poor from the center of the city to its edges and suburbs, where they remain today (the opposite of most postwar American cities—Manhattan and San Francisco are exceptions—abandoned to the poor when the middle class flocked to the suburbs). Other efforts were made to civilize the "wilderness" of the city in the nineteenth century: streetlights, house numbers, sidewalks, regularly posted street names, maps, guidebooks, increased policing, and the registration, prosecution, or both of prostitutes.

The real complaint against Haussmann seems to be twofold. The first is that in tearing down so much of the old city, he obliterated the delicate interlace of mind and architecture, the mental map walkers carried with them and the geographical correlatives to their memories and associations. In a poem about walking through one of Haussmann's construction sites near the Louvre, Baudelaire complained

> Paris is changing! but nothing within my melancholy
> has shifted! New palaces, scaffoldings, piles of stone
> Old neighborhoods—everything has become allegory for me
> and my dear memories are heavier than stones.

CASTIGLIONE   *The patisseries, though! Several on every block, it seemed. The window displays were*

Baudelaire being Baudelaire, the poem ends "in the forest of my exiled soul." And the brothers Edmond and Jules Goncourt wrote in their journal on November 18, 1860, "My Paris, the Paris in which I was born, the Paris of the manners of 1830 to 1848, is vanishing, both materially and morally. . . . I feel like a man merely passing through Paris, a traveller. I am foreign to that which is to come, to that which is, and a stranger to these new boulevards that go straight on, without meandering, without the adventures of perspective. . . ."

The second complaint is that with his broad, straight avenues, Haussmann turned the wilderness into a formal garden. The new boulevards continued a project begun two centuries before by André Le Nôtre, who went on to design the vast gardens of Versailles for Louis XIV. It was Le Nôtre who had designed the gardens of the Tuileries and the garden-boulevard of the Champs-Élysées extending west from the Tuileries to the Étoile, where Napoleon later placed his Arc de Triomphe. Most of these designs of Le Nôtre were outside the city walls and thus outside the economic life of the city, but the city expanded to absorb them. Thus the boulevards that Le Nôtre built for pleasure alone in the 1660s were developed for pleasure and industry by Haussmann in the 1860s (and these long axes had been widely emulated long before; Washington, D.C., is one of the cities that derives from this imperial geometry). Haussmann was as much an aesthete as Le Nôtre; he annoyed his emperor by leveling hills and taking other pains to make his streets utterly straight, opening up the long vistas that now seem so characteristic of Paris. It is a great irony that though the English garden had triumphed and gardens had become "natural"—irregular, asymmetrical, full of serpentine rather than straight lines—a formal French garden had been hacked out of the wilds of Paris.

The damp, intimate, claustrophobic, secretive, narrow, curving streets with their cobblestones sinuous like the scales of a snake had given way to ceremonial public space, space full of light, air, business, and reason. And if the old city had so often been compared to a forest, it may have been because it was an organic accretion of independent gestures by many creatures, rather than the implementation of a master plan made by one; it had not been designed but grown. No map had dictated that meandering organic form. And many hated

like pastry erotica. I discovered I particularly liked a vanilla custard pastry cut into segments like a

the change: "For the promenaders, what necessity was there to walk from the Madeleine to the Étoile by the shortest route? On the contrary, the promenaders like to prolong their walk, which is why they walk the same alley three or four times in succession," wrote Adolphe Thiers. Walking in the wilderness is one kind of pleasure, demanding daring, knowledge, strength—for savages, detectives, women in men's clothes; walking in a garden is a far milder one. Haussmann's boulevards made far more of the city a promenade and far more of its citizens promenaders. The arcades began their long decay as the streets bloomed with boutiques and the grand department stores were born—and during the Commune of 1871 the barricades of street revolutionaries were built across the great boulevards.

It wasn't Baudelaire who had first drawn Benjamin's attention to the arcades and to the possibility of configuring walking as a cultural act, but Benjamin's contemporaries—fellow Berliner and friend Franz Hessel and the surrealist writer Louis Aragon. He found Aragon's 1926 book *Paysan de Paris (Paris Peasant)* so exhilarating that "evenings in bed I could not read more than a few words of it before my heartbeat got so strong I had to put the book down. . . . And in fact the first notes of the *Passagenwerk* [or Arcades Project] come from this time. Then came the Berlin years, in which the best part of my friendship with Hessel was nourished by the Passagen-project in frequent conversations." In his Berlin essay, Benjamin describes Hessel as one of those guides who had introduced him to the city, and Hessel himself had written about walking Berlin (and, with Benjamin, worked on a translation of Marcel Proust's *Remembrance of Things Past*, a novel that with its themes of memory, walks, chance encounters, and Parisian salons fits neatly between the two bodies of French literature Benjamin took on). It is these twentieth-century writers and artists who best fit the descriptions of the nineteenth-century flâneur.

Aragon's *Paysan de Paris* is one of a trio of surrealist books published in the late 1920s; the other two are André Breton's *Nadja* and Philippe Soupault's *Last Nights of Paris*. All three are first-person narratives about a man wandering in Paris, give very specific place names and descriptions of places, and make prostitutes one of their main

---

wheel of Brie. Walking along the street, Eiffel Tower in sight ahead, eating the pastry out of my hand,

---

destinations. Surrealism prized dreams, the free associations of an unconscious or unself-conscious mind, startling juxtapositions, chance and coincidence, and the poetic possibilities of everyday life. Wandering around a city was an ideal way to engage with all these qualities. Breton wrote, "I still recall the extraordinary role that Aragon played in our daily strolls through Paris. The localities that we passed through in his company, even the most colourless ones, were positively transformed by a spellbinding romantic inventiveness that never faltered and that needed only a street-turning or a shop-window to inspire a fresh outpouring."

Paris, which had been stripped of its mystery by Haussmann, had recovered it to serve once again as a kind of muse to its poets. Both *Nadja* and *Last Nights* are organized around the pursuit of an enigmatic young woman met through a chance encounter, and it is this pursuit that gives the books their narratives. Such encounters are a staple of city-walking literature: Bretonne follows women with beautiful feet; Whitman eyeballs men in Manhattan; both Nerval and Baudelaire wrote poems about a passing glimpse of a woman who could have been their great love. Breton "spoke to this unknown woman, though I must admit that I expected the worst." Soupault's nameless narrator stalked his subject like a detective and came to know the underworld she and her associates inhabited, though this sordid realm of the ambitious, the demented, and the murderous neither explains nor fully dispels her fascination. Aragon's book, the least conventional of the three, has no narrative and is organized, like Benjamin's *One-Way Street*, around geography: it explores a few Parisian places—the first of which is the passage de l'Opéra, a shopping arcade already slated for destruction when Aragon wrote about it. (It was, tidily enough, torn down to make way for the expansion of the boulevard Haussmann.) *Paysan de Paris* demonstrated how rich a subject the city itself was for wandering, on foot and in the imagination.

Aragon made the city itself his subject, but Breton and Soupault pursued women who were embodiments of the city: Nadja and Georgette. Soupault writes of his protagonist spying on Georgette as she takes a customer to a hotel near the Pont Neuf and returns to the streets. Afterward, "Georgette resumed her stroll about Paris, through the mazes of the night. She went on, dispelling sorrow, solitude or tribulation. Then more than ever did she display her strange power: that of transfiguring the night. Thanks to her, who was no more than one of the hundred thousands, the Parisian night became a mysterious domain, a great and marvelous country, full of flowers, of birds, of glances and of stars, a hope launched into space. . . . That night, as we were pursuing, or more exactly, tracking Georgette, I saw Paris for the first time. It was surely not the same city. It lifted itself above the mists, rotating like the earth on its axis, more feminine than usual. And Georgette herself became a city." Once again and yet more deliriously, Paris is a wilderness, bedroom, and book to be read by walking. The protagonist—nameless, without a profession, the perfect flâneur at last—has taken up Bretonne's task of exploring the night, but by pursuing a single woman entangled with a single crime, a murder whose aftermath they both witnessed. The protagonist is a detective on the trail of crime and aesthetic experience, and Georgette embodies both.

Later Georgette tells him she took up her profession because she and her brother needed to live, and "Everything is so simple when one knows all the streets as I do, and all the people who move in them. They are all seeking something without seeming to do so." Like Nadja, she is a flâneuse, one who has made of the street a sort of residence. While Last Nights of Paris is a novel, Nadja is based on Breton's encounters with a real woman, and to underscore his book's nonfictionality, he reproduces photographs of people (though not of pseudonymous Nadja), places, drawings, and letters in the pages of his narrative. On one of their dates, Nadja leads him to the place Dauphine at the west end of the Île de la Cité, and he writes, "Whenever I happen to be there, I feel the desire to go somewhere else gradually ebbing out of me, I have to struggle against myself to get free from a gentle, over-insistent, and finally, crushing embrace."

Thirty years later, in his Pont Neuf, Breton, in the words of one critic, "famously proposes a detailed 'interpretation' of the topography of central Paris according to which the geographical and architectural layout of the Île de la Cité, and the bend of the Seine where it is situated, are seen to make up the body of a recumbent woman whose vagina is located in the place Dauphine, 'with its triangular, slightly curvilinear form bisected by a slit separating two wooded spaces.'"

---

take long walks along the Champs-Élysées to the Étoile, and exercise became a kind of punishment. As

Breton spends the night in a hotel with Nadja, and Soupault's narrator
hires Georgette for sex, but in these tales eroticism is not focused on
bodily intimacy in bed but diffused throughout the city, and noctural
walking rather than copulating is the means by which they bask in this
charged atmosphere. The women they pursue are most themselves,
most enchanting, and most at home on the streets, as though the profession of streetwalker was at last truly to walk the streets (no longer
victims of or refugees from the streets, as so many earlier heroines had
been). Nadja and Georgette are, like most surrealist representations of
women, too burdened with being incarnations of Woman—degraded
and exalted, muse and whore, city incarnate—to be individual women,
and this is most evident in their magical strolls through the city, strolls
that lure the narrators to follow these sirens on a chase that is also an
homage to and tour of Paris. The love of a citizen for his city and the
lust of a man for a passerby has become one passion. And the consummation of this passion is on the streets and on foot. Walking has
become sex. Benjamin concurred in this transformation of city into
female body, walking into copulating, when he concluded his passage
about Paris as labyrinth, "Nor is it to be denied that I penetrated to its
innermost place, the Minotaur's chamber, with the only difference
being that this mythological monster had three heads: those of the
occupants of the small brothel on rue de la Harpe, in which, summoning my last reserves of strength . . . I set my foot." Paris is a labyrinth whose center is a brothel, and in this labyrinth it is the arrival, not
the consummation, that seems to count, and the foot that seems to be
the crucial anatomical detail.

Djuna Barnes wrote a sort of coda to these books in her 1936
Nightwood, where once again the erotic love of an enchanted mad-
woman mingles with the fascinations of Paris and the night. The
heroine of Barnes's great lesbian novel, Robin Vote, walks the streets
"rapt and confused," abandoning her lover Nora Flood and directing
"her steps toward that night life that was a known measure between
Nora and the cafés. Her meditations, during this walk, were a part of
the pleasure she expected to find when the walk came to an end. . . .
Her thoughts were in themselves a form of locomotion." A cross-
dressing Irish doctor who frequents the pissoirs of the boulevards
explains the night in a long soliloquy to Nora, and Barnes must have

known what she was doing when she housed this Dr. O'Connor on the rue Servandoni by the place Saint-Sulpice, the same small street in which Dumas had housed one of his Three Musketeers and Hugo had settled *Les Misérables'* hero Jean Valjean. Such a density of literature had accumulated in Paris by the time of *Nightwood* that one pictures characters from centuries of literature crossing paths constantly, crowding each other, a Metro car full of heroines, a promenade populated by the protagonists of novels, a rioting mob of minor characters. Parisian writers always gave the street address of their characters, as though all readers knew Paris so well that only a real location in the streets would breathe life into a character, as though histories and stories themselves had taken up residence throughout the city.

Walter Benjamin described himself as "a man who has, with great difficulty, pried open the jaws of a crocodile and set up housekeeping there." He managed to live most of his life drifting about like a minor character in the literature he preferred. Perhaps it was French literature that led him to his death, for he delayed leaving Paris until it was too late. Boys' adventure books and the chronicles of the explorers would have better prepared him for his last years in the shadow of the Third Reich. When war broke out in September of 1939, he was rounded up with other German men in France and marched to a camp in Nevers, more than a hundred miles to the south. Now plump and afflicted with heart trouble that even on the streets of Paris had made him stop every few minutes, he collapsed several times on the march, but revived enough during his nearly three months of internment in the camp to teach courses in philosophy for a fee of a few cigarettes. His release secured by the P.E.N. club, he returned to Paris, where he continued to work on the *Arcades Project*, tried to secure a visa, and wrote the piercingly lyrical "Theses on the Philosophy of History." After the Nazi occupation of France, he fled south and with several others walked the steep route over the Pyrenees into Port Bou, Spain. He carried a heavy briefcase with him containing, he said, a manuscript more precious than his life, and in a steep vineyard he was so overcome that his companions had to support him on the walk. "No one knew the path," wrote a Frau Gurland who went with him. "We

---

or 14 kilometers every day and looking carefully in the seine to see if there is any place where I could

had to climb part of the way on all fours." In Spain the authorities demanded an exit visa from France and refused to honor the entrance visa for the United States Benjamin's friends had finally secured. In despair at his circumstances and the prospect of having to walk back over the mountains, he took an overdose of morphine in Spain and died on September 26, 1940—"whereupon the border officials, upon whom this suicide had made an impression," writes Hannah Arendt, "allowed his companions to proceed to Portugal." His briefcase vanished.

In the same essay, Arendt, who had lived in Paris herself in the 1960s, wrote, "In Paris a stranger feels at home because he can inhabit the city the way he lives in his own four walls. And just as one inhabits an apartment, and makes it comfortable, by living in it instead of just using it for sleeping, eating, and working, so one inhabits a city by strolling through it without aim or purpose, with one's stay secured by the countless cafés which line the streets and past which the life of the city, the flow of pedestrians moves along. To this day Paris is the only one among the large cities which can be comfortably covered on foot, and more than any other city it is dependent for its liveliness on people who pass by in the streets, so that the modern automobile traffic endangers its very existence not only for technical reasons." When I ran away to Paris at the end of the 1970s, the city was still more or less a walker's paradise, if you discounted the petty lecheries and rudeness of some of its men, and I was so poor and so young that I walked everywhere, for hours, and in and out of the museums (which are free to people under eighteen). Now I know that even I was living in a Paris that was disappearing. The vast void on the Right Bank was the site where the great Les Halles markets had recently been eradicated, but I didn't know that the spiral-walled pissoirs like little labyrinths for the mystery of male privilege were vanishing too, that traffic lights would come to the crooked old streets of the Latin Quarter and illuminated plastic signs for fast food would mar the old walls, that the old hulk on the quai d'Orsay was to become a flashy new museum, that the Tuileries' and Luxembourg's metal chairs with their spiral arms and perforated circular seats (in much the same aesthetic vein as the pissoirs) would be replaced by more rectilinear and less beautiful chairs painted the same green. It was nothing like the

throw Bloom in with a 50 lb. weight fixed to his feet."—RICHARD ELLMANN, *JAMES JOYCE   Fantine,*

transformations Parisians experienced during the Revolution, or during Haussmannization or at many other times, but this small register of changes has made me too the possessor of a lost city, and perhaps Paris is always a lost city, a city full of things that only live in imagination. Most dismaying of all when I returned recently was the change Arendt had foreseen: the dominance of the streets by cars. Cars had returned Paris's streets to the dirty and dangerous state in which they once had been, in the days when Rousseau was run over by a coach and walking the streets was a feat. To compensate for the automotive apotheosis, cars are banished on Sundays from certain streets and quays so that people might once again promenade there, as they always have in the gardens and on the wide sidewalks of the boulevards (and as I write, efforts are being made to take back more space—notably the great expanse of the place de la Concorde, which has in recent decades become a congested traffic circle).

One glory remains to Paris, that of possessing the chief theorists of walking, among them Guy DeBord in the 1950s, Michel de Certeau in the 1970s, and Jean Christophe Bailly in the 1990s. DeBord addressed the political and cultural meanings of cities' architecture and spatial arrangements; deciphering and reworking those meanings was one of the tasks of the Situationist Internationale he cofounded and whose principal documents he wrote. "Psychogeography," he declared in 1955, was a discipline that "could set for itself the study of the precise laws and specific effects of the geographical environment, consciously organized or not, on the emotions and behaviors of individuals." He decried the apotheosis of the automobile in that essay and elsewhere, for psychogeographies were best perceived afoot: "The sudden change of ambience in a street within the space of a few meters; the evident division of the city into zones of distinct psychic atmospheres; the path of least resistance which is automatically followed in aimless strolls (and which has no relation to the contour of the ground)" were among the subtleties he charted, proposing "the introduction of psychogeographic maps, or even the introduction of alterations" to "clarify certain wanderings that express not subordination to randomness but complete insubordination to habitual influences (influences generally categorized as tourism, that popular drug as repugnant as sports or buying on credit)." Another of DeBord's

---

in those labyrinths of the hill of the Pantheon, where so many ties are knotted and unloosed, long fled from

---

pugnacious treatises was the "Theory of the Dérive" (*dérive* is French for *drifting*), "a technique of transient passage through varied ambiances. . . . In a dérive one or more persons during a certain period drop their usual motives for movement and action, their relations, their work and leisure activities, and let themselves be drawn by the attractions of the terrain and the encounters they find there." That flâneury seemed to DeBord a radical new idea all his own is somewhat comic, as are his authoritarian prescriptions for subversion—but his ideas for making urban walking yet more conscious an experiment are serious. "The point," writes Greil Marcus, who has studied Situationism, "was to encounter the unknown as a facet of the known, astonishment on the terrain of boredom, innocence in the face of experience. So you can walk up the street without thinking, letting your mind drift, letting your legs, with their internal memory, carry you up and down and around turns, attending to a map of your own thoughts, the physical town replaced by an imaginary city." The Situationists' combination of cultural means and revolutionary ends has been influential, nowhere more so than in Paris's 1968 student uprising, when Situationist slogans were painted on the walls.

De Certeau and Bailly are far more mild, though they see futures as dark as DeBord's. The former devotes a chapter of his *Practice of Everyday Life* to urban walking. Walkers are "practitioners of the city," for the city is made to be walked, he wrote. A city is a language, a repository of possibilities, and walking is the act of speaking that language, of selecting from those possibilities. Just as language limits what can be said, architecture limits where one can walk, but the walker invents other ways to go, "since the crossing, drifting away, or improvisation of walking privilege, transform or abandon spatial elements." Further, he adds, "the walking of passers-by offers a series of turns (tours) and detours that can be compared to 'turns of phrase' or 'stylistic figures.'" De Certeau's metaphor suggests a frightening possibility: that if the city is a language spoken by walkers, then a postpedestrian city not only has fallen silent but risks becoming a dead language, one whose colloquial phrases, jokes, and curses will vanish, even if its formal grammar survives. Bailly lives in this car-choked Paris and documents this decline. In the words of an interpreter, he states that the social and imaginative function of cities "is under threat

---

Tholomyès, but in such a way as always to meet him again. There is a way of avoiding a person which

from the tyranny of bad architecture, soulless planning and indifference to the basic unit of urban language, the street, and the 'ruissellement de paroles' (stream of words), the endless stories, which animate it. Keeping the street and the city alive depends on understanding their grammar and generating the new utterances on which they thrive. And for Bailly, the principal agency of this process is walking, what he calls the 'grammaire generative de jambes' (generative grammar of the legs)." Bailly speaks of Paris as a collection of stories, a memory of itself made by the walkers of the streets. Should walking erode, the collection may become unread or unreadable.

Chapter 13

CITIZENS OF THE STREETS:

Parties, Processions, and

Revolutions

I turned all the way around to see that it was his wings that had made the angel just behind me look so odd out of the corner of my eye. At least, he was dressed as an angel, and various space aliens, tarts, disco kings, and two-legged beasts were all streaming down the street in the same direction, toward Castro Street, as they do every Halloween. The night before I had taken my bike down to the foot of Market Street to ride in Critical Mass, the group ride that is both a protest of the lack of safe space for bicyclists and a festive seizure of that space. Several hundred bicyclists riding together filled the streets, as they have the last Friday of every month since the event began here in 1992. (Cyclists stage Critical Masses around the world, from Geneva to Sydney to Jerusalem to Philadelphia.) Some of the more righteous bicyclists had taken to wearing T-shirts that say "One Less Car," so a trio of runners accompanied us wearing "One Less Bike" shirts, and in honor of the impending holiday some of the cyclists had donned masks or costumes.

Halloween in the Castro is a similarly hybrid event, both celebration and, at least in its origins, political statement—for asserting a queer identity is a bold political statement in itself. Asserting such an identity festively subverts the long tradition of sexuality being secret

---

shape for him, thanks to one of the enormous antitheses that were indispensable to his inspiration. In Hugo

---

and homosexuality being shameful—and in dreary times joy itself is insurrectionary, as community is in times of isolation. Nowadays, the Castro's Halloween street party is a magnet for a lot of straight people as well, but everyone seems to operate under the aegis of tolerance, campiness, and shameless staring in this event that is nothing more than a few thousand people milling along several blocks of shutdown streets. Nothing is sold, no one is in charge, and everyone is both spectacle and spectator. Earlier Halloween night, several hundred people had marched from Castro Street to the Hall of Justice to protest and mourn the murder of a young gay man in Wyoming, a pretty routine demonstration for San Francisco and for the Castro, which is both a temple of consumerism and home base for a politically active community.

November 2, Día de los Muertos, the Day of the Dead, was celebrated on Twenty-fourth Street in the Mission District. As always, the Aztec dancers—barefoot, spinning and stamping, clad in loincloths, leg rattles, and four-foot-long feather plumes—led the parade. They were followed by participants who bore altars on long poles—a Virgin of Guadalupe atop one and an Aztec god on the other. Behind the altars walked people carrying huge crosses draped in tissue paper, people with faces painted as skulls, people carrying candles, perhaps a thousand participants in all. Unlike bigger parades, this one was made up almost entirely of participants, with only a few onlookers from the windows of their homes. Perhaps it is better described as a procession, for a procession is a participants' journey, while a parade is a performance with audience. Walking together through the streets felt very different than did milling around on Halloween; there was a more tender, melancholic mood about this festival of death and a delicate but satisfying sense of camaraderie in the air that might have come from nothing more than sharing the same space and same purpose while moving together in the same direction. It was as though in aligning our bodies we had somehow aligned our hearts. At Twenty-fifth and Mission another procession invaded ours, a louder one chanting against the impending execution of a death-row inmate, and though it was annoying to be demonstrated at as though we were the executioners, it was useful to be reminded of the reality of death. The bakeries stayed open late selling *pan de muerto*—sweet bread baked

---

the crowd enters literature as an object of contemplation. The surging ocean is its model, and the thinker

---

into human figures—and the holiday was a fine hybrid of Christian
and indigenous Mexican tradition, revised and metamorphosed at the
hands of San Francisco's many cultures. Like Halloween, the Day of
the Dead is a liminal festival, celebrating the threshholds between life
and death, the time in which everything is possible and identity itself
is in flux, and these two holidays have become thresholds across
which different factions of the city meet and the boundaries between
strangers drop.

The great German artist Joseph Beuys used to recite, as a maxim
and manifesto, the phrase "Everyone an artist." I used to think it meant
that he thought everyone should make art, but now I wonder if he
wasn't speaking to a more basic possibility: that everyone could
become a participant rather than a member of the audience, that everyone could become a producer rather than a consumer of meaning
(the same idea lies behind punk culture's DIY—do it yourself—
credo). This is the highest ideal of democracy—that everyone can
participate in making their own life and the life of the community—
and the street is democracy's greatest arena, the place where ordinary
people can speak, unsegregated by walls, unmediated by those with
more power. It's not a coincidence that _media_ and _mediate_ have the same
root; direct political action in real public space may be the only way
to engage in unmediated communication with strangers, as well as a
way to reach media audiences by literally making news. Processions
and street parties are among the pleasant manifestations of democracy, and even the most solipsistic and hedonistic expressions
keep the populace bold and the avenues open for more overtly political uses. Parades, demonstrations, protests, uprisings, and urban
revolutions are all about members of the public moving through
public space for expressive and political rather than merely practical
reasons. In this, they are part of the cultural history of walking.

Public marches mingle the language of the pilgrimage, in which
one walks to demonstrate one's commitment, with the strike's picket
line, in which one demonstrates the strength of one's group and one's
persistence by pacing back and forth, and the festival, in which the
boundaries between strangers recede. Walking becomes testifying.
Many marches arrive at rally points, but the rallies generally turn
participants back into audiences for a few select speakers; I myself

---

who reflects on this eternal spectacle is the true explorer of the crowd in which he loses himself as he loses

---

have often been deeply moved by walking through the streets en masse and deeply bored by the events after arrival. Most parades and processions are commemorative, and this moving through the space of the city to commemorate other times knits together time and place, memory and possibility, city and citizen, into a vital whole, a ceremonial space in which history can be made. The past becomes the foundation on which the future will be built, and those who honor no past may never make a future. Even the most innocuous parades have an agenda: Saint Patrick's Day parades go back more than two hundred years in New York, and they demonstrate the religious convictions, ethnic pride, and strength of a once-marginal community, as do the much more glittering Chinese New Year's Day parade in San Francisco and colossal Gay Pride parades around the continent. Military parades have always been shows of strength and incitements to tribal pride or citizen intimidation. In Northern Ireland, Orangemen have used their marches celebrating past Protestant victories to symbolically invade Catholic neighborhoods, while Catholics have made the funerals of the slain into massive political processions.

On ordinary days we each walk alone or with a companion or two on the sidewalks, and the streets are used for transit and for commerce. On extraordinary days—on the holidays that are anniversaries of historic and religious events and on the days we make history ourselves—we walk together, and the whole street is for stamping out the meaning of the day. Walking, which can be prayer, sex, communion with the land, or musing, becomes speech in these demonstrations and uprisings, and a lot of history has been written with the feet of citizens walking through their cities. Such walking is a bodily demonstration of political or cultural conviction and one of the most universally available forms of public expression. It could be called marching, in that it is common movement toward a common goal, but the participants have not surrendered their individuality as have those soldiers whose lockstep signifies that they have become interchangeable units under an absolute authority. Instead they signify the possibility of common ground between people who have not ceased to be different from each other, people who have at last become the public. When bodily movement becomes a form of speech, then the distinctions between words and deeds, between representations and actions, begin to blur, and so marches can themselves be liminal, another form of walking into the realm of the representational and symbolic—and sometimes, into history.

Only citizens familiar with their city as both symbolic and practical territory, able to come together on foot and accustomed to walking about their city, can revolt. Few remember that "the right of the people peaceably to assemble" is listed in the First Amendment of the U.S. Constitution, along with freedom of the press, of speech, and of religion, as critical to a democracy. While the other rights are easily recognized, the elimination of the possibility of such assemblies through urban design, automotive dependence, and other factors is hard to trace and seldom framed as a civil rights issue. But when public spaces are eliminated, so ultimately is the public; the individual has ceased to be a citizen capable of experiencing and acting in common with fellow citizens. Citizenship is predicated on the sense of having something in common with strangers, just as democracy is built upon trust in strangers. And public space is the space we share with strangers, the unsegregated zone. In these communal events, that abstraction the public becomes real and tangible. Los Angeles has had tremendous riots—Watts in 1965 and the Rodney King uprising in 1992—but little effective history of protest. It is so diffuse, so centerless, that it possesses neither symbolic space in which to act, nor a pedestrian scale in which to participate as the public (save for a few relict and re-created pedestrian shopping streets). San Francisco, on the other hand, has functioned like the "Paris of the West" it was once called, breeding a regular menu of parades, processions, protests, demonstrations, marches, and other public activities in its central spaces. San Francisco, however, is not a capital, as Paris is, so it is not situated to shake the nation and the national government.

Paris is the great city of walkers. And it is the great city of revolution. Those two facts are often written about as though they are unrelated, but they are vitally linked. Historian Eric Hobsbawm once speculated on "the ideal city for riot and insurrection." It should, he concluded, "be densely populated and not too large in area. Essentially it should still be possible to traverse it on foot. . . . In the ideal insurrectionary city

into the streets, arm in arm, continuing the topics of the day, or roaming far and wide until a late hour,

the authorities—the rich, the aristocracy, the government or local administration—will therefore be as intermingled with the central concentration of the poor as possible." All the cities of revolution are old-fashioned cities: their stone and cement are soaked with meanings, with histories, with memories that make the city a theater in which every act echoes the past and makes a future, and power is still visible at the center of things. They are pedestrian cities whose inhabitants are confident in their movements, familiar with the crucial geography. Paris is all these things, and it has had major revolutions and insurrections in 1789, 1830, 1848, 1871, and 1968, and in recent times, myriad protests and strikes.

Hobsbawm addresses Haussmann's reshaping of Paris when he writes, "Urban reconstruction, however, had another and probably unintended effect on potential rebellions, for the new and wide avenues provided an ideal location for what became an increasingly important aspect of popular movements, the mass demonstration or rather procession. The more systematic these rings and cartwheels of boulevards, the more effectively isolated those were from the surrounding inhabited area, the easier it became to turn such assemblies into ritual marches rather than preliminaries to riot." In Paris itself, it seems that the saturation of ceremonial, symbolic, and public space makes the people there peculiarly susceptible to revolution. That is to say, the French are a people for whom a parade is an army if it marches like one, for whom the government falls if they believe it has, and this seems to be because they have a capital where the representational and the real are so interfused and because their imaginations too dwell in public, engaged with public issues, public dreams. "I take my desires for reality, because I believe in the reality of my desires," said graffiti on the Sorbonne in the student-led uprising of May 1968. That uprising captured its most crucial territory, the national imagination, and it was on this territory as well as the Latin Quarter and the strike sites around France that they came within a hairsbreadth of toppling Europe's strongest government. "The difference between rebellion at Columbia and rebellion at the Sorbonne is that life in Manhattan went on as before, while in Paris every section of society was set on fire, in the space of a few days," wrote Mavis Gallant, who was there in the streets of the Latin Quarter. "The collective hallucination

seeking amid the wild lights and shadows of the populous city, that infinity of mental excitement which was that life can change, quite suddenly and for the better. It still strikes me as a noble desire."

Everyone knows how the French Revolution began. On July 11, 1789, Louis XVI dismissed the popular minister Jacques Necker, further stirring up his already turbulent capital. Parisians must have been imagining an armed revolt, for 6,000 of them spontaneously assembled to storm the Invalides and seize the rifles stored there, then went on to conquer the Bastille across the river for more military supplies, with results still celebrated in parades and festivals throughout France every July 14, Bastille Day. Life did change, suddenly and, in the long run, for the better. The liberation of that medieval fortress-prison symbolically ended centuries of despotism but the revolution didn't really begin until the march of the market women three months later. The revolution's intellectual origins lay in the ideals of liberty and justice prompted in part by Enlightenment philosophers such as Thomas Paine, Rousseau, and Voltaire, but it also had bodily origins. In the summer of 1788 a devastating hailstorm had wiped out much of the harvest across France, and in 1789 the people felt the effects. Bread rose in price and became scarce, ordinary people often began standing in line at the bakeries at 4 A.M. in the hope of buying a loaf that day, and the poor began to become the hungry. Bodily causes had bodily effects; it was to be a revolution not merely of ideas but of bodies liberated, starving, marching, dancing, rioting, decapitated, on the stage of Parisian streets and squares. Revolutions are always politics made bodily, politics when actions become the usual form of speech. Britain and France had had food and tax riots before, but nothing quite like this combination of hunger for food and for ideals.

In the heady days after the fall of the Bastille, the market women and *poissardes*, or fishwives, had grown accustomed to marching together, and they must have first felt their common desires and collective strength during the religious processions they went on that season. At least one local was alarmed "at the discipline, pageantry, and magnitude of the almost daily processions of market women, laundresses, tradesmen, and workers of different districts that, during August and September, wound up the rue Saint-Jacques to the newly built church of Sainte-Genevieve [patroness of Paris] for thanksgiving services." Simon Schama points out that on the feast-day of Saint

----

quiet observation can afford.—EDGAR ALLAN POE, "THE MURDERS IN THE RUE MORGUE"   *In cities*

Louis, August 25, the market women of Paris traditionally went to Versailles to present the queen with bouquets. It is as though having learned the form of the procession, they could give it new content: having marched to pay homage to church and state, they were ready to march to demand terms.

On the morning of October 5, 1789, a girl took a drum to the central markets of Les Halles, while in the insurrectionary faubourg Saint-Antoine a woman compelled a local cleric to ring the church bells in his church. Drum and bells gathered a crowd. The women—now numbering in the thousands—chose a hero of the Bastille to lead them, Stanislas-Marie Maillard, who found himself constantly preaching moderation to his followers. Though made up mostly of poor working women—fishwives, market women, laundresses, portresses—the crowd included some women of means and a few noted revolutionaries, such as Theroigne de Mericourt, known as Theroigne the Amazon. (Prostitutes and men dressed as women loomed large in contemporary accounts of the march, but this seems to have been because many believed "respectable" women were incapable of such insurrection.) The women insisted on moving straight through the Tuileries, still the gardens of the king, and when a guard pulled his sword on one of the women in the lead, Maillard came to her defense—but "she delivered such a blow with her broom to the crossed swords of the men that they were both disarmed." They continued on chanting "Bread and to Versailles!" Later that day the marquis de Lafayette, hero of the American Revolution, led an army of about 20,000 national guards after them in equivocal support.

By early evening they were at the National Assembly in Versailles, demanding that this new governing body deal with the food shortage, and a few women were taken before the king to make their case. Before midnight the crowd was at the palace gates; and early in the morning the crowd came inside. It was a gory arrival—after a guardsman shot a young woman, the crowd decapitated two guards and rushed the royal apartments looking for the hated queen, Marie Antoinette. That day, the terrified royal family was forced to return to Paris with the jubilant, exhausted, victorious crowd. At the head of the long procession—Lafayette estimated it at 60,000—came the royal family in a carriage surrounded by women carrying branches of laurel, followed by the National Guard, escorting wagonloads of

men cannot be prevented from concerting together, and awakening a mutual excitement which prompts

wheat and flour. At the rear, writes one historian, marched more women, "their decorated branches amidst the gleaming iron of pikes and musket barrels giving the impression, as one observer thought, of 'a walking forest.' It was still raining, and the roads were ankle deep in mud, yet they all seemed content, even cheerful." They shouted to passersby, "Here come the Baker, the Baker's Wife, and the Baker's Little Boy." The king in Paris was a very different entity than the king in Versailles. There the once absolute power of the French monarchy ebbed away, and he became a constitutional monarch, then a prisoner, and within a few years a victim of the guillotine as the revolution spiraled down into factions and bloodbaths.

History is often described as though it were made up entirely of negotiations in closed spaces and wars in open ones—of talking and fighting, of politicians and warriors. Earlier events of that revolution— the birth of the National Assembly and the storming of the Bastille— correspond to these versions. Yet the market women had managed to make history as ordinary citizens engaged in ordinary gestures. During the walk of the thousands of women to Versailles, they had overcome the weight of the past in which they had been deferential to all the usual authorities, while the traumas of the future were yet unforeseen. They had one day in which the world was with them, they feared nothing, armies followed in their wake, and they were not grist for history's mill but the grinders. Like mass marchers everywhere, they displayed a collective power—the power at the very least to withdraw their support and at the most to revolt violently—but they managed to start the revolution largely as marchers. They carried branches as well as muskets—for muskets operate in the realm of the real, but branches in that of the symbolic.

This intertwining of religious festivity, huge gatherings in public squares, and mass marches would appear again on the two hundredth anniversary of the beginning of the French Revolution. The revolutionary year began inauspiciously with government tanks literally crushing the student democracy movement in Beijing's Tiananmen Square, but across Europe Communist governments had lost their appetite for or their confidence in violent repression. Violence itself had become a far less casual tool than it had been before Gandhi spread his doctrine of nonviolence, human rights had become far more established, and media had made events around the world more visible. The American civil rights movement had demonstrated its effectiveness in the West, and peace movements and nonviolent direct-action tactics had become a global language of citizen resistance. As Hobsbawm points out, marching down the boulevard had largely replaced rioting in the quarter. Throughout Eastern Europe the insurrectionaries made it clear that nonviolence was part of their ideology. The revolution in Poland worked the way nonviolent changes are supposed to—slowly, with lots of outside political pressure and inside political negotiation, culminating in the free election of June 4, 1989—and all the revolutions benefited from Mikhail Gorbachev's shrewd dismantling of the Soviet Union. But in Hungary, East Germany, and Czechoslovakia, history was made in the streets, and their old cities accommodated public gatherings beautifully.

It was, reported Timothy Garton Ash, a funeral held thirty-one years late for Imre Nagy, executed for his part in the unsuccessful 1956 revolt, that started the revolution in Hungary. On June 16, two hundred thousand people marched in a gathering that would have been violently crushed in previous years. In the exhilaration of having recovered their history and their voice, dissidents stepped up their efforts, and on October 23, the new Hungarian Republic was born. East Germany was next. Repressive measures were at first stepped up—students on their way home from school and employees returning for work were arrested just for being in the vicinity of disturbances in East Berlin: even the everyday freedom to walk about had become criminalized (as, with curfews and bans on assembly, it often is in turbulent times or under repressive regimes). But Leipzig's Nikolaikirche had long held Monday-evening "prayers for peace" followed by demonstrations on adjacent Karl-Marx-Platz, and there the numbers began to grow. On October 2, fifteen to twenty thousand gathered at that square by the church in the largest spontaneous demonstration in East Germany since 1953, and by October 30, nearly half a million people marched. "From that time forward," writes Ash, "the people acted and the Party reacted." On November 4 a million people gathered in East Berlin's Alexanderplatz, carrying flags, banners, and posters,

inhabitants are members; their populace exercise a prodigious influence upon the magistrates, and

and on November 9 the Berlin Wall fell. A friend who was there told me it fell because so many people showed up when a false report circulated that the wall was down that they made it into a real event— the guards lost their nerve and let them through. It became true because enough people were there to make it true. Once again people were writing history with their feet.

Czechoslovakia's "Velvet Revolution" was the most marvelous of them all, and the last (Romania's Christmastime violence was something else altogether). In January of that magic year, playwright Václav Havel had been imprisoned for participating in a twentieth-anniversary commemoration of a student who had burned himself to death in Prague's heart, Wenceslas Square, in protest of the crushed "Prague Spring" revolution of 1968. November 17, 1989, was the anniversary of another Czech student martyr, killed by Nazis during the occupation, and this commemorative procession was far larger and far bolder than that of January. The crowd marched from Charles University, and when the official itinerary was over at dusk, they lit candles, produced flowers, and continued on through the streets, singing and chanting antigovernment slogans—the past once again becoming an occasion to address the present. At Wenceslas Square, policemen surrounded them and began clubbing anyone within reach. Marchers stampeded down side streets, where some slipped away or were taken into nearby homes, but many were injured. False accounts that one student had joined the ranks of student martyrs infuriated the nation. Afterward came spontaneous marches, strikes, and gatherings in Wenceslas Square—really a kilometer-long, immensely wide boulevard in the heart of the city—with hundreds of thousands of participants. Behind the scenes, in the Magic Lantern Theater, the recently released Havel brought together all the opposition groups into a political force to make something pragmatic of the power being taken in the streets (the Czech opposition was called the Civic Forum; the Slovak equivalent was called the Public Against Violence).

Czechoslovakians had begun to live in public, gathering every day in Wenceslas Square and proceeding down adjoining Národní Avenue, getting their news from other participants, making and reading posters and signs, creating altars of flowers and candles—reclaiming the street as public space whose meaning would be determined by the

---

frequently execute their own wishes without the intervention of public officers.—ALEXIS DE

public. "Prague," reported one journalist, "seemed hypnotized, caught in a magical trance. It had never ceased to be one of Europe's most beautiful cities, but for two long decades a cloud of repressive sadness had enveloped the Gothic and baroque towers. Now it vanished. The crowds were calm, confident and civilized. Each day, people assembled after work at 4pm, filing politely, patiently and purposefully into Wenceslas Square. . . . The city burst with color: posters were plastered on walls, on shop windows, on any inch of free space. After each mass rally, the crowd sang the National Anthem." Four days later the country's two most famous dissidents—Havel and the hero of 1968, Alexander Dubc̆ek—appeared on a balcony above the square, the latter in his first public appearance after twenty-one years of enforced silence. Dubc̆ek said at this time, "The government is telling us that the street is not the place for things to be solved, but I say the street was and is the place. The voice of the street must be heard."

The revolution that began by remembering a student peaked by celebrating a saint. Saint Agnes of Bohemia, great-granddaughter of the saintly Wenceslas, had been canonized a few weeks earlier. Prague's archbishop, a supporter of the opposition, held an outdoor mass for hundreds of thousands in the snow a few days after Dubc̆ek reappeared. Like the Hungarians, the Czechoslovakians had wrested their future free by remembering the heroes and martyrs of the past, for by December 10 there was a new government. Michael Kukral, a young American geographer who was there throughout the Velvet Revolution, wrote, "The time of massive and daily street demonstrations was over after November 27th, and thus, the entire character of the revolution metamorphosed. I did not awaken the next morning to find myself transformed into a giant bug, but I did feel a sense of sadness knowing that I will probably never again experience the momentum, spontaneity, and exhilaration of these past ten days."

Nineteen-eighty-nine was the year of the squares—of Tiananmen Square, of the Alexanderplatz, of Karl-Marx-Platz, of Wenceslas Square—and of the people who rediscovered the power of the public in such places. Tiananmen Square serves as a reminder that marches,

---

TOQUEVILLE, DEMOCRACY IN AMERICA   *We are nothing if we walk alone, we are everything when we*

protests, and seizures of public space don't always produce the desired results. But many other struggles lie somewhere in between the Velvet Revolution and the bloodbaths of repression, and the 1980s were a decade of great political activism: in the colossal antinuclear movements in Kazakhstan, Britain, Germany, and the United States, in the myriad marches against U.S. intervention in Central America, in the students around the world who urged their universities to divest from South Africa and helped topple the apartheid regime there, in the queer parades increasing through the decade and the radical AIDS activists at the end of the decade, in the populist movements that took to the streets of the Philippines and many other countries.

A few years earlier another insurrection found a square for its stage. The saga of the Mothers of the Plaza de Mayo began when these women started to notice each other at the police stations and government offices, making the same fruitless inquiries after children who had been "disappeared" by agents of the brutal military junta that seized power in 1976. "Secrecy," writes Marguerite Guzman Bouvard, "was a hallmark of the junta's Dirty War. . . . In Argentina the abductions were carried out beneath a veneer of normalcy so that there would be no outcry, so that the terrible reality would remain submerged and elusive even to the families of the abducted." Mostly homemakers with little education and no political experience, these women came to realize that they had to make the secret public, and they pursued their cause with a stunning lack of regard for their own safety. On April 30, 1977, fourteen mothers went to the Plaza de Mayo in the center of Buenos Aires. It was the place where Argentinean independence had been proclaimed in 1810 and where Juan Perón had given his populist speeches, a plaza at the heart of the country. Sitting there was, a policeman shouted, tantamount to holding an illegal meeting, and so they began walking around the obelisk in the center of the plaza.

There and then, wrote a Frenchman, the generals lost their first battle and the Mothers found their identity. It was the plaza that gave them their name, and their walks there every Friday that made them famous. "Much later," writes Bouvard, "they described their walks as marches, not as walking, because they felt that they were marching toward a goal and not just circling aimlessly. As the Fridays succeeded

----

walk together in step with other dignified feet.—SUBCOMMANDANTE MARCOS, 1995   *She's gravely*

----

one another and the numbers of Mothers marching around the plaza increased, the police began to take notice. Vanloads of policemen would arrive, take names, and force the Mothers to leave." Attacked with dogs and clubs, arrested and interrogated, they kept returning to perform this simple act of remembrance for so many years that it became ritual and history and made the name of the plaza known around the world. They marched carrying photographs of those children mounted like political placards on sticks or hung around their neck, and wearing white kerchiefs embroidered with the names of their disappeared children and the dates of their disappearances (later they were embroidered instead, "Bring Them Back Alive").

"They tell me that, while they are marching they feel very close to their children," wrote the poet Marjorie Agosin, who walked with them. "And the truth is, in the plaza where forgetting is not allowed, memory recovers its meaning." For years these women taking the national trauma on a walk were the most public opposition to the regime. By 1980 they had created a network of mothers around the country, and in 1981 they began the first of their annual twenty-four-hour marches to celebrate Human Rights Day (they also joined religious processions around the country). "By this time the Mothers were no longer alone during their marches; the Plaza was swarming with journalists from abroad who had come to cover the strange phenomenon of middle-aged woman marching in defiance of a state of siege." When the military junta fell in 1983, the Mothers were honored guests at the inauguration of the newly elected president, but they kept up their weekly walks counterclockwise around the obelisk in the Plaza de Mayo, and the thousands who had been afraid before joined them. They still walk counterclockwise around the tall obelisk every Thursday.

There are many ways to measure the effectiveness of protest. There's its impact on the wider public, directly and through the media, and there's its impact on the government—on its audiences. But what's often forgotten is its impact on the protesters, who themselves suddenly become the public in literal public space, no longer an audience but a force. I had a taste, once, of this public life during the first weeks

---

ill now from a hunger strike, and she's told the generals that she'll gladly leave. But she said, "I want all

of the Gulf War, of living there more intensely than in San Francisco's
many annual marches and parades before and since. Not much was
written then or has been since about the huge protests all over the
country in January 1991—surrounding Philadelphia's Independence
Hall, gathering in Lafayette Park across from the White House, occupying the Washington State and Texas legislatures, shutting down
the Brooklyn Bridge, covering Seattle in posters and demonstrations,
holding "gas-pump protests" across the South. But there was, amid the
fear and more deferential versions of patriotism, a huge outcry that
continued for weeks in San Francisco. I don't mean to suggest that we
had the courage of the Mothers of the Plaza de Mayo or the impact
of the people of Prague, only that we too lived for a while in public.
The whole strategy of that war—its speed, its colossal censorship, its
reliance on high-tech weaponry, its very limited ground combat—
was organized to defeat opposition at home by limiting information
and U.S. casualties, which suggests that protest and popular opinion
were so strong a force that the war (and the little wars like it since)
was a preemptive strike against them.

We went out into the streets anyway, and the very space of the
city was transformed. Before the first bombs dropped, people began
to gather spontaneously, to march together, to make bonfires out of
the old Christmas trees put out on the streets, to organize rituals and
gatherings, to plaster the city with posters that seemed to make the
very walls break their silence with calls for specific actions and caustic
commentaries on the meaning of the war. Many of the demonstrations here, as elsewhere, instinctively headed for the traffic arteries—
bridges, highways—or for the power points—the federal building,
the stock exchange—and shut them down. There were protests
almost daily into February. The city was being remade as a place
whose center did not belong to business or to cars, but to pedestrians
moving down the street in this most bodily form of free speech. The
streets were no longer antechambers to the interiors of homes,
schools, offices, shops, but a colossal amphitheater. I wonder now if
anyone has ever protested or paraded simply because such occasions
provide the only time when American city streets are a perfect place
to be a pedestrian, safe from assault by cars and strangers if not,

political prisoners released, and I want to walk to the airport." It's twenty miles from her house to the

occasionally, police. From the middle of the street, the sky is wider and the shop windows are opaque.

The Saturday evening before the war began, I ditched my car and walked in the boisterous march that coalesced spontaneously, drawing people out of bars and cafés and homes. I marched in the well-organized protest the day before the war broke out with a few thousand others. I joined more thousands the afternoon the war broke out to march again through the dark and our own horror to the Federal Building. The next morning I blockaded Highway 101 with the group of activists I spent much of the war with, until the highway patrol began clubbing away and broke one man's leg, and later that morning I walked with twenty or thirty others again down the city streets into the financial and commercial district. On the weekend after the war broke out, I walked with 200,000 others who gathered to protest the war with banners and placards, puppets and chants. For those weeks my life seemed to be one continuous procession through the trans-formed city. Private concerns and personal fears faded away in the in-cendiary spirit of the time. The streets were our streets, and all our fear was for others. There were mutterings about using nuclear weapons and suggestions that Israel might be drawn into a conflict that seemed as though it could spread like wildfire into a worldwide conflagration. The horror about what was happening far away and the strength of the incendiary resistance inside us and around us generated extraordinary feeling. I have never felt anything as intensely as I did that war except for the most passionate love and the most mourned deaths (and it was a war with plenty of deaths, though few were of Americans until the effects of the war's toxic materials began to materialize).

The afternoon of the first day of the war, I got caught up in a police sweep and spent a few hours sitting down for a change, hand-cuffed in a bus near the center of activity, looking out the window, and in an odd truce, listening with the policemen to an arrested journalist's shortwave radio broadcasting the war. Missiles were being fired on Israel, and the radio said the inhabitants of Tel Aviv were all in sealed rooms wearing gas masks. That image stuck with me, of a war in which civilians lost sight of the world and of each others' faces and, from behind their hideous masks, lost even the ability to speak. Most

----

airport. And she has so rallied the spirit of her people that the generals rightly fear that the whole country

----

Americans weren't much better off, voiceless in front of televisions running the same uninformative footage of the censored war over and over again. In living on the streets we were refusing to consume the meaning of that war and instead producing our own meaning, on our streets and in our hearts if not in our government and media.

In those moments of moving through the streets with people who share one's beliefs comes the rare and magical possibility of a kind of populist communion—perhaps some find it in churches, armies, and sports teams, but churches are not so urgent, and armies and teams are driven by less noble dreams. At such times it is as though the still small pool of one's own identity has been overrun by a great flood, bringing its own grand collective desires and resentments, scouring out that pool so thoroughly that one no longer feels fear or sees the reflections of oneself but is carried along on that insurrectionary surge. These moments when individuals find others who share their dreams, when fear is overwhelmed by idealism or by outrage, when people feel a strength that surprises them, are moments in which they become heroes—for what are heroes but those so motivated by ideals that fear cannot sway them, those who speak for us, those who have power for good? A person who feels this all the time may become a fanatic or at least an annoyance, but a person who never feels it is condemned to cynicism and isolation. In those moments everyone becomes a visionary, everyone becomes a hero.

Histories of revolutions and uprisings are full of stories of generosity and trust between strangers, of incidents of extraordinary courage, of transcendence of the petty concerns of everyday life. In 1793, Victor Hugo's novel of revolution, he wrote, "People lived in public: they ate at tables spread outside the doors; women seated on the steps of the churches made lint as they sang the 'Marseillaise.' Park Monceaux and the Luxembourg Gardens were parade-grounds. . . . Everything was terrible and no one was frightened. . . . Nobody seemed to have leisure: all the world was in a hurry." At the beginning of the Spanish Civil War, George Orwell wrote of Barcelona's transformation, "The revolutionary posters were everywhere, flaming from the walls in clean reds and blues that made the few remaining

would turn out to cheer her if she took that twenty-mile walk.—PAUL MONETTE, ON AUNG SAN SUU

advertisements look like daubs of mud. Down the Ramblas, the wide central artery of the town where crowds of people streamed constantly to and fro, the loud-speakers were bellowing revolutionary songs all day and far into the night. . . . Above all there was a belief in the revolution and the future, a feeling of having suddenly emerged into an era of equality and freedom." To use a Situationist word, there seems to be a psychogeography of insurrection in which life is lived in public and is about public issues, as manifested by the central ritual of the march, the volubility of strangers and of walls, the throngs in streets and plazas, and the intoxicating atmosphere of potential freedom that means the imagination has already been liberated. "Revolutionary moments are carnivals in which the individual life celebrates its unification with a regenerated society," writes Situationist Raoul Vaneigem.

But nobody remains heroic forever. It is the nature of revolutions to subside, which is not the same thing as to fail. A revolution is a lightning bolt showing us new possibilities and illuminating the darkness of our old arrangements so that we will never see them quite the same way again. People rise up for an absolute freedom, a freedom they will only find in their hopes and their acts at the height of that revolution. Sometimes they may have overthrown a dictator, but other dictators will arise and bring with them other ways of intimidating or enslaving the populace. Sometimes everyone will have a vote at last, food and justice will be adequate if not ideal, but ordinary traffic will return to the streets, the posters will fade, revolutionaries will go back to being housewives or students or garbage collectors, and the heart will become private again. On the first anniversary of the storming of the Bastille came the Fête de la Federation, a national festival of dances, visits, parades, and overflowing joy, and it was the spontaneous participation of all classes of Parisians in readying the Champ de Mars for their fête, rather than the fete itself, that was most exhilarating. A year later, on July 12, 1791, there was a military parade commemorating Voltaire, and the people who had participated in history ferociously and then joyously had become spectators again.

"Resistance is the secret of joy," proclaimed the pamphlet someone from Reclaim the Streets handed me in the middle of a Birmingham

KYI IN BURMA *The Orthodox forbid women to pray together in groups, to pray out loud, and to hold*

street, in the midst of one of their street parties. Reclaim the Streets was founded in London in May 1995 with the understanding that if the twin forces of privatized space and globalized economies are alienating us from each other and from local culture, the reclaiming of public space for public life and public festival is one way to resist both. The very act of revolting—happily and communally and in the middle of the street—was no longer a means to an end but victory itself. Imagined thus, the difference between revolutions and festivals becomes even less distinct, for in a world of dreary isolation festivals are inherently revolutionary. The RTS party in Birmingham three years later was intended as a counterpoint to the Group of Eight meeting that weekend, in which leaders from the world's top economic powers would make the world's future without consulting citizens or the poorer nations. Hundreds of thousands gathered by the group Christian Aid formed a human chain around the central city to demand that third-world debt be forgiven. Reclaim the Streets wasn't asking, but taking what they wanted.

There was a glorious moment when trumpeters blew a sort of pedestrian charge, and the thousands who'd come for this Global Street Party surged out of the bus station into Birmingham's main street. People quickly shimmied up light poles and hung banners: "Beneath the Tarmac the Grass," said one about sixty feet long, copping a line from May '68 in Paris, and "Stop the Car/Free the City" said another. Once people settled in, the great spirit of the move forward subsided into a fairly standard party of mostly young and scruffy people, dancing, mingling, stripping down in the steamy heat, not notably different from, say, Halloween in the Castro, except that it was illegal and obstructionist. Walking and marching are communal in spirit in ways that mingling after arrival is not. It wasn't, an RTS activist told me later, one of their great street parties, nothing compared to their three-day street party with the striking Liverpool dockworkers, or the rave-style protest of an intrusive new highway near London that included giant puppets wearing hoop skirts beneath which hid jackhammer operators putting holes in the overpass that were then planted with trees, or RTS spinoff the Revolutionary Pedestrian Front's pranks at an Alfa Romeo promotional event, or the taking over of Trafalgar Square. Perhaps some of the other places

a Torah. These women gathered to pray at the Western Wall in Jerusalem, and were assaulted by

where sister street parties were held that day—Ankara, Berlin, Bogotá, Dublin, Istanbul, Madrid, Prague, Seattle, Turin, Vancouver, Zagreb— lived up to the glorious rhetoric of Reclaim the Streets' publications. Though Reclaim the Streets may not have fulfilled its goal, it has set a new one for every street action—now every parade, every march, every festival, can be regarded as a triumph over alienation, a reclaiming of the space of the city, of public space and public life, an opportunity to walk together in what is no longer a journey but already an arrival.

the Orthodox men, their case is still pending in the Israeli Supreme Court.—MARTHA SHELLEY,

---

WALKING AFTER MIDNIGHT:
Women, Sex, and Public Space

Caroline Wyburgh, age nineteen, went "walking out" with a sailor in Chatham, England, in 1870. Walking had long been an established part of courtship. It was free. It gave the lovers a semiprivate space in which to court, whether in a park, a plaza, a boulevard, or a byway (and such rustic landscape features as lovers' lanes gave them private space in which to do more). Perhaps, in the same way that marching together affirms and generates solidarity between a group, this delicate act of marching the rhythms of their strides aligns two people emotionally and bodily; perhaps they first feel themselves a pair by moving together through the evening, the street, the world. As a way of doing that something closest to doing nothing, strolling together allows them to bask in each other's presence, obliged neither to converse continually nor to do something so engaging as to prevent them from conversing. And in Britain the term "walking out together" sometimes meant something explicitly sexual, but more often expressed that an ongoing connection had been established, akin to the modern American phrase "going steady." In James Joyce's novella *The Dead*, the husband who has just discovered that his wife had a suitor in her youth asks if she loved that now-dead boy, and she replies, devastatingly, "I used to go out walking with him."

    Caroline Wyburgh, age nineteen, was seen walking with her soldier, and because of it she was dragged from her bed late one night

---

---

by a police inspector. The Contagious Diseases Acts in effect at that time gave police in barracks towns the power to arrest anyone they suspected of being a prostitute. Merely walking about in the wrong time or place could put a woman under suspicion, and the law allowed any woman so accused or suspected to be arrested. If the arrested woman refused to undergo a medical examination, she could be sentenced to months in jail; but the painful and humiliating medical examination constituted punishment too; and if she was found to be infected, she was confined to a medical prison. Guilty till proven innocent, she could not escape unscathed. Wyburgh supported herself and her mother by washing doorsteps and basements, and her mother, fearing the loss of their income for so long, tried to persuade her to submit to the examination rather than to serve the three-month prison sentence. She refused, and so the officers of the law strapped her to a bed for four days. On the fifth day she agreed to be examined, but her willingness failed her after she was taken to the surgery, straitjacketed, thrust onto an examining couch with her feet strapped apart, and held down by an assistant who planted an elbow on her chest. She struggled, rolled off the couch with her ankles still strapped in, and severely injured herself. But the surgeon laughed, for his instruments of inspection had deflowered her, and blood poured between her legs. "You have been telling the truth," he said. "You are not a bad girl."

The soldier was never named, arrested, inspected, or otherwise drawn into the legal system, and men have usually had an easier time walking down the street than have women. Women have routinely been punished and intimidated for attempting that most simple of freedoms, talking a walk, because their walking and indeed their very beings have been construed as inevitably, continually sexual in those societies concerned with controlling women's sexuality. Throughout the history of walking I have been tracing, the principal figures— whether of peripatetic philosophers, flâneurs, or mountaineers—have been men, and it is time to look at why women were not out walking too.

"Being born a woman is my awful tragedy," wrote Sylvia Plath in her journal when she too was nineteen. "Yes, my consuming desire to mingle with road crews, sailors and soldiers, barroom regulars—to be part of a scene, anonymous, listening, recording—all is spoiled by the fact that I am a girl, a female always in danger of assault and battery.

---

the problem. "There is no proof it affects the vision at all," she said. "And if women see less well at

My consuming interest in men and their lives is often misconstructed as a desire to seduce them, or as an invitation to intimacy. Yes, God, I want to talk to everybody I can as deeply as I can. I want to be able to sleep in an open field, to travel west, to walk freely at night." Plath seems to have been interested in men for the very reason she was unable to investigate them—because their greater freedom made their lives more interesting to a young woman just setting out on her own. There are three prerequisites to taking a walk—that is, to going out into the world to walk for pleasure. One must have free time, a place to go, and a body unhindered by illness or social restraints. Free time has many variables, but most public places at most times have not been as welcoming and as safe for women. Legal measures, social mores subscribed to by both men and women, the threat implicit in sexual harassment, and rape itself have all limited women's ability to walk where and when they wished. (Women's clothes and bodily confinements—high heels, tight or fragile shoes, corsets and girdles, very full or narrow skirts, easily damaged fabrics, veils that obscure vision—are part of the social mores that have handicapped women as effectively as laws and fears.)

Women's presence in public becomes with startling frequency an invasion of their private parts, sometimes literally, sometimes verbally. Even the English language is rife with words and phrases that sexualize women's walking. Among the terms for prostitutes are streetwalkers, women of the streets, women on the town, and public women (and of course phrases such as a public man, man about town, or man of the streets mean very different things than do their equivalents attached to women). A woman who has violated sexual convention can be said to be strolling, roaming, wandering, straying—all terms that imply that women's travel is inevitably sexual or that their sexuality is transgressive when it travels. Had a group of women called themselves the Sunday Tramps, as did a group of Leslie Stephen's male friends, the monicker would have implied not that they went walking but that they engaged in something salacious on Sundays. Of course women's walking is often construed as performance rather than transport, with the implication that women walk not to see but to be seen, not for their own experience but for that of a male audience, which means that they are asking for whatever attention they

---

night, well, no good Muslim woman should be out at night unaccompanied."—JAN GOODWIN, PRICE

---

receive. Much has been written about how women walk, as erotic assessment—from the seventeenth-century miss whose "feet beneath her petticoat / like little mice, stole in and out" to Marilyn Monroe's wiggle—and as instruction on the right way to walk. Less has been written about where we walk.

Other categories of people have had their freedom of movement limited, but limitations based on race, class, religion, ethnicity, and sexual orientation are local and variable compared to those placed on women, which have profoundly shaped the identities of both genders over the millennia in most parts of the world. There are biological and psychological explanations for these states of affairs, but the social and political circumstances seem most relevant. How far back can one go? In Middle Assyria (circa the seventeenth to eleventh centuries B.C.), women were divided into two categories. Wives and widows "who go out onto the street" may not have their heads uncovered, said the law; prostitutes and slave girls, contrarily, must not have their heads covered. Those who illicitly wore a veil could be given fifty lashes or have pitch poured over their heads. The historian Gerda Lerner comments, "Domestic women, sexually serving one man and under his protection, are here designated as 'respectable' by being veiled; women not under one man's protection and sexual control are designated as 'public women,' hence unveiled. . . . This pattern of enforced visible discrimination recurs throughout historical time in the myriad regulations which place 'disreputable women' in certain districts or certain houses marked with clearly identifiable signs or which force them to register with the authorities and carry identification cards." Of course "respectable" women have been equally regulated, but more by social constraints than legal ones. Many things are remarkable about the appearance of this law, whose ordering of the world seems to have prevailed ever since. It makes women's sexuality a public rather than a private matter. It equates visibility with sexual accessibility, and it requires a material barrier rather than a woman's morality or will to make her inaccessible to passersby. It separates women into two publicly recognized castes based on sexual conduct but allows men, whose sexuality remains private, access to both castes. Membership in the respectable caste comes at the cost of consignment to private life; membership in the

---

OF HONOR: MUSLIM WOMEN LIFT THE VEIL OF SILENCE ON THE ISLAMIC WORLD   I could not walk out

---

caste with spatial and sexual freedom comes at the cost of social respect. Either way, the law makes it virtually impossible to be a respected public female figure, and ever since, women's sexuality has
been public business.

Homer's Odysseus travels the world and sleeps around. Odysseus's wife Penelope stays dutifully at home, rebuffing the suitors she
lacks the authority to reject outright. Travel, whether local or global,
has remained a largely masculine prerogative ever since, with women
often the destination, the prize, or keepers of the hearth. By the
fifth century B.C. in Greece, these radically different roles were defined as those of the interior and exterior, the private and the public
spheres. Athenian women, writes Richard Sennett, "were confined to
houses because of their supposed physiological defects." He quotes
Pericles concluding his funeral oration with advice to the women of
Athens—"The greatest glory of a woman is to be least talked about
by men, whether they are praising you or criticizing you"—and Xenophon telling wives, "Your business will be to stay indoors." Women
in ancient Greece lived far from the celebrated public spaces and
public life of the cities. Throughout much of the Western world into
the present, women have remained relatively housebound, not only
by law in some countries even now, but by custom and fear in others.
The usual theory for this control of women is that in cultures where
patrilineal descent is important for inheritance and identity, controlling women's sexuality has been the means of ensuring paternity.
(Anyone who thinks such matters are archaic or irrelevant need
merely remember the anatomist-evolutionist Owen Lovejoy, discussed in chapter 3, attempting to naturalize this social order by the-
orizing that female monogamy and immobility were important for
our species long before we became human.) But there are many other
factors pertaining to the creation of a dominant gender whose privi-
leges include controlling and defining the female sexuality often
viewed as chaotic, threatening, and subversive—a sort of wild nature
to be subdued by masculine culture.

Architectural historian Mark Wiggins writes, "In Greek thought
women lack the internal self-control credited to men as the very mark
of their masculinity. This self-control is no more than the mainte-
nance of secure boundaries. These internal boundaries . . . cannot be

---

alone, without giving suspicion to the whole family, should I be watched, and seen to meet a man—judge

maintained by a woman because her fluid sexuality endlessly over-
flows and disrupts them. And more than this she endlessly disrupts
the boundaries of others, that is, men. . . . In these terms the role of
architecture is explicitly the control of sexuality, or more precisely,
women's sexuality, the chastity of the girl, the fidelity of the wife. . . .
While the house protects the children from the elements, its primary
role is to protect the father's genealogical claims by isolating women
from other men." Thus, women's sexuality is controlled via the regulation of public and private space. In order to keep women "private,"
or sexually accessible to one man and inaccessible to all others, her
whole life would be consigned to the private space of the home that
served as a sort of masonry veil.

Prostitutes have been more regulated than any other women, as
though the social constraints they had escaped pursued them as laws.
(Prostitutes' customers, of course, have almost never been regulated
in any way, either by law or by social condemnation: think of Walter
Benjamin and André Breton a few chapters ago, who managed to
write about their relations with prostitutes without fear of losing their
status as public intellectuals or marriageable men.) Throughout the
nineteenth century, many European governments attempted to reg-
ulate prostitution by limiting the circumstances in which it could be
carried out, and this often became a limitation of the circumstances
in which any woman could walk. Nineteenth-century women were
often portrayed as too frail and pure for the mire of urban life and
compromised for being out at all if they didn't have a specific purpose.
Thus women legitimized their presence by shopping—proving they
were not for purchase by purchasing—and stores have long provided
safe semipublic havens in which to roam. One of the arguments about
why women could not be flâneurs was that they were, as either com-
modities or consumers, incapable of being sufficiently detached from
the commerce of city life. Once the stores closed, so did much of
their opportunity to wander (which was hardest on working women,
for whom the evening was their only free time). In Germany the vice
squad persecuted women who were out alone in the evening, and a
Berlin doctor commented, "The young men strolling on the streets

---

of the consequence!—MARY WORTLEY MONTAGU, 1712   *How beautiful are thy feet in shoes,*

think only that a woman of good reputation does not allow herself to be seen in the evening." Public visibility and independence were still equated, as they had been three thousand years earlier, with sexual disreputability; women's sexuality could still be defined by geographical as well as temporal locale. Think of Dorothy Wordsworth and her fictional sister Elizabeth Bennet upbraided for going out walking in the country, or Edith Wharton's New York heroine in _The House of Mirth_ risking her social status at the beginning of the novel to walk into a man's house unchaperoned for a cup of tea and ruining that status for good by being seen to leave another man's house in the evening (while the law controls "disreputable women," "respectable women" often patrol each other).

By the 1870s in France, Belgium, Germany, and Italy, prostitutes were only allowed to solicit at certain times. France was particularly cynical in its regulation of prostitution; the practice was licensed, and both the licensing and the banning of unlicensed sexual commerce allowed the police to control women. Any woman could be arrested for soliciting merely because she appeared in the times and places associated with the sex industry, while known prostitutes could be arrested for appearing in any other time or place—women had been divided into diurnal and nocturnal species. One prostitute was arrested for "shopping in Les Halles at nine o'clock in the morning and was charged with speaking to a man (the stall holder), and with being off the beat stipulated on her registration license." By that time, the Police des Moeurs, or Morals Police, could arrest working-class women for anything or nothing, and they would sometimes round up groups of female passersby on the boulevards to meet their quotas. At first watching the women get arrested was a masculine pastime, but by 1876 the abuses became so extreme that boulevardiers sometimes tried to interfere and got arrested themselves. The mostly young, mostly poor, unmarried women and girl children arrested were seldom found innocent; many were incarcerated behind the high walls of Saint Lazare prison, where they lived in dire circumstances, cold, malnourished, unwashed, overworked, and forbidden to speak. They were released when they agreed to register as prostitutes, while women who ran away from licensed brothels were given the choice of either returning to the brothel or being sent to Saint Lazare—thus women

---

O prince's daughter! the joints of thy thighs are like jewels.—SONG OF SOLOMON 7:1   _"You're wearing_

were forced into rather than out of prostitution. Many committed suicide rather than face arrest. The great champion of human rights for prostitutes Josephine Butler visited Saint Lazare in the 1870s: "I asked what the crime was for which the greater number were in prison and was told it was for walking in streets which are forbidden, and at hours which are forbidden!"

Butler, a well-educated, upper-class woman who grew up amid progressives, was the most effective opponent of Britain's Contagious Diseases Acts passed in the 1860s. A devout Christian, she opposed the laws both because they put the state in the business of regulating prostitution and thus, implicitly, of condoning it and because they enforced a double standard. Women could be punished by incarceration or by the inspections dubbed "surgical rape" for the slightest suspicion of being a prostitute, and a woman found to carry a venereal disease was confined and treated, while men were left free to continue spreading it (similar measures have been considered and sometimes carried out in regards to prostitutes and AIDS in recent years). The law had been passed to protect the health of the army, whose soldiers had a much higher incidence of such diseases than the general public; it seems to have been based on a cynical recognition that the health, freedom, and civil rights of men were of greater value to the state than those of women. Many more extreme abuses than that of Caroline Wyburgh were carried out, and at least one woman— a widowed mother of three—was hounded into suicide. Going out walking had become evidence of sexual activity, and sexual activity on the part of women had been criminalized. Though the laws in the United States were never quite so bad, similar circumstances sometimes prevailed. In 1895 a young working-class New Yorker named Lizzie Schauer was arrested as a prostitute because she was out alone after dark and had stopped to ask directions of two men. Though she was in fact on her way to her aunt's house on the Lower East Side, the act and the time were interpreted as signs that she was soliciting. Only after a medical examination proved she was a "good girl" was she released. Had she not been a virgin, she might well have been found guilty of a crime compounded of the twin acts of having been sexual and of walking alone in the evening.

Though protecting respectable women from vice had long been

---

flat 'eels. Are you a Lesbian?"—LONDON PASSERBY TO A STREETWALKER   *Her body is perfectly*

---

one rationale for state regulation and prosecution of prostitution, the eminently respectable Butler took on the formidable task of protecting women from the state, for which she was vilified and chased by mobs (often hired by brothel owners). On one occasion the mob caught her and she was badly beaten and smeared with dirt and excrement, her hair and clothes torn; on another, a prostitute she came across as she fled a mob led her through a labyrinth of back streets and empty warehouses to safety. Of course she herself had transgressed by moving into the public sphere of political discourse and challenging the sexual conduct of men, and she was decried by one member of Parliament as "worse than prostitutes." As she lay dying in 1906, far more women were moving into that sphere and meeting with similar treatment. The women's suffrage movement in the United States and Britain, after decades of quiet and ineffectual effort to gain the vote for women, became militant in the first decade of the twentieth century, with an extraordinary campaign of marches, demonstrations, and public meetings— the now-usual forms of outdoor politicking available to those denied entrée to the system. These demonstrations were met with an unusual degree of violence—by the police in Britain, and by crowds of soldiers and other men in the United States. Union activists, religious nonconformists, and others had been met by violence before, but some of the things that happened to the suffragettes were unique. In Britain archaic laws were invoked to criminalize the women's public gatherings, and current laws that gave all citizens the right to petition the government were violated. In both the United States and Britain these women arrested for exercising their right to be and to speak in public went on hunger strike, demanding they be recognized as political prisoners. Both governments responded by force-feeding the prisoners, and the agonizing procedure—which involved restraining the woman, forcing a tube down her nostrils to her stomach, and pumping in food— became a new form of institutional rape. Once again women who had attempted to participate in public life by walking down the street were locked up and found the privacy of their bodily interiors violated by the state.

But women won the vote, and in recent decades most of this strange duet between public space and private parts has been not between women and the government but between women and men.

---

balanced, she holds herself straight, and yet nothing suggests a ramrod. She takes steps of medium length,

---

Feminism has largely addressed and achieved reforms of interactions indoors—in the home, the workplace, the schools, and the political system. Yet access to public space, urban and rural, for social, political, practical, and cultural purposes is an important part of everyday life, one limited for women by their fear of violence and harassment. The routine harassment women experience ensures, in the words of one scholar of the subject, "that women will not feel at ease, that we will remember our role as sexual beings, available to, accessible to men. It is a reminder that we are not to consider ourselves equals, participating in public life with our own right to go where we like when we like, to pursue our own projects with a sense of security." Both men and women may be assaulted for economic reasons, and both have been incited by crime stories in the news to fear cities, strangers, the young, the poor, and uncontrolled spaces. But women are the primary targets of sexualized violence, which they encounter in suburban and rural as well as urban spaces, from men of all ages and income levels, and the possibility of such violence is implicit in the more insulting and aggressive propositions, comments, leers, and intimidations that are part of ordinary life for women in public places. Fear of rape puts many women in their place—indoors, intimidated, dependent yet again on material barriers and protectors rather than their own will to safeguard their sexuality. Two-thirds of American women are afraid to walk alone in their own neighborhoods at night, according to one poll, and another reported that half of British women were afraid to go out after dark alone and 40 percent were "very worried" about being raped.

Like Caroline Wyburgh and Sylvia Plath, I was nineteen when I first felt the full force of this lack of freedom. I had grown up on the suburban edge of the country in the days before children were closely supervised and I went to town or to the hills at will, and at seventeen I ran away to Paris, where the men who often propositioned and occasionally grabbed me in the streets seemed more annoying than terrifying. At nineteen, I moved to a poor San Francisco neighborhood with less street life than the gay neighborhood I had moved from and discovered that at night the day's constant threats were more likely to be carried out. Of course it wasn't only poor neighborhoods and nighttime in which I was threatened. I was, for example, followed near

and like all people who move and dance well, walks from the hip, not the knee. On no account does

Fisherman's Wharf one afternoon by a well-dressed man who mur-
mured a long stream of vile sexual proposals to me; when I turned
around and told him off, he recoiled in genuine shock at my profanity,
told me I had no right to speak to him like that, and threatened to kill
me. Only the earnestness of his death threat made the incident stand
out from hundreds of others more or less like it. It was the most devastating discovery of my life that I had no real right to life, liberty, and
pursuit of happiness out-of-doors, that the world was full of strangers
who seemed to hate me and wish to harm me for no reason other than
my gender, that sex so readily became violence, and that hardly
anyone else considered it a public issue rather than a private problem.
I was advised to stay indoors at night, to wear baggy clothes, to cover
or cut my hair, to try to look like a man, to move to someplace more
expensive, to take taxis, to buy a car, to move in groups, to get a man
to escort me—all modern versions of Greek walls and Assyrian veils,
all asserting it was my responsibility to control my own and men's
behavior rather than society's to ensure my freedom. I realized that
many women had been so successfully socialized to know their place
that they had chosen more conservative, gregarious lives without realizing why. The very desire to walk alone had been extinguished in
them—but it had not in me.

The constant threats and the few incidents of real terror transformed me. Still, I stayed where I was, became more adept at navi-
gating the dangers of the street, and became less of a target as I grew
older. Almost all my interactions nowadays with passersby are civil,
and some are delightful. Young women receive the brunt of such harassment, I think, not because they are more beautiful but because
they are less sure of their rights and boundaries (though such un-
sureness manifested as naïveté and timidity are often part of what is
considered beauty). The years of harassment received in youth constitute an education in the limits of one's life, even long after the daily
lessons stop. Sociologist June Larkin got a group of Canadian teenagers to keep track of their sexual harassment in public and found
they were leaving the less dramatic incidents out because, as one said,
"If I wrote down every little thing that happened on the street, it
would take up too much time." Having met so many predators, I
learned to think like prey, as have most women, though fear is far

---

she swing her arms, nor does she rest a hand on her hip! Nor, when walking, does she wave her

more minor an element of my everyday awareness than it was when I
was in my twenties.

The movements for women's rights often came out of the movements
for racial justice. The first great women's convention at Seneca Falls,
New York, was organized by abolitionists Elizabeth Cady Stanton
and Lucretia Mott out of anger over the discrimination they faced
even while trying to fight against slavery—they had attended the
World's Anti-Slavery Convention in London, only to find that the
male-dominated organization would not seat any female delegates.
"Stanton and Mott," writes one historian, "began to see similarities
between their own circumscribed status and that of slaves." Josephine
Butler and the English suffrage leader Emmeline Pankhurst also came
from abolitionist families, and in recent years some of the most
original and important feminists have been black women—bell hooks,
Michelle Wallace, June Jordan—who address both race and gender.

  When I wrote of the gay poets of New York, I left out the
Harlem-born James Baldwin, because for him Manhattan was not a
deliciously liberatory place where he could lose himself, as it was for
Whitman and Ginsberg. It threatened instead never to let him forget
himself, whether it was the policemen near the Public Library telling
him to stay uptown, the pimps on uptown Fifth Avenue trying to re-
cruit him, when he was a boy, to become one of the dangers, or the
people in his own neighborhood keeping track of him as do people in
small towns. He wrote about walking the city as a black man rather
than a gay one, though he was both; his race limited his roaming until
he moved to Paris. Black men nowadays are seen as working-class
women were a century ago: as a criminal category when in public, so
that the law often actively interferes with their freedom of movement.
In 1983 an African-American man, Edward Lawson, won a Supreme
Court case challenging a California statute that "required persons
who loiter or wander on the streets to provide a credible and reliable
identification and to account for their presence when requested by a
peace officer." Lawson, who, the *New York Times* reported, "liked to
walk and was often stopped late at night in residential areas," had
been arrested fifteen times for refusing to identify himself under this

---

hands about in gesticulation.—EMILY POST, ETIQUETTE   *"I call them limousine shoes," she says of*

---

statute criminalizing walking. An athletic man with tidy dreadlocks, he used to dance at the same nightclub I did then.

But in public space, racism has often been easier to recognize than sexism and far more likely to become an issue. Late in the 1980s two young black men died for being in "the wrong place at the wrong time." Michael Griffith was chased by a gang of hostile white men in Howard Beach, ran out into traffic to escape their persecution, and was killed by a car. Like Griffith, Yusef Hawkins was bludgeoned to death for being a black man in another white neighborhood, this time in Brooklyn. An enormous outcry arose over these two cases; people rightly understood that these young men's civil rights had been stripped from them when they were attacked for walking down the street. Not long after Griffith and Hawkins died, the "Central Park Jogger Case" erupted. A woman who went running in the park was raped and beaten so badly her skull was crushed, bones broken, and most of her blood lost. Her survival and recovery from a coma was considered miraculous. Five teenage boys from Harlem who had been in the park were charged with the crime and imprisoned—and in 2002 the real assailant was found and the five exonerated.

The Central Park Jogger Case was discussed in startlingly different terms. Considerable public outrage had been expressed that the two murdered men had been denied the basic liberty to roam the city, and the crimes were universally recognized as racially motivated. But in a careful study of the Central Park case, Helen Benedict wrote, "Throughout the case, even up to the start of the trial, the white and black press kept running articles trying to analyze why the youths had committed this heinous crime. . . . They looked for answers in race, drugs, class, and in the ghetto's 'culture of violence.'" The reasons proferred, she concludes, "were woefully inadequate as an explanation . . . because the press never looked at the most glaring reason of all for rape: society's attitude toward women." Portraying it as a case about race—the assailants were Latino and black—rather than gender failed to make an issue at all of violence against women. And almost no one at all discussed the Central Park case as a civil rights issue—as part of a pattern of infringements on women's right to roam the city (women of color rarely show up in crime reportage

---

the pointy toes that will make the rounds this fall. "One woman wrote to me, 'My boyfriend wants me

at all, apparently since they lack men's status as citizens and white women's titillating appeal as victims). A decade after Bensonhurst and Central Park, the gruesome lynching of a black man in Texas has been greeted with outrage as a hate crime and an infringement on the civil rights of people of color, as has the brutal death of a young gay man in Wyoming—for gays and lesbians are also frequent targets of violence that "teaches them their place" or punishes them for their nonconformity. But similar murders motivated by gender, though they fill the newspapers and take the lives of thousands of women every year, are not contextualized as anything but isolated incidents that don't require social reform or national soul-searching.

The geography of race and gender are different, for a racial group may monopolize a whole region, while gender compartmentalizes in local ways. Many people of color find the whiter parts of rural America unwelcoming, to say the least, even in the places where a white woman might feel safe (white supremacists seem to arise from or flock to some of the most scenic parts of the country). Evelyn C. White writes that when she first tried to explore rural Oregon, memories of southern lynchings "could leave me speechless and paralyzed with the heart-stopping fear that swept over me as when I crossed paths with loggers near the McKenzie River or whenever I visited the outdoors." In Britain the photographer Ingrid Pollard made a series of wry portraits of herself in the Lake District, where she apparently went to try to feel like Wordsworth and felt nervous instead. Nature romanticism, she seemed to be saying, is not available to people of her color. But many white women too feel nervous in any isolated situation, and some have personal experience to draw upon. When she was young, the great climber and mountaineer Gwen Moffat went to the beautiful Isle of Skye off Scotland's west coast to climb by herself. After a drunken neighbor broke into her bedroom in the middle of the night, she cabled for a man to join her and recounts, "Had I been older and more mature, I could have coped with life on my own, but living as I did I laid myself open to all kinds of advances and speculations. Ordinary, conventional men thought this way of life an open invitation and I couldn't face the resentment which I knew they felt when they were rebuffed."

---

to wear high heels, but they hurt my feet.' I said, 'Tell him you'll wear the shoes if he provides

---

* * *

Women have been enthusiastic participants in pilgrimages, walking clubs, parades, processions, and revolutions, in part because in an already defined activity their presence is less likely to be read as sexual invitation, in part because companions have been women's best guarantee of public safety. In revolutions the importance of public issues seems to set aside private matters temporarily, and women have found great freedom during them (and some revolutionaries, such as Emma Goldman, have made sexuality one of the fronts on which they sought freedom). But walking alone also has enormous spiritual, cultural, and political resonance. It has been a major part of meditation, prayer, and religious exploration. It has been a mode of contemplation and composition, from Aristotle's peripatetics to the roaming poets of New York and Paris. It has supplied writers, artists, political theorists, and others with the encounters and experiences that inspired their work, as well as the space in which to imagine it, and it is impossible to know what would have become of many of the great male minds had they been unable to move at will through the world. Picture Aristotle confined to the house, Muir in full skirts. Even in times when women could walk by day, the night—the melancholic, poetic, intoxicating carnival of city nights—was likely to be off limits to them, unless they had become "women of the night." If walking is a primary cultural act and a crucial way of being in the world, those who have been unable to walk out as far as their feet would take them have been denied not merely exercise or recreation but a vast portion of their humanity.

Women from Jane Austen to Sylvia Plath have found other, narrower subjects for their art. Some have broken out into the larger world—Peace Pilgrim (in middle age), George Sand (in men's clothes), Emma Goldman, Josephine Butler, Gwen Moffat, come to mind—but many more must have been silenced altogether. Virginia Woolf's famous *Room of One's Own* is often recalled as though it were literally a plea for women to have home offices, but it in fact deals with economics, education, and access to public space as equally necessary to making art. To prove her point, she invents the blighted life of Shakespeare's equally talented sister, and asks of this Judith Shakespeare,

---

curb-to-curb service.'"—HARPER'S BAZAAR, 1997   *After the [Korean] city gates were closed . . .*

"Could she even get her dinner in a tavern or roam the streets at midnight?"

Sarah Schulman wrote a novel that is, like Woolf's essay, a commentary on the circumscription of women's freedom. Titled *Girls, Visions and Everything* after a phrase from Jack Kerouac's *On the Road*, it is among other things an investigation into how useful Kerouac's credo is for a young lesbian writer, Lila Futuransky. "The trick," thinks Futuransky, "was to identify with Jack Kerouac instead of with the women he fucks along the way," for like Odysseus, Kerouac was a traveling man in a landscape of immobile women. She explores the charms of Lower East Side Manhattan in the mid-1980s as he did America in the 1950s, and among "the things she loved best" was to "walk the streets for hours with nowhere to go but where she ended up." But as the novel progresses, her world becomes more intimate rather than more open: she falls in love and the possibility of a free life in public space recedes.

Near the end of the novel, she and her lover go out for an evening walk in Washington Square Park and come back to eat ice cream together in front of her apartment building when they overhear a man in a group of men: "That's gay liberation. They think they can do whatever they want whenever they want it." They have been, like lovers since time immemorial, walking out together. Like Lizzie Schauer, arrested in the Lower East Side ninety years earlier for walking alone, their venture into public space threatens to become an invasion of their private lives and their bodies:

"Lila didn't want to go upstairs, because she didn't want them to see where she lived. They started walking slowly away, but the men followed.

"'Come on you cunt. I bet you've got a nice pussy, you suck each other's pussy, right? I'll show you a cock that you'll never forget. . . .'

"For Lila, this was a completely normal though unnecessary part of daily life. As a result she had learned docility, to keep quiet and do a shuffle, to avoid having her ass kicked in. . . . Lila walked in the streets like someone who had always walked in the streets and for whom it was natural and rich. She walked with the illusion that she was safe and that the illusion would somehow keep her that way. Yet, that particular night as she went out for cigarettes, Lila walked uneasily,

---

the city was turned over to women, who were then free to walk abroad. They strolled and chatted in

her mind wandering until it stopped of its own accord on the simple fact that she was not safe. She could be physically hurt at any time and felt, for a fleeting moment that she would be. She sat on the trunk of a '74 Chevy and accepted that this world was not hers. Even on her own block."

Part IV

PAST THE END
OF THE *R*OAD

Chapter 15

AEROBIC SISYPHUS AND THE
SUBURBANIZED PSYCHE

Freedom to walk is not of much use without someplace to go. There is a sort of golden age of walking that began late in the eighteenth century and, I fear, expired some decades ago, a flawed age more golden for some than others, but still impressive for its creation of places in which to walk and its valuation of recreational walking. This age peaked around the turn of the twentieth century, when North Americans and Europeans were as likely to make a date for a walk as for a drink or meal, walking was often a sort of sacrament and a routine recreation, and walking clubs were flourishing. At that time, nineteenth-century urban innovations such as sidewalks and sewers were improving cities not yet menaced by twentieth-century speedups, and rural developments such as national parks and mountaineering were in first bloom. Thus far this book has surveyed pedestrian life in rural and urban spaces, and the history of walking is a history of cities and countryside, with a few towns and mountains thrown in for good measure. Perhaps 1970, when the U.S. Census showed that the majority of Americans were—for the first time in the history of any nation—suburban, is a good date for this golden age's tombstone. Suburbs are bereft of the natural glories and civic pleasures of those older spaces, and suburbanization has radically changed the scale and texture of everyday life, usually in ways inimical to getting about on foot. This transformation has happened in the mind as well as on the

---

Walking as a form of transport in modern middle-class Euro-American life is essentially obsolete. It is

ground. Ordinary Americans now perceive, value, and use time, space, and their own bodies in radically different ways than they did before. Walking still covers the ground between cars and buildings and the short distances within the latter, but walking as a cultural activity, as a pleasure, as travel, as a way of getting around, is fading, and with it goes an ancient and profound relationship between body, world, and imagination. Perhaps walking is best imagined as an "indicator species," to use an ecologist's term. An indicator species signifies the health of an ecosystem, and its endangerment or diminishment can be an early warning sign of systemic trouble. Walking is an indicator species for various kinds of freedoms and pleasures: free time, free and alluring space, and unhindered bodies.

I. SUBURBIA

In _Crabgrass Frontier: The Suburbanization of the United States_, Kenneth Jackson outlines what he calls "the walking city" that preceded the development of middle-class suburbs: it was densely populated; it had "a clear distinction between city and country," often by means of walls or some other abrupt periphery; its economic and social functions were intermingled (and "factories were almost non-existent" because "production took place in the small shops of artisans"); people rarely lived far from work; and the wealthy tended to live in the center of the city. His walking city and my golden age find their end in the suburbs, and the history of suburbia is the history of fragmentation.

Middle-class suburban homes were first built outside London in the late eighteenth century, writes Robert Fishman in another history of suburbs, _Bourgeois Utopias_, so that pious merchants could separate family life from work. Cities themselves were looked upon askance by these upper-middle-class evangelical Christians: cards, balls, theaters, street fairs, pleasure gardens, taverns were all condemned as immoral. At the same time the modern cult of the home as a consecrated space apart from the world began, with the wife-mother as a priestess who was, incidentally, confined to her temple. This first suburban community of wealthy merchant families who shared each other's values sounds, in Fishman's account, paradisiacal and, like most paradises,

---

the rare individual who commutes to work on foot. Walking is usually linked with leisure. . . . One

---

dull: a place of spacious freestanding houses, with little for their residents to do outside the home and garden. These villas were miniaturized English country estates, and like such estates they aspired to a kind of social self-sufficiency. However, a whole community of farmworkers, gamekeepers, servants, guests, and extended families had inhabited the estate, which usually encompassed working farms and had thus been a place of production, while the suburban home housed little more than the nuclear family and was to become more and more a site only of consumption. Too, the estate was on a scale that permitted walking without leaving the grounds; the suburban home was not, but suburbia would eat up the countryside and diffuse the urban anyway.

It was in Manchester, during the industrial revolution, that the suburb came into its own. The suburb is a product of that revolution, radiating outward from Manchester and the north Midlands, which has so thoroughly fragmented modern life. Work and home had never been very separate until the factory system came of age and the poor became wage-earning employees. Those jobs, of course, fragmented work itself as craftsmanship was broken down into unskilled repetitive gestures in attendance on machines. Early commentators deplored how factory work destroyed family life, taking individuals out of the home and making family members strangers to each other during their prodigiously long workdays. Home for factory workers was little more than a place to recuperate for the next day's work, and the industrial system made them far poorer and unhealthier than they had been as independent artisans. In the 1830s Manchester's manufacturers began to build the first large-scale suburbs to escape the city they had created and to enhance family life for their class. Unlike the London evangelicals, they were fleeing not temptation but ugliness and danger—industrial pollution, the bad air and sanitation of a poorly designed city, and the sight and threat of their miserable workforce.

"The decision to suburbanize had two great consequences," says Fishman. "First the core emptied of residents as the middle class left and workers were pushed out by the conversion of their rooms in the back streets to offices. . . . Visitors were surprised to find an urban core that was totally quiet and empty after business hours. The

---

Irishwoman made a similar observation: "Just think, the two most important forms of transport early this

central business district was born. Meanwhile, the once peripheral factories were now enclosed by a suburban belt, which separated them from the now-distant rural fields. The grounds of the suburban villas were enclosed by walls, and even the tree-lined streets on which they stood were often forbidden except to the residents and their guests. One group of workers attempted to keep open a once-rural footpath that now ran through the grounds of a factory owner's suburban villa. . . . Mr. Jones responded with iron gates and ditches." Fishman's picture shows a world where the fertile mix of urban life in the "walking city" has been separated out into its sterile constituent elements.

The workers responded by fleeing to the fields on Sundays and, eventually, fighting for access to the remaining rural landscape in which to walk, climb, cycle, and breathe (as chapter 10 chronicles). The middle class responded by continuing to develop and dwell in suburbs. Men commuted to work and women to shop by private carriages, then horse-drawn omnibuses (which, in Manchester, were priced too high to accommodate the poor), and eventually trains. In fleeing the poor and the city, they had left behind pedestrian scale. One could walk in the suburbs, but there was seldom anyplace to go on foot in these homogenous expanses of quiet residential streets behind whose walls dwelt families more or less like each other. The twentieth-century American suburb reached a kind of apotheosis of fragmentation when proliferating cars made it possible to place people farther than ever from work, stores, public transit, schools, and social life. The modern suburb as described by Philip Langdon is antithetical to the walking city: "Offices are kept separate from retailing. The housing is frequently divided into mutually exclusive tracts... with further subdivision by economic status. Manufacturing, no matter how clean and quiet—today's industries are rarely the noisy, smoke-belching mills of urban memory—is kept away from residential areas or excluded from the community entirely. Street layouts in new developments enforce apartness. To unlock the rigid geographic segregation, an individual needs to obtain a key—which is a motor vehicle. For obvious reasons those keys are not issued to those under sixteen, the very population for whom the suburbs are

---

century are now highly specialized hobbies!"—NANCY LOUISE FREY, *PILGRIM STORIES: ON AND OFF THE*

---

supposedly most intended. These keys are also denied to some of the elderly who can no longer drive."

Getting a license and a car is a profound rite of passage for modern suburban teenagers; before the car, the child is either stranded at home or dependent upon chauffeuring parents. Jane Holtz Kay, in her book on the impact of cars, *Asphalt Nation*, writes of a study that compared the lives of ten-year-olds in a walkable Vermont small town and an unwalkable southern California suburb. The California children watched four times as much television, because the outdoor world offered them few adventures and destinations. And a recent study of the effects of television on Baltimore adults concluded that the more local news television, with its massive emphasis on sensational crime stories, locals watched, the more fearful they were. Staying home to watch TV discouraged them from going out. That *Los Angeles Times* advertisement for an electronic encyclopedia I cited at the beginning of this book—"You used to walk across town in the pouring rain to use our encyclopedias. We're pretty confident we can get your kid to click and drag"—may describe the options open to a child who no longer has a library within walking distance and may not be allowed to walk far alone anyway (walking to school, which was for generations the great formative first foray alone into the world, is likewise becoming a less common experience). Television, telephones, home computers, and the Internet complete the privatization of everyday life that suburbs began and cars enhanced. They make it less necessary to go out into the world and thus accommodate retreat from rather than resistance to the deterioration of public space and social conditions.

These American suburbs are built car-scale, with a diffuseness the unenhanced human body is inadequate to cope with, and just as gardens, sidewalks, arcades, and wilderness trails are a kind of infrastructure for walking, so modern suburbs, highways, and parking lots are an infrastructure for driving. Cars made possible the development of the great Los Angelean sprawls of the American West, those places not exactly suburbs because there is no urbanity to which they are subsidiary. Cities like Albuquerque, Phoenix, Houston, and Denver

---

ROAD TO SANTIAGO   *"People don't walk in Texas. Only Mexicans."*—CHARACTER IN EDNA

may or may not have a dense urban core floating somewhere in their bellies like a half-digested snack, but most of their space is too diffuse to be well served by public transit (if it exists) or to be traversed on foot. In these sprawls, people are no longer expected to walk, and they seldom do. There are many reasons why. Suburban sprawls generally make dull places to walk, and a large subdivision can become numbingly repetitive at three miles an hour instead of thirty or sixty. Many suburbs were designed with curving streets and cul-de-sacs that vastly expand distances: Langdon gives an example of an Irvine, California, subdivision where in order to reach a destination a quarter mile away as the crow flies the traveler must walk or drive more than a mile. Too, when walking is not an ordinary activity, a lone walker may feel ill at ease about doing something unexpected and isolated.

Walking can become a sign of powerlessness or low status, and new urban and suburban design disdains walkers. Many places have replaced downtown with shopping malls inaccessible by any means but cars, or by building cities that never had downtowns, buildings meant to be entered through parking garages rather than front doors. In Yucca Valley, the town near Joshua Tree National Park, all the businesses are strung out along several miles of highway, and crosswalks and traffic lights are rare: though, for example, my bank and food store are only a few blocks apart, they are on opposite sides of the highway, and a car is the only safe, direct way to travel between them. Throughout California more than 1,000 crosswalks have been removed in recent years, more than 150 of them in traffic-clogged Silicon Valley, apparently in the spirit of the L.A. planners who proclaimed in the early 1960s, "The pedestrian remains the largest single obstacle to free traffic movement." Many parts of these western sprawl-cities were built without sidewalks altogether, in both rich and poor neighborhoods, further signaling that walking has come to an end by design. Lars Eigner, who during a homeless and largely penniless phase of his life in the 1980s hitchhiked with his dog Lizbeth between Texas and southern California, wrote eloquently about his experiences, and one of the worst came about when a driver dropped him off in the wrong part of town: "South Tucson simply has no sidewalks. I thought at first this was merely in keeping with the general wretchedness of the place, but eventually it seemed to me that the

FERBER'S *GIANT*   *A black performance artist, Keith Antar Mason, told me recently that he is now*

public policy in Tucson is to impede pedestrians as much as possible. In particular, I could find no way to walk to the main part of town in the north except in the traffic lanes of narrow highway ramps. I could not believe this at first, and Lizbeth and I spent several hours wandering on the south bank of the dry gash that divides Tucson as I looked for a walkway."

Even in the best places, pedestrian space is continually eroding: in the winter of 1997–98, New York mayor Rudolph Giuliani decided that pedestrians were interfering with traffic (one could just as well have said, in this city where so many still travel and take care of their business on foot, that cars interfere with traffic). The mayor ordered the police to start citing jaywalkers and fenced in the sidewalks of some of the busiest corners of the city. New Yorkers, to their eternal glory, rebelled by staging demonstrations at the barriers and jaywalking more. In San Francisco, faster and denser traffic, shorter walk lights, and more belligerent drivers intimidate and occasionally mangle pedestrians. Here 41 percent of all traffic fatalities are pedestrians killed by cars, and more than a thousand walkers are injured every year. In Atlanta, the figures are 80 pedestrians killed per year and more than 1,300 injured. In Giuliani's New York, almost twice as many people are killed by cars as are murdered by strangers—285 versus 150 in 1997. Walking the city is not now an attractive prospect for those unequipped to dodge and dash.

Geographer Richard Walker defines urbanity as "that elusive combination of density, public life, cosmopolitan mixing, and free expression." Urbanity and automobiles are antithetical in many ways, for a city of drivers is only a dysfunctional suburb of people shuttling from private interior to private interior. Cars have encouraged the diffusion and privatization of space, as shopping malls replace shopping streets, public buildings become islands in a sea of asphalt, civic design lapses into traffic engineering, and people mingle far less freely and frequently. The street is public space in which First Amendment rights of speech and assembly apply, while the mall is not. The democratic and liberatory possibilities of people gathered together in public don't exist in places where they don't have space in which to gather. Perhaps it was meant that way. As Fishman argues, the suburbs were a refuge— first from the sin and then from the ugliness and anger of the city and

working increasingly in the only public spaces for African-Americans that are supported actively by

its poor. In postwar America "white flight" sent middleclass whites to the suburbs from multiracial cities, and in the new sprawl-cities of the West and suburbs around the country a fear of crime that often seems to be a broader fear of difference is further eliminating public space and pedestrian possibilities. Political engagement may be one of the things suburbs have zoned out.

Early on in the development of the American suburbs, the porch, an important feature for small-town social life, was replaced at the front of the home by the blind maw of the garage (and the sociologist Dean McCannell tells me some new homes have pseudo-porches that make them look sweetly old-fashioned but are actually too shallow to sit on). More recent developments have been more radical in their retreat from communal space: we are in a new era of walls, guards, and security systems, and of architecture, design, and technology intended to eliminate or nullify public space. This withdrawal from shared space seems, like that of the Manchester merchants a century and a half ago, intended to buffer the affluent from the consequences of economic inequity and resentment outside the gates; it is the alternative to social justice. The new architecture and urban design of segregation could be called Calvinist: they reflect a desire to live in a world of predestination rather than chance, to strip the world of its wide-open possibilities and replace them with freedom of choice in the marketplace. "Anyone who has tried to take a stroll at dusk through a neighborhood patrolled by armed security guards and signposted with death threats quickly realizes how merely notional, if not utterly obsolete, is the old idea of 'freedom of the city,'" writes Mike Davis of the nicer suburbs of Los Angeles. And Kierkegaard long ago exclaimed, "It is extremely regrettable and demoralizing that robbers and the elite agree on just one thing—living in hiding."

If there was a golden age of walking, it arose from a desire to travel through the open spaces of the world unarmored by vehicles, unafraid to mingle with different kinds of people. It emerged in a time when cities and countryside grew safer and desire to experience that world was high. Suburbia abandoned the space of the city without returning to the country, and in recent years a second wave of that impulse has beefed up this segregation with neighborhoods of high-priced bunkers. But even more importantly, the disappearance of

---

government—the prisons.—NORMAN KLEIN, THE HISTORY OF FORGETTING   *The stationary cycle*

---

pedestrian space has transformed perception of the relationship between bodies and spaces. Something very odd has happened to the
very state of embodiment, of being corporeal, in recent decades.

II. THE DISEMBODIMENT OF EVERYDAY LIFE

The spaces in which people live have changed dramatically, but so
have the ways they imagine and experience that space. I found a
strange passage in a 1998 *Life* magazine celebrating momentous events
over the past thousand years. Accompanying a picture of a train was
this text: "For most of human history, all land transport depended on
a single mode of propulsion—feet. Whether the traveller relied on his
own extremities or those of another creature, the drawbacks were the
same, low cruising speed, vulnerability to weather, the need to stop
for food and rest. But on September 15, 1830, foot power began its
long slide toward obsolescence. As brass bands played, a million
Britons gathered between Liverpool and Manchester to witness the
inauguration of the world's first fully steam-driven railway. . . . Despite
the death of a member of Parliament who was run down by the train
at the opening ceremony, the Liverpool and Manchester inspired a
rash of track-laying round the world." The train was, like the factory
and the suburb, part of the apparatus of the industrial revolution; just
as factories mechanically sped up production, so trains sped up distribution of goods, and then of travelers.

Life magazine's assumptions are interesting; nature as biological
and meteorological factors is a drawback rather than an occasional
inconvenience; progress consists of the transcendence of time, space,
and nature by the train and later the car, airplane, and electronic communications. Eating, resting, moving, experiencing the weather, are
primary experiences of being embodied; to view them as negative is
to condemn biology and the life of the senses, and the passage does
exactly that in its most lurid statement, that "foot power began its
long slide toward obsolescence." Perhaps this is why neither *Life* nor
the crowd apparently mourned the squashed Parliamentarian. In a
way, the train mangled not just that one man's body, but all bodies in
the places it transformed, by severing human perception, expectation,

---

and the treadmill both have slot machines attached, allowing casino customers to sweat and bet at the

and action from the organic world in which our bodies exist. Alienation from nature is usually depicted as estrangement from natural spaces. But the sensing, breathing, living, moving body can be a primary experience of nature too: new technologies and spaces can bring about alienation from both body and space.

In his brilliant *The Railway Journey: The Industrialization of Time and Space in the Nineteenth Century*, Wolfgang Schivelbusch explores the ways trains changed their passengers' perceptions. Early railroad travelers, he writes, characterized this new technology's effects as the elimination of time and space, and to transcend time and space is to begin to transcend the material world altogether—to become disembodied. Disembodiment, however convenient, has side effects. "The speed and mathematical directness with which the railroad proceeds through the terrain destroy the close relationship between the traveller and the travelled space," Schivelbusch writes. "The train was experienced as a projectile, and travelling on it as being shot through the landscape—thus losing control of one's senses. . . . The traveller who sat inside that projectile ceased to be a traveller and became, as noted in a popular metaphor of the century, a parcel." Our own perceptions have sped up since, but trains were then dizzyingly fast. Earlier forms of land travel had intimately engaged travelers with their surroundings, but the railroad moved too fast for nineteenth-century minds to relate visually to the trees, hills, and buildings whipping by. The spatial and sensual engagement with the terrain between here and there began to evaporate. Instead, the two places were separated only by an ever-shortening amount of time. Speed did not make travel more interesting, Schivelbusch writes, but duller; like the suburb, it put its inhabitants in a kind of spatial limbo. People began to read on the train, to sleep, to knit, to complain of boredom. Cars and airplanes have vastly augmented this transformation, and watching a movie on a jetliner 35,000 feet above the earth may be the ultimate disconnection of space, time, and experience. "From the elimination of the physical effort of walking to the sensorimotor loss induced by the first fast transport, we have finally achieved states bordering on sensory deprivation," writes Paul Virilio. "The loss of the thrills of the old voyage is now compensated for by the showing of a film on a central screen."

The *Life* writers may be right. Bodies are not obsolete by any

----

same time. . . . "People are going crazy about this," said Kathy Harris, president of the Fitness Gaming

----

objective standard, but they increasingly are perceived as too slow, frail, and unreliable for our expectations and desires—as parcels to be transported by mechanical means (though of course many steep, rough, or narrow spaces can only be traversed on foot, and many remote parts of the world can't be reached by any other means; it takes a built environment, with tracks, graded roads, landing strips, and energy sources, to accommodate motor transport). A body regarded as adequate to cross continents, like John Muir's or William Wordsworth's or Peace Pilgrim's, is experienced very differently than a body inadequate to go out for the evening under its own power. In a sense the car has become a prosthetic, and though prosthetics are usually for injured or missing limbs, the auto-prosthetic is for a conceptually impaired body or a body impaired by the creation of a world that is no longer human in scale. In one of the *Alien* movies the actress Sigourney Weaver lurches along in a sort of mechanized body armor that wraps around her limbs and magnifies her movements. It makes her bigger, fiercer, stronger, able to battle with monsters, and it seems strange and futuristic. But this is only because the relationship between the body and the prosthetic machine is so explicit here, the latter so obviously an extension of the former. In fact, from the first clasped stick and improvised carrier, tools have extended the body's strength, skill, and reach to a remarkable degree. We live in a world where our hands and feet can direct a ton of metal to go faster than the fastest land animal, where we can speak across thousands of miles, blow holes in things with no muscular exertion but the squeeze of a forefinger.

It is the unaugmented body that is rare now, and that body has begun to atrophy as both a muscular and a sensory organism. In the century and a half since the railroad seemed to go too fast to be interesting, perceptions and expectations have sped up, so that many now identify with the speed of the machine and look with frustration or alienation at the speed and ability of the body. The world is no longer on the scale of our bodies, but on that of our machines, and many need—or think they need—the machines to navigate that space quickly enough. Of course, like most "time-saving" technologies, mechanized transit more often produces changed expectations than free time; and modern Americans have significantly less time than they did three decades ago.

Corporation in Fairfax, Va. . . . Ms. Harris pointed out that the machines were wired so "you can't

---

To put it another way, just as the increased speed of factory production did not decrease working hours, so the increased speed of transportation binds people to more diffuse locales rather than liberating them from travel time (many Californians, for example, now spend three or four hours driving to and from work each day). The decline of walking is about the lack of space in which to walk, but it is also about the lack of time—the disappearance of that musing, unstructured space in which so much thinking, courting, daydreaming, and seeing has transpired. Machines have sped up, and lives have kept pace with them.

The suburbs made walking ineffective transportation within their expanses, but the suburbanization of the American mind has made walking increasingly rare even when it is effective. Walking is no longer, so to speak, how many people think. Even in San Francisco, very much a "walking city" by Jackson's criteria, people have brought this suburbanized consciousness to their local travel, or so my observations seem to indicate. I routinely see people drive and take the bus remarkably short distances, often distances that could be covered more quickly by foot. During one of my city's public transit crises, a commuter declared he could _walk_ downtown in the time it took the streetcar, as though walking was some kind of damning comparison—but he had apparently been traveling from a destination so near downtown he could've walked every day in less than half an hour, and walking was one transit option the newspaper coverage never proposed (obvious things could be said about bicycling here, were this not a book about walking). Once I made my friend Maria—a surfer, biker, and world traveler—walk the half mile from her house to the bars on Sixteenth Street, and she was startlingly pleased to realize how close they were, for it had never occurred to her before that they were accessible by foot. Last Christmas season, the parking lot of the hip outdoor equipment store in Berkeley was full of drivers idling their engines and waiting for a parking space, while the streets around were full of such spaces. Shoppers weren't apparently willing to walk two blocks to buy their outdoor gear (and since then I have noticed that nowadays drivers often wait for a close parking spot rather than walk in from the farther reaches of the lot). People have a kind of

gamble unless you're pedaling and you can't pedal unless you're gambling." The company's motto: "Put

mental radius of how far they are willing to go on foot that seems to be shrinking; in defining neighborhoods and shopping districts, planners say it is about a quarter mile, the distance that can be walked in five minutes, but sometimes it hardly seems to be fifty yards from car to building.

Of course the people idling their engines at the outdoor equipment store may have been there to buy hiking boots, workout clothes, climbing ropes—equipment for the special circumstances in which people will walk. The body has ceased to be a utilitarian entity for many Americans, but it is still a recreational one, and this means that people have abandoned the everyday spaces—the distance from home to work, stores, friends—but created new recreational sites that are most often reached by car: malls, parks, gyms. Parks, from pleasure gardens to wilderness preserves, have long accommodated bodily recreation, but the gyms that have proliferated wildly in the past couple of decades represent something radically new. If walking is an indicator species, the gym is a kind of wildlife preserve for bodily exertion. A preserve protects species whose habitat is vanishing elsewhere, and the gym (and home gym) accommodates the survival of bodies after the abandonment of the original sites of bodily exertion.

III. THE TREADMILL

The suburb rationalized and isolated family life as the factory did manufacturing work, and the gym rationalizes and isolates not merely exercise but nowadays even each muscle group, the heart rate, the "burn zone" of most inefficient calorie use. Somehow all this history comes back to the era of the industrial revolution in England. "The Tread-Mill," writes James Hardie in his little book of 1823 on the subject, "was, in the year 1818, invented by Mr. William Cubitt, of Ipswich, and erected in the House of Correction at Brixton, near London." The original treadmill was a large wheel with sprockets that served as steps that several prisoners trod for set periods. It was meant to rationalize prisoners' psyches, but it was already an exercise machine. Their bodily exertion was sometimes used to power grain mills

---

your heart into gambling."—NEW YORK TIMES   We've all heard of that future, and it sounds pretty

---

or other machinery, but it was the exertion, not the production, that was the point of the treadmill. "It is its monotonous steadiness and not its severity, which constitutes its terror, and frequently breaks down the obstinate spirit," Hardie wrote of the treadmill's effect in the American prison he oversaw. He added, however, that "the opinions of the medical officers in attendance at the various prisons, concur in declaring that the general health of the prisoners has, in no degree suffered injury, but that, on the contrary, the labor has, in this respect, been productive of considerable benefit." His own prison of Bellevue on New York's East River included 81 male and 101 female vagrants, as well as 109 male and 37 female convicts, and 14 female "maniacs." Vagrancy—wandering without apparent resources or purpose—was and sometimes still is a crime, and doing time on the treadmill was perfect punishment for it.

Repetitive labor has been punitive since the gods of Greek myth sentenced Sisyphus—who had, Robert Graves tells us, "always lived by robbery and often murdered unsuspecting travellers"—to his famous fate of pushing a boulder uphill. "As soon as he has almost reached the summit, he is forced back by the weight of the shameless stone, which bounces to the very bottom once more; where he wearily retrieves it and must begin all over again, though sweat bathes his limbs." It is hard to say if Sisyphus is the first weight lifter or the first treadmiller, but easy to recognize the ancient attitude to repetitive bodily exertion without practical results. Throughout most of human history and outside the first world nowadays, food has been relatively scarce and physical exertion abundant; only when the status of these two things is reversed does "exercise" make sense. Though physical training was part of ancient Greek citizens' education, it had social and cultural dimensions missing from modern workouts and Sisyphean punishments, and while walking as exercise had long been an aristocratic activity, industrial workers' enthusiasm for hiking, particularly in Britain, Austria, and Germany, suggests that it was far more than a way to make the blood circulate or calories burn. Under the heading "Alienation," Eduardo Galeano wrote a brief essay about fishermen in a remote village of the Dominican Republic puzzling over an advertisement for a rowing machine not very long ago. "Indoors? They use it indoors? Without water? They row without water?

----
lonely. In the next century, the line of thinking goes, everyone will work at home, shop at home, watch
----

And without fish? And without the sun? And without the sky?" they exclaimed, telling the resident alien who has shown them the picture that they like everything about their work *but* the rowing. When he explained that the machine was for exercise, they said "Ah. And exercise—what's that?" Suntans famously became status symbols when most of the poor had moved indoors from the farm to the factory, so that browned skin indicated leisure time rather than work time. That muscles have become status symbols signifies that most jobs no longer call upon bodily strength: like tans, they are an aesthetic of the obsolete.

The gym is the interior space that compensates for the disappearance of outside and a stopgap measure in the erosion of bodies. The gym is a factory for the production of muscles or of fitness, and most of them look like factories: the stark industrial space, the gleam of metal machines, the isolated figures each absorbed in his or her own repetitive task (and like muscles, factory aesthetics may evoke nostalgia). The industrial revolution institutionalized and fragmented labor; the gym is now doing the same thing, often in the same place, for leisure. Some gyms actually are born-again industrial sites. The Chelsea Piers in Manhattan were built in the first decade of this century for ocean liners—for the work of longshoremen, stevedores, and clerks, and for the travel of emigrants and elites. They now house a sports center with indoor track, weight machines, pool, climbing gym, and most peculiarly, a four-story golf driving range, destinations in themselves rather than points of arrival and departure. An elevator takes golfers to their stalls, where all the gestures of golf—walking, carrying, gazing, situating, removing, communicating, retrieving or following the ball—have vanished with the landscape of the golf course. Nothing remains but the single arc of a drive: four tiers of solitary stationary figures making the same gesture, the sharp sound of balls being hit, the dull thud of their landing, and the miniaturized armored-car vehicles that go through the green artificial-grass war zone to scoop up the balls and feed them into the mechanism that automatically pops up another ball as each one is hit. Britain has specialized in the conversion of industrial sites into climbing gyms. Among them are a former electrical substation in London, the Warehouse on Gloucester's Severn River waterfront, the Forge in Sheffield

---

movies at home and communicate with all their friends through videophones and e-mail. It's as if science

---

on one side of the Peak District, an early factory in downtown Birmingham, and, according to a surveyor friend, a "six-story former cotton mill near Leeds" I couldn't locate (not to mention a desanctified church in Bristol). It was in some of these buildings that the industrial revolution was born, with the Manchester and Leeds textile mills, Sheffield's iron- and steelworks, the innumerable manufactories of "the workshop of the world" that Birmingham once was. Climbing gyms are likewise established in converted industrial buildings in the United States, or at least in those cities old enough to have once had industrial-revolution architecture. In those buildings abandoned because goods are now made elsewhere and First World work grows ever more cerebral, people now go for recreation, reversing the inclinations of their factory-worker predecessors to go out—to the outskirts of town or at least out-of-doors—in their free time. (In defense of climbing gyms, it should be said they allow people to polish skills and, during foul weather, to stay fit; for some the gym has only augmented the opportunities, not replaced the mountain, though for others the unpredictabilities and splendors of real rock have become dispensible, annoying—or unknown.)

And whereas the industrial revolution's bodies had to adapt to the machines, with terrible consequences of pain, injury, and deformity, exercise machines are adapted to the body. Marx said history happens the first time as tragedy, the second as farce; bodily labor here happens the first time around as productive labor and the second as leisuretime consumption. The deepest sign of transformation is not merely that this activity is no longer productive, that the straining of the arms no longer moves wood or pumps water. It is that the straining of the muscles can require a gym membership, workout gear, special equipment, trainers and instructors, a whole panoply of accompanying expenditures, in this industry of consumption, and the resulting muscles may not be useful or used for any practical purpose. "Efficiency" in exercise means that consumption of calories takes place at the maximum rate, exactly the opposite of what workers aim for, and while exertion for work is about how the body shapes the world, exertion for exercise is about how the body shapes the body. I do not mean to denigrate the users of gyms—I have sometimes been one

---

and culture have progressed for one purpose only: to keep us from ever having to get out of our

myself—only to remark on their strangeness. In a world where manual labor has disappeared, the gym is among the most available and efficient compensations. Yet there is something perplexing about this semipublic performance. I used to try to imagine, as I worked out on one or another weight machine, that this motion was rowing, this one pumping water, this one lifting bales or sacks. The everyday acts of the farm had been reprised as empty gestures, for there was no water to pump, no buckets to lift. I am not nostalgic for peasant or farmworker life, but I cannot avoid being struck by how odd it is that we reprise those gestures for other reasons. What exactly is the nature of the transformation in which machines now pump our water but we go to other machines to engage in the act of pumping, not for the sake of water but for the sake of our bodies, bodies theoretically liberated by machine technology? Has something been lost when the relationship between our muscles and our world vanishes, when the water is managed by one machine and the muscles by another in two unconnected processes?

The body that used to have the status of a work animal now has the status of a pet: it does not provide real transport, as a horse might have; instead, the body is exercised as one might walk a dog. Thus the body, a recreational rather than utilitarian entity, doesn't work, but works out. The barbell is only abstracted and quantified materiality to shift around—what used to be a sack of onions or a barrel of beer is now a metal ingot—and the weight machine makes simpler the act of resisting gravity in various directions for the sake of health, beauty, and relaxation. The most perverse of all the devices in the gym is the treadmill (and its steeper cousin, the Stairmaster). Perverse, because I can understand simulating farm labor, since the activities of rural life are not often available—but simulating walking suggests that space itself has disappeared. That is, the weights simulate the objects of work, but the treadmill and Stairmaster simulate the surfaces on which walking takes place. That bodily labor, real or simulated, should be dull and repetitive is one thing; that the multifaceted experience of moving through the world should be made so is another. I remember evenings strolling by Manhattan's many glass-walled second-floor gyms full of rows of treadmillers looking as though they were trying to leap

---

pajamas.—SAN FRANCISCO CHRONICLE   *Some people walk with both eyes focused on their goal: the*

---

through the glass to their destruction, saved only by the Sisyphean contraption that keeps them from going anywhere at all—though probably they didn't see the plummet before them, only their own reflection in the glass.

I went out the other day, a gloriously sunny winter afternoon, to visit a home-exercise equipment store and en route walked by the University of San Francisco gym, where treadmillers were likewise at work in the plate-glass windows, most of them reading the newspaper (three blocks from Golden Gate Park, where other people were running and cycling, while tourists and Eastern European emigrés were walking). The muscular young man in the store told me that people buy home treadmills because they allow them to exercise after work when it might be too dark for them to go out safely, to exercise in private where the neighbors will not see them sweating, to keep an eye on the kids, and to use their scarce time most efficiently, and because it is a low-impact activity good for people with running injuries. I have a friend who uses a treadmill when it's painfully cold outside in Chicago, and another who uses a no-impact machine whose footpads rise and fall with her steps because she has an injured hamstring (injured by driving cars designed for larger people, not by running). But a third friend's father lives two miles from a very attractive Florida beach, she tells me, full of low-impact sand, but he will not walk there and uses a home treadmill instead.

The treadmill is a corollary to the suburb and the autotropolis: a device with which to go nowhere in places where there is now nowhere to go. Or no desire to go: the treadmill also accommodates the automobilized and suburbanized mind more comfortable in climate-controlled indoor space than outdoors, more comfortable with quantifiable and clearly defined activity than with the seamless engagement of mind, body, and terrain to be found walking out-of-doors. The treadmill seems to be one of many devices that accommodate a retreat from the world, and I fear that such accommodation disinclines people to participate in making that world habitable or to participate in it at all. It too could be called Calvinist technology, in that it provides accurate numerical assessments of the speed, "distance" covered, and even heart rate, and it eliminates the unpredictable and unforeseeable from the routine—no encounters with acquaintances or strangers, no

---

highest mountain peak in the range, the fifty-mile marker, the finish line. They stay motivated by anticipating

---

sudden revelatory sights around a bend. On the treadmill, walking is no longer contemplating, courting, or exploring. Walking is the alternate movement of the lower limbs.

Unlike the prison treadmills, of the 1820s, the modern treadmill does not produce mechanical power but consumes it. The new treadmills have two-horsepower engines. Once, a person might have hitched two horses to a carriage to go out into the world without walking; now she might plug in a two-horsepower motor to walk without going out into the world. Somewhere unseen but wired to the home is a whole electrical infrastructure of power generation and distribution transforming the landscape and ecology of the world—a network of electrical cables, meters, workers, of coal mines or oil wells feeding power plants or of hydropower dams on rivers. Somewhere else is a factory making treadmills, though factory work is a minority experience in the United States nowadays. So the treadmill requires far more economic and ecological interconnection than does taking a walk, but it makes far fewer experiential connections. Most treadmillers read or otherwise distract themselves. *Prevention* magazine recommends watching TV while treadmilling and gives instructions on how treadmill users can adapt their routines to walking about outside when spring comes (with the implication that the treadmill, not the walk, is the primary experience). The *New York Times* reports that people have begun taking treadmill classes, like the stationary bicycling classes that have become so popular, to mitigate the loneliness of the long-distance treadmiller. For like factory labor, treadmill time is dull—it was the monotony that was supposed to reform prisoners. Among the features of the Precor Cardiologic Treadmill, says its glossy brochure, are "5 programmed courses" that "vary in distance, time and incline. . . .The Interactive Weight Loss course maintains your heart rate within your optimum weight loss zone by adjusting workload," while "custom courses allow you to easily create and store personalized programs of up to 8 miles, with variations as small as 1/10th mile increments." It's the custom courses that most amaze me; users can create an itinerary like a walking tour over varied terrain, only the terrain is a revolving rubber belt on a platform about six feet long. Long ago when railroads began to erode the experience of space, journeys began to be spoken of in terms of time rather than

---

the end of the journey. Since I tend to be easily distracted, I travel somewhat differently—one step at a time,

distance (and a modern Angeleno will say that Beverly Hills is twenty minutes from Hollywood rather than so many miles). The treadmill completes this transformation by allowing travel to be measured entirely by time, bodily exertion, and mechanical motion. Space—as landscape, terrain, spectacle, experience—has vanished.

THE SHAPE OF A WALK

The disembodiment of everyday life I have been tracing is a majority experience, part of automobilization and suburbanization. But walking has sometimes been, at least since the late eighteenth century, an act of resistance to the mainstream. It stood out when its pace was out of keeping with the time—which is why so much of this history of walking is a First World, after-the-industrial-revolution history, about when walking ceased to be part of the continuum of experience and instead became something consciously chosen. In many ways, walking culture was a reaction against the speed and alienation of the industrial revolution. It may be countercultures and subcultures that will continue to walk in resistance to the postindustrial, postmodern loss of space, time, and embodiment. Most of these cultures draw from ancient practices—of peripatetic philosophers, of poets composing afoot, of pilgrims and practitioners of Buddhist walking meditation—or old ones, such as hiking and flâneury. But one new realm of walking opened up in the 1960s, walking as art.

Artists, of course, have walked. In the nineteenth century the development of photography and spread of plein-air painting made walking an important means for image makers—but once they found their view, they stopped traipsing around, and more importantly, their images stopped the view forever. There are countless wonderful paintings of walkers, from Chinese prints in which tiny hermits stray amid the heights to, for example, Thomas Gainsborough's *Morning Walk* or

---

minutes to ten hours. More often they're less definite. . . . Trapped by our concepts and languages and

Gustave Caillebotte's *Paris Street, Rainy Day*, with its umbrella-carrying citizens going wherever they please on the Parisian cobblestones. But the aristocratic young couple in *The Morning Walk* are forever frozen with their best foot forward. Among all the works that come to mind, only the nineteenth-century Japanese printmaker Hiroshige's *Fifty-three Views on the Tokuida Road* seem to suggest walking rather than stopping; they are, like stations of the cross, sequenced to reprise a journey, this time a 312-mile journey from Edo (now Tokyo) to Kyoto, which most then made on foot, as they do in the prints. It is a road movie from when roads were for walkers and movies were woodblock prints.

Language is like a road; it cannot be perceived all at once because it unfolds in time, whether heard or read. This narrative or temporal element has made writing and walking resemble each other in ways art and walking do not—until the 1960s, when everything changed and anything became possible under the wide umbrella of visual art. Every revolution has many parents. One godfather for this one is the abstract expressionist painter Jackson Pollock, at least as one of his offspring portrayed him. Allan Kaprow, himself an important performance and interdisciplinary artist, wrote in 1958 that Pollock shifted the emphasis from the painting as an aesthetic object to a "diaristic gesture." The gesture was primary, the painting secondary, a mere souvenir of that gesture which was now its subject. Kaprow's analysis becomes an exuberant, prophetic manifesto as he weighs the consequences of what the older artist had done: "Pollock's near destruction of this tradition may well be a return to the point where art was more actively involved in ritual, magic, and life than we have known it in our recent past. . . . Pollock, as I see him, left us at the point where we must become preoccupied with and even dazzled by the space and object of our everyday life, either our bodies, clothes, rooms, or, if need be, the vastness of Forty-second Street. Not satisfied with the suggestion through paint of our other senses, we shall utilize the specific substances of sight, sound, movements, people, odors, touch."

For the artists who took up the invitation Kaprow outlined, art ceased to be a craft-based discipline of making objects and become a kind of unbounded investigation into the relationship between ideas, acts, and the material world. At a time when the institutions of galleries and museums and the objects made for them seemed mor-

the utter predictability of our five senses, we often forget to wonder what we're missing as we hurry

ibund, this new conceptual and dematerialized art sought a new arena and a new immediacy for artmaking. Art objects might be only the evidence of such an investigation or props and prompts for the viewers' own investigations, while artists could model themselves after scientists, shamans, detectives, or philosophers as they expanded the possible repertoire of gestures far beyond that of the painter at his canvas. Artists' bodies themselves became a medium for performances, and as art historian Kristine Stiles writes, "Emphasizing the body as art, these artists amplified the role of process over product and shifted from representational objects to presentational modes of action." In retrospect, it seems as though these artists were remaking the world, act by act, object by object, starting with the simplest substances, shapes, gestures. One such gesture—an ordinary one from which the extraordinary can be derived—is walking.

Lucy Lippard, who has been writing subversive histories of art for more than thirty years, traces the parentage for walking as a fine art to sculpture, not performance. She focuses on Carl Andre's 1966 sculpture *Lever* and his 1968 *Joint*, the former made of bricks lined up to extend from one room to another so that the viewer has to travel, the other a similar line but this time of hay bales in a meadow traversing a far greater distance. "My idea of a piece of sculpture is a road," Andre wrote then. "That is, a road doesn't reveal itself at any particular point or from any particular point. Roads appear and disappear. . . . We don't have a single point of view for a road at all, except a moving one, moving along it." Andre's minimal sculptures, like Chinese scrolls, reveal themselves over time in response to the movement of the looker; they incorporate travel into their form. "By incorporating an oriental notion of multiple viewpoints and both implied movement and direct intervention in the landscape, Andre set the scene for a subgenre of dematerialized sculpture which is simply, and not so simply, *walking,*" concludes Lippard.

Other artists had already built roads of sorts: Carolee Schneemann built a labyrinth for friends to walk through out of a fallen tree of heaven and other tornado debris in her Illinois backyard in the summer of 1960, before she moved to New York and became one of the most radical artists in the burgeoning field of performance art. Kaprow himself was building environments for audiences and performers to

---

along toward goals we may not even have chosen. I became a tracker by default, not design, when my

---

move through and participate in by the early 1960s. The same year Andre built *Joint*, Patricia Johanson built *Stephen Long*. As Lippard describes it, "A 1,600-foot wooden trail painted in gradated pastels and laid out along an abandoned railroad track in Buskirk, New York, it was intended to take color and light beyond traditional impressionism by adding the elements of distance and time taken to perceive it." In the American West, even longer lines were being drawn, though they were no longer necessarily related to walking: Michael Heizer made his "motorcycle drawings" by using said vehicle to draw in the desert; Walter de Maria commissioned a bulldozer to make similarly grandiose earth art on Nevada's arid surface, with lines that could be seen whole from a airplane or perceived over time and partially from the ground; but perhaps Robert Smithson's 1,500-foot-long *Spiral Jetty*, a rough but walkable path of rock and earth curling into the Great Salt Lake, was human in scale. Though its first inhabitants walked for millennia, the American West is often perceived as inimical to pedestrian scale; the earth art, as it came to be called, made there often seemed to echo the massive development projects of the conquest of the West, the railroads, dams, canals, mines.

England, on the other hand, has never ceased to be pedestrian in scale, and its landscape is not available for much further conquest, so artists there must use a lighter touch. The contemporary artist most dedicated to exploring walking as an artistic medium is the Englishman Richard Long. Much of what he has done since was already present in his early *Line Made by Walking* of 1967. The black-and-white photograph shows a path in grass running straight up the center to the trees at the far end of the meadow. As the title makes clear, Long had drawn the line with his feet. It was both more ambitious and more modest than conventional art: ambitious in scale, in making his mark upon the world itself; modest in that the gesture was such an ordinary one, and the resultant work was literally down to earth, underfoot. Like that of many other artists who emerged at the time, Long's work was ambiguous: was *A Line Made by Walking* a performance of which the line was a residual trace, or a sculpture—the line—of which the photograph was documentation, or was the photograph the work of art, or all of these?

Walking became Long's medium. His exhibited art since has

consisted of works on paper documenting his walks, photographs of further marks in the landscape made in the course of those walks, and other sculptures made indoors that reference his outdoor activities. Sometimes the walk is represented by a photograph with text, or a map, or by text alone. On the maps the route of the walk is drawn in to suggest that walking is drawing on a grand scale, that his walking is to the land itself as his pen is to the map, and he often walks straight lines, circles, squares, spirals. Similarly, his sculptures in the landscape are usually made by rearranging (without relocating) rocks and sticks into lines and circles, a reductive geometry that evokes everything— cyclical and linear time, the finite and the infinite, roads and routines— and says nothing. Yet other works lay out those lines, circles, and labyrinths of sticks, stones, or mud on the gallery floor. But walking in the landscape is always primary to the work. One magnificent early sculpture uniting these approaches was titled *A Line the Length of a Straight Walk from the Bottom to the Top of Silbury Hill*. With boots dipped in mud he had walked the distance not as a straight line but as a spiral on the gallery floor, so that the muddy path both represented the route he had taken elsewhere and became a new route indoors, evidence of and an invitation to walk. It plays with the concreteness of experience— the walk and its location (Silbury Hill is an ancient earthwork of unknown religious significance in southern England)—and with the abstractions of language and measurement in which that walk is described. The experience cannot be reduced to a place name and a length, but even this scant information is enough to start the imagination going. "A walk expresses space and freedom and the knowledge of it can live in the imagination of anyone, and that is another space too," Long wrote years later.

In some ways Long's works resemble travel writing, but rather than tell us how he felt, what he ate, and other such details, his brief texts and uninhabited images leave most of the journey up to the viewer's imagination, and this is one of the things that distinguishes such contemporary art, that it asks the viewer to do a great deal of work, to interpret the ambiguous, imagine the unseen. It gives us not a walk nor even a representation of a walk, only the idea of a walk and an evocation of its location (the map) or one of its views (the photograph). Formal and quantifiable aspects are emphasized: geometry,

often puzzling patterns somewhere, anywhere—to their source or end or simply to some midpoint in

measurement, number, duration. There is, for example, Long's piece—
a drawing of a squared-off spiral—captioned "A Thousand Miles A
Thousand Hours A Clockwise Walk in England Summer 1974." It
plays with correlations between time and space without showing or
telling us anything further of the walk but the nation and the year,
plays with what can be measured and what cannot. Yet it is enough to
know that in 1974, as life seemed to get more complicated, crowded,
and cynical, someone found the time and space to engage in such an
arduous and apparently satisfying encounter with the land in quest
of alignments between geography, body, and time. Then there was
the map with inset text, " A Six Day Walk Over All Roads, Lanes
and Double Tracks Inside a Six-Mile Wide Circle Centred on the
Giant of Cerne Abbas," with the routes he had walked radiating like
arteries out from the figure Long had placed at the heart of his walk.
Another inset portrayed that figure—a 2,000-year-old chalk outline
of a 180-foot-tall figure with a club and an erect penis on a Dorset
hillside.

Long likes places where nothing seems to have broken the connection to the ancient past, so buildings, people, and other traces of
the present or recent past rarely appear. His work revises the British
tradition of country walks while representing its most enchanting and
problematic faces. He has gone to Australia, the Himalayas, and the
Bolivian Andes to make his work, and the idea that all these places can
be assimilated into a thoroughly English experience smacks of colo-
nialism or at least high-handed tourism. It raises once again the perils
of forgetting that the rural walk is a culturally specific practice, and
though it may be a civil, gentle thing in itself, imposing its values elsewhere is not. But while the literary art of the rural walk bogged down
in convention, sentimentality, and autobiographical chatter, Long's art
is austere, almost silent, and entirely new in its emphasis on the walk
itself as having shape, and this is less a cultural legacy than a creative
reassessment. His work is breathtakingly beautiful at times, and its
insistence that the simple gesture of walking can tie the walker to the
surface of the earth, can measure the route as the route measures the
walker, can draw on a grand scale almost without leaving a trace, can
be art, is profound and elegant. Long's friend and contemporary
Hamish Fulton has also made walking his art, and his photographs-

---

between. But when I began tracking lost people, what had begun as an eccentric habit—following

with-text pieces are almost indistinguishable from the other peripatetic Englishman's. But Fulton emphasizes a more spiritual-emotional
side to his walking, more often choosing sacred sites and pilgrimage
routes, and he makes no sculptures in the gallery or marks in the land.

There have been other kinds of walking artists. Probably the first
artist to have made walking into performance art is a little-known
emigré from Dutch Surinam, Stanley Brouwn. In 1960 he asked
strangers on the street to draw him directions to locations around
town and exhibited the results as a vernacular art of encounters or a
collection of drawings; later he held a conceptual exhibition of "all the
shoe-shops in Amsterdam" which would've called for viewers to take
a walking tour; installed in a gallery signposts pointing out cities
around the world and inviting viewers to take the first few steps
toward Khartoum or Ottawa; spent a whole day in 1972 counting his
steps; and otherwise explored the everyday world of urban walking.
The magisterial German performance artist and sculptor Joseph
Beuys, who often imbued simple acts with profound meaning, did one
performance where he simply swept up after a political parade and
another where he walked through one of the bogs he loved. This
1971 *Bog Action* was documented in photographs that show him
walking, sometimes with only his head and trademark fedora visible
above water.

The New York performance artist Vito Acconci did his *Following
Piece* over twenty-three days in 1969; like much conceptual art of the
time, it played with the intersection between arbitrary rules and
random phenomena by choosing a stranger and following him or her
until he or she entered a building. Sophie Calle, a French photographer whose works arise from interactions and encounters, later revised Acconci's performance with two of her own, documented in
photographs and text. *Suite Venitienne* recounts how she met a man at a
party in Paris and surreptitiously followed him to Venice, where she
tailed him like a detective until he confronted her; years later she had
her mother hire an actual detective to do the same to her in Paris, and
incorporated the detective's photographs of her into her own artwork
as a kind of commissioned portraiture. These pieces explored the
city's potential for suspicion, curiosity, and surveillance arising from
the connections and disconnections between strangers on the street.

---

footprints on the ground—quickly matured into an avocation. . . . I now commonly walk toward a

In 1985 and 1986, the Palestinian-British artist Mona Hatoum used the street as a performance space, stenciling footprints containing the word *unemployed* down streets in Sheffield, as if to make visible the sad secrets of passersby in that economically devastated city, and performing two different walking acts in Brixton, a working-class outpost of London.

Of all the performances involving walking, the most dramatic, ambitious, and extreme was Marina Abramović and Ulay's 1988 *Great Wall Walk*. Radical performance artists from the Communist east—she from Yugoslavia, he from East Germany—they began to collaborate in 1976 on a series of what they called "relation works." They were interested in testing both their own and the audiences' physical and psychic boundaries with performances that threatened danger, pain, transgression, boredom; they were also interested in symbolically uniting the genders into an ideal whole; and they were increasingly influenced by shamanistic, alchemical, Tibetan Buddhist, and other esoteric traditions. Their work calls to mind what Gary Snyder described as the Chinese tradition of the "'four dignities'—Standing, Lying, Sitting, and Walking. They are 'dignities' in that they are ways of being fully ourselves, at home in our bodies, in their fundamental modes," or Vipassana Buddhism's similar emphasis on meditating in these four postures. In their first piece, *Relation in Space*, they walked rapidly from opposite walls of a room toward each other until they collided, again and again. In 1977's *Imponderabilia* they stood nude and motionless in the doorway of a museum so that visitors had to decide who to face as they slipped sideways between them. In 1980's *Rest Energy*, they stood together while she held a bow and he held the arrow notched on the taut bowstring, pointing at her heart; their balanced tension and stillness prolonged this moment and stabilized its danger. That same year, they went to the Australian outback hoping to communicate with aboriginal people, who ignored them. They stayed and spent months of a scorching desert summer practicing sitting without moving, learning "immobility, silence and watchfulness" from the desert. Afterward, they found the locals more communicative. From this experience came their *Nightsea Crossing* performance in Sydney, Toronto, Berlin, and other cities: while remaining silent and fasting

---

single goal: to meet the person at the other end of the tracks.—HANNAH NYALA, *POINT LAST SEEN   In*

twenty-four hours a day, they spent several hours each day on successive days in a museum or public space sitting motionless, facing
each other across a table, living sculptures displaying a kind of fero-
cious commitment.

"When I went to Tibet and the Aborigines I was also introduced
to some Sufi rituals. I saw that all these cultures pushed the body to
the physical extreme in order to make a mental jump, to eliminate the
fear of death, the fear of pain, and of all the bodily limitations we live
with," Abramović later said. "Performance was the form enabling me
to jump to that other space and dimension." The *Great Wall Walk* was
planned at the height of her collaboration with Ulay. They intended
to walk toward each other from opposite ends of the 4,000-kilometer
wall, meet, and marry. Years afterward, when they had finally cleared
the bureaucratic hurdles set up by the Chinese government, their relationship had so changed that the walk became instead the end of
their collaboration and relationship. In 1988 they spent three months
walking toward each other from 2,400 miles away, embraced at the
center, and went their separate ways.

The Great Wall, built to keep marauding nomads out of China, is
one of the world's great emblems of the desire to define and secure
self or nation by sealing its boundaries. For these two raised behind
the Iron Curtain, this transformation of a wall separating north from
south into a road linking east to west is full of political ironies and
symbolic meanings. After all, walls divide and roads connect. Their
performance could be read as a symbolic meeting of East and West,
male and female, the architecture of sequestration and of connection.
Too, the artists believed the wall had been, in the words of Thomas
McEvilley, the critic who has most closely followed their work,
"mapped out over the millennia by feng shui experts, so if you followed the wall exactly you would be touching the serpent-power lines
that bind together the surface of the earth." The book on the project
records, "On March 30, 1988 Marina Abramović and Ulay began
their walk over the Great Wall from opposite ends. Marina embarked
from the east, by the sea. Ulay started far to the west, in the Gobi
desert. On June 27, to the blare of horns, they met up in a mountain
pass near Shenmu in Shaanxi Province, in the midst of Buddhist,

Confucian and Taoist temples." McEvilley points out that this last performance also expanded upon their first, in which they strode toward each other until they collided.

Both artists have a section in this book in which sparse words and evocative photographs give a sense of their experience, functioning like Richard Long's photograph-and-text pieces to evoke carefully chosen fragments of a complex experience. In between the two texts McEvilley's essay revealed another face of the walk: its entanglement with endless layers of bureaucracy throughout the journey. Like Tolstoy's Princess Marya wishing to be a pilgrim on the road, Abramović and Ulay seem to have set out with an image of themselves walking alone in a clear, uncluttered space and state of mind, but McEvilley describes the minivans that took them to lodgings every night, the handlers, translators, and officials that bustled around them, ensuring they met the government's requirements and attempting to slow them down so they would spend more time and thus money in each province, the quarrel Ulay got into at a dance hall, the way schedules, rules, and geography had fragmented Ulay's walk (while Abramović made sure she started each morning where she stopped the night before, declaring, "I walk every fucking centimeter of the wall."). The wall was crumbling in many sections, calling for climbing as much as walking, and atop it the wind was often overwhelming. The walk had, in McEvilley's version, become another kind of performance, like a record-seeking one, in which the official goal is realized only at the cost of countless unofficial distractions and annoyances. But perhaps the two artists who had worked so long on their powers of concentration were able to shut out the surrounding clutter from their time on the wall. Their texts and images speak of the essence of walking, of the basic simplicity of the act amplified by the ancient emptiness of the desert around them. Like Long's pieces, theirs seem a gift to viewers of the assurance that a primeval purity of bodily encounter with the earth is still possible and that the human presence so crowded and dominating elsewhere is still small when measured against the immensity of lonely places. "It took a great number of days before, for the first time, I felt the right pace," Ulay wrote. "When mind and body harmonized in the rhythmical sway of walking."

Afterward, Abramović began to make sculptures that invited

----

hour by plane. Today the capital is only several minutes away from anywhere else . . . —PAUL VIRILIO,

----

viewers to participate in the basic acts her performances had explored. She set geodes, chunks of crystal, and polished stones into
wooden chairs or on pedestals and mounts where they could be sat
with or stood under—furniture for contemplation and for encounters
with the elemental forces she believes the stones hold. The most
spectacular of the sculptures were several pairs of amethyst shoes included in a big survey of her work at the Irish Museum of Modern Art
in 1995. I had arrived there at the end of a long walk from downtown
Dublin to find that the museum was housed in an elegant old military
hospital, and the walk and the building's history seemed preparation
for the shoes—great rough chunks of translucent mottled purple that
had been hollowed out and polished inside, like a fairy-tale version of
the wooden shoes European peasants once wore. Viewers were invited to put them on and close their eyes, and with them on I realized
my feet were, in a sense, inside the earth itself, and though it was possible to walk, it was difficult to do so. I closed my eyes and saw strange
colors, and the shoes seemed like fixed points around which the hospital, Dublin, Ireland, Europe, revolved, shoes not to travel in but to
realize you might already be there. Later I read that they were made
for walking meditation, to heighten awareness of every step. They
were titled *Shoes for Departure.*

Kaprow's 1958 prophesy is fulfilled by all these walking artists: "They
will discover out of ordinary things the meaning of ordinariness.
They will not try to make them extraordinary but will only state their
real meaning. But out of this they will devise the extraordinary."
Walking as art calls attention to the simplest aspects of the act: the
way rural walking measures the body and the earth against each
other, the way urban walking elicits unpredictable social encounters.
And to the most complex: the rich potential relations between thinking
and the body; the way one person's act can be an invitation to another's imagination; the way every gesture can be imagined as a brief
and invisible sculpture; the way walking reshapes the world by
mapping it, treading paths into it, encountering it; the way each act
reflects and reinvents the culture in which it takes place.

SPEED AND POLITICS   *An automobile which cuts out the use value from your feet . . . . I was recently*

LAS VEGAS, OR THE LONGEST
DISTANCE BETWEEN TWO POINTS

I would have preferred to step out into the Peak District. I had been looking for a last tour of the sites of walking's history, and that locale seemed to have everything. I had envisioned starting in the hedge maze at the magnificent estate of Chatsworth, then wandering through the surrounding formal gardens into the Capability Brown–landscaped later gardens. From there I could go into the wilder reaches of the Peak, toward Kinder Scout, where the great right-of-way battles were fought, and past the famous gritstone climbs where "the working-class revolution in climbing" took place, then head for bordering Manchester with its formative suburbs or Sheffield with its industrial ruins and climbing gym in a former forge. Or I could begin with the industrial cities and work my way into the country and then to the garden and the maze. But all these picturesque schemes came to an end with the sneaking suspicion that proving that it was still possible to walk in Britain didn't count for much at all. Even Britain's industrial wastelands signify the pale northern European past, and it wasn't pedestrianism's past but its prognosis that I wanted to inspect. So one December morning I stepped out of Pat's van onto Fremont Street in downtown Las Vegas, and he set off to spend the day climbing the boulders and cliffs at Red Rocks.

Down most of Vegas's east-west avenues straight as latitude lines you can see the thirteen-mile-long escarpment of Red Rocks and, behind its ruddy sandstone domes and pillars, the ten-thousand-foot-high gray peaks of the Spring Range. One of the least celebrated aspects of this arid, amnesiac boomtown is its spectacular setting, with mountain ranges on three sides and glorious desert light, but Las Vegas has never been about nature appreciation. *Las vegas* means "the meadows" and makes it clear the Spanish got to this Southern Paiute oasis before the Anglos did, but the oasis didn't become a town until the twentieth century—until 1905, when the railroad from Los Angeles to Salt Lake City decided to put in a station here. Las Vegas remained a town for drifters and tourists long after the oasis was sucked dry. It lacked the mineral resources of much of the state and only began to flourish when gambling became legal in Nevada in 1931, while Hoover Dam was being built on the Colorado River, thirty miles to the southeast. In 1951 the Nevada Test Site was established sixty miles to the northwest, and in the decades since more than a thousand nuclear weapons have been detonated on its premises (until 1963, most of the tests were above ground, and there are some startling photographs of mushroom clouds rising up above the casinos' towering signs). Las Vegas is bracketed by these colossal monuments to the ambition to dominate rivers, atoms, wars, and to some extent, the world. It may be, however, a much smaller but more pervasive invention that most shaped this city in the Mojave Desert: air conditioning, which has much to do with the recent American mass migration south and west to places where many will spend all summer in climate-controlled interiors. Often portrayed as exceptional, Las Vegas is instead emblematic, an extreme version of new kinds of places being built around the United States and the world.

Las Vegas's downtown was built around the railroad station: visitors were expected to get off the train and walk to the casinos and hotels of downtown's compact Glitter Gulch area around Fremont Street. As cars came to supersede trains for American travelers, the focus shifted: in 1941 the first casino-hotel complex went in along what was then the highway to Los Angeles, Highway 91, and is now the Las Vegas Strip. Long ago, after falling asleep in a car headed for the annual antinuclear gathering at the Nevada Test Site, I woke up when we came to a halt at a traffic light on the Strip to see a jungle of neon vines and flowers and words dancing, bubbling, exploding. I still

---

somebody could just walk! He can jog perhaps in the morning, but he can't walk anywhere! The world

---

remember the shock of that spectacle after the blackness of the desert, heavenly and hellish in equal measure. In the 1950s, cultural geographer J. B. Jackson described the then-new phenomenon of roadside strips as another world, a world built for strangers and motorists. "The effectiveness of this architecture is finally a matter of what that other world is: whether it is one that you have been dreaming about or not. And it is here that you begin to discover the real vitality of this new other-directed architecture along our highways: it is creating a dream environment for our leisure that is totally unlike the dream environment of a generation ago. It is creating and at the same time reflecting a new public taste." That taste, he said, was for something wholly new, something that dismissed earlier Eurocentric aping-one's-betters notions of recreation and taste, something adapted to cars and to the new futuristic and tropical fantasies of those cars' inhabitants: "those streamlined facades, flamboyant entrances and deliberately bizarre color effects, those cheerfully self-assertive masses of color and light and movement that clash so roughly with the old and traditional." This vernacular architecture invented for automotive America was celebrated in the famous 1972 architectural manifesto *Learning from Las Vegas*.

In recent years, however, something wholly unexpected has happened on the Strip. Like those islands where an introduced species reproduces so successfully that its teeming hordes devastate their surroundings and starve en masse, the Strip has attracted so many cars that its eight lanes of traffic are in continual gridlock. Its fabulous neon signs were made to be seen while driving past at a good clip, as are big signs fronting mediocre buildings on every commercial strip, but this Strip of Strips has instead in the last several years become a brand-new outpost of pedestrian life. The once-scattered casinos on the Strip have grown together into a boulevard of fantasies and lures, and tourists can now stow their car in one casino's behemoth parking lot and wander the Strip on foot for days, and they do, by the millions—more than 30 million a year, upward of 200,000 at once on the busiest weekends. Even in August, when it was about 100 degrees Fahrenheit after dark, I have seen the throngs stream back and forth on the Strip, slowly—though not much more slowly than the cars. Casino architecture itself has undergone radical changes since the

---

has become inaccessible because we drive there.—IVAN ILLICH, *WHOLE EARTH REVIEW*   *"After several*

prescient 1966 Caesars Palace played up fantasy architecture over neon signage and the 1989 Mirage presented the first facade specifically designed as a pedestrian spectacle. It seemed to me that if walking could suddenly revive in this most inhospitable and unlikely place, it had some kind of a future, and that by walking the Strip myself I might find out what that future was.

Fremont Street's old-fashioned glitter has suffered by comparison with the Strip's new fantasy environments, so it has been redesigned as a sort of cyber-arcade. Its central blocks have been closed to cars so that pedestrians can mill around freely, and up above the resurfaced street is set a high, arched roof on which laser shows are beamed by night, so that what was once sky is now a kind of giant television screen. It's still a sad half-abandoned place in daylight, and it didn't take me long to tour it and wander south on Las Vegas Boulevard, which would eventually become the Strip. Before it becomes the Strip, the boulevard is a skid row of motels, shabby apartments, and sad souvenir, pornography, and pawn shops, the ugly backside of the gambling, tourism, and entertainment industries. A homeless black man huddled in a brown blanket at a bus stop looked at me walking by, and I looked at an Asian couple across the street coming out of one of the tiny wedding chapels, him in a dark suit, her in a chalk-white dress, so impersonally perfect they could have fallen off a colossal wedding cake. Here each enterprise seemed to stand on its own; the wedding chapel unintimidated by the sex shops, the fanciest casinos by the ruins and vacant lots around them. There weren't many people on the sidewalk with me in this sagging section between the two official versions of Vegas.

Farther on I came to the old El Rancho hotel, burned out and boarded up. The desert and the West had been romanticized by many of the earlier casinos: the Dunes, the Sands, the Sahara, and the Desert Inn on the Strip and the Pioneer Club, the Golden Nugget, the Frontier Club, and the Hotel Apache on Fremont Street, but more recent casinos have thrown regional pride to the winds and summoned up anyplace else, the less like the Mojave the better. The Sands is being replaced by the Venetian, complete with canals. I realized later that my walk was an attempt to find a continuity of experience here, the spatial continuity walking usually provides, but the

---

hours, I begin to feel something new, something never before experienced. I strongly sense, with my

place would defeat me with its discontinuities of light and fantasy. It defeated me another way too. Las Vegas, which had a population of 5 at the beginning of the twentieth century, 8,500 in 1940, and about half a million in the 1980s, when the casinos seemed to stand alone in the sweep of creosote bushes and yuccas, now has about 1.25 million residents and is the fastest growing city in the country. The glamorous Strip is surrounded by a colossal sprawl of trailer parks, golf courses, gated communities, and generic subdivisions—one of Vegas's abundant ironies is that it has a pedestrian oasis in the heart of the ultimate car suburb. I had wanted to walk from the Strip to the desert to connect the two, and I called the local cartographic company for recommendations about routes, since all my maps were long out of date. They told me that the city was growing so fast they put out a new map every month and recommended some of the shortest routes between the southern Strip and the city edge, but Pat and I drove along them and saw they were alarming places for a solitary walker— a mix of warehouses, light industrial sites, dusty lots, and walled homes from which an aura of abandonment emanated, only occasionally interrupted by a car or grimy hobo. So I stuck to the pedestrian oasis and found that I could mentally move the great casinos like chess pieces from the flat board of the desert: ten years back the fantasy casinos were gone; twenty years back the casinos were scattered and there were almost no pedestrians; fifty years back there were only a few isolated outposts; and a century ago only a tiny whiskey-saloon hamlet disturbed the pale earth spreading in every direction.

Under the tired pavilion in front of the Stardust, an old French couple asked me for directions to the Mirage. I watched them walk slowly away from the old vehicular American fantasyland of glittering futures and toward the new nostalgic fantasyland at the Strip's heart and followed them south myself. The scattered walkers began to become crowds as I traveled south. The bride and groom I had seen come out of the wedding chapel showed up again walking down the boulevard near me, she with a delicate quilted jacket over her wedding dress and spike heels. Tourists come here from around the affluent world (and employees from some of the less affluent ones, notably Central America). Another of Vegas's ironies is that it is one of the

---

whole self, that I am moving from one place to another. . . . I am not passing through space, as one does

---

world's most visited cities, but few will notice the actual city. In, for example, Barcelona or Katmandu, tourists come to see the locals in their natural habitat, but in Vegas the locals appear largely as employees and entertainers in the anywhere-but-here habitat built for tourists. Tourism itself is one of the last major outposts of walking. It has always been an amateur activity, one not requiring special skills or equipment, one eating up free time and feeding visual curiosity. To satisfy curiosity you must be willing to seem naive, to engage, to explore, to stare and be stared at, and people nowadays seem more willing or able to enter that state elsewhere than at home. What is often taken as the pleasure of another place may be simply that of the different sense of time, space, and sensory stimulation available anywhere one goes slowly.

The Frontier was the first casino I went inside. For six and a half years visitors could watch an outdoor floor show there, a round-the-clock picket of workers—maids, cocktail waitresses, busboys—fighting the union-busting new owners, testifying with their feet and their signs day and night, in summer's withering heat and winter's storms. During those years 101 babies were born and 17 people died among the Frontier strikers, and none of them crossed the picket line. It became the great union battle of the decade, a national inspiration to labor activists. In 1992 the AFL-CIO organized a related event, the Desert Solidarity March. Union activists and strikers walked three hundred miles from the Frontier across the desert to the courthouse in Los Angeles in a show of their willingness to suffer and prove their commitment. Vegas filmmaker Amie Williams commented, while showing me rushes from her documentary on the Frontier strike, that the union is like an American religion of family and solidarity: it has a credo, "An injury to one is an injury to all"—and in the Desert Solidarity March it got its pilgrimage. In Amie's footage, people who looked like they didn't walk much at all straggled in a line alongside old Route 66, bared their feet for bandaging in the evenings, and got up and did it again the next day. A carpenter and union representative named Homer, a bearded man who looked like a biker, testified to the pilgrimage's miracle: in the middle of a rainstorm, a sunspot followed them, and they stayed dry, and he sounded as enthused as one of the children of Israel for whom the Red Sea parted. Finally the union-busting family that had

---

in a car or airplane. I feel I am in a place, actually, in an infinite number of places. I am not in an

bought the Frontier was forced to sell, and on January 31, 1997, the new owners invited the union back in. Those who had spent six exhilarating years walking the line went back to mixing drinks and making beds. There was nothing of that struggle to see inside the Frontier, just the usual supernova of dizzily patterned carpet, jingling slots, flashing lights, mirrors, staff moving briskly and visitors milling slowly in the twilight that suffuses every casino. They are modern mazes, made to get lost in, with their windowless expanses full of odd angles, view-obscuring banks of slot machines, and other distractions designed, like those in malls and department stores, to prolong visitors' encounters with the temptations that might make them open their wallets before they find the well-concealed exits. Many casinos have "people movers"—moving ramps like those in airports—but here they are all inward-bound. Finding the way out is up to you.

Wandering and gambling have some things in common; they are both activities in which anticipation can be more delicious than arrival, desire more reliable than satisfaction. To put one foot in front of one another or one's cards on the table is to entertain chance, but gambling has become a highly predictable science for the casinos, and they and the law enforcement of Las Vegas are trying to control the odds on walking down the Strip too. The Strip is a true boulevard. It is exposed to the weather, open to its surroundings, a public space in which those glorious freedoms granted by the First Amendment can be exercised, but a mighty effort is being made to take them away, so that the Strip will instead be a sort of amusement park or mall, a space in which we can be consumers but not citizens. Next to the Frontier is the Fashion Show Shopping Center, where leafletters hang out together, forming one of the Strip's many subcultures. Many are un-documented Central Americans, Amie Williams said, and the leaflets most often pertain to sex (though Vegas has a huge sex industry, cus-tomers are sought largely by advertisements, not by street hustling; the dozens of clusters of newsstands along the Strip contain very few newspapers and a veritable library of brochures, cards, and leaflets with color photographs advertising an army of "private dancers" and "escort services"). Since the women themselves are largely invisible, the visibility of the ads has come under assault. The county passed an

---

undifferentiated space—what one feels in many modern places that, really are non-places; they are

ordinance making "off-premises canvassing" in the "resort district" a misdemeanor. The director of the Nevada outpost of the American Civil Liberties Union, Gary Peck, spoke to me of "the almost transparent paradox that Vegas markets itself as anything goes—sex, alcohol, gambling—and on the other hand has this almost obsessive attempt at thorough control of public space, advertising on billboards, in the airport, panhandling, free speech." The ACLU's fight over the handbill ordinance had reached the Federal Court of Appeals, and other issues kept cropping up. Earlier that year, petition-gathers were harassed and a pastor and four companions were arrested for proselytizing in the Fremont Street Experience on charges of "blocking a sidewalk," though the now-pedestrian arcade is one vast sidewalk it would take dozens to obstruct.

The casinos and the county, Peck told me, are trying to privatize the very sidewalks, to give themselves more muscle for prosecuting or removing anyone engaging in First Amendment activities—speaking about religion, sex, politics, economics—or otherwise ruffling the smooth experience visitors are supposed to have (similarly, Tucson has recently looked into privatizing sidewalks by leasing them for one dollar to businesses, to allow them to drive out the homeless). Peck worries that if they succeed in taking away the ancient "freedom of the city" at the sidewalk level, it will set a precedent for the rest of the country, malling what were once genuine public spaces, making cities into theme parks. "The theme park," writes Michael Sorkin, "presents its happy regulated version of pleasure—all those artfully hoodwinking forms—as a substitute for the democratic urban realm, and it does so appealingly by stripping troubled urbanity of its sting, of the presence of the poor, of crime, of dirt, of work." The Mirage has already posted a little sign on one of its lawns: "This sidewalk is the private property of the Mirage Casino Hotel upon which an easement has been granted to facilitate pedestrian movement. Anyone found loitering or otherwise impeding pedestrian movement is subject to arrest for trespass," and signs up and down the strip say, "Resort District: No Obstructive Uses Permitted on Public Sidewalks." The signs are there not to protect the freedom of movement of pedestrians, but to restrict what those pedestrians can do or see.

---

simply repetitions of concepts—the concept of hospital space, shopping space, mall space, airport

* * *

I was hot and weary from the four miles or so I'd gone from Fremont
Street, for it was a warm day and the air was stale with exhaust. Distance is deceptive on the Strip: the major intersections are about a
mile apart, but the new casinos with their twenty-or-thirty-story
hotel towers tend to look closer because their scale doesn't register.
Treasure Island is the first of the new theme-park casinos one reaches
from the north, and one of the most fantastic—named not after a
place or period like the rest, but after a boys' book about pirate life in
the South Seas. With a facade of fake rock and picturesque building
fronts behind its lagoon of palm trees and pirate ships, it resembles a
hotel-resort version of Disneyland's Pirates of the Caribbean ride. But
it was the adjoining Mirage that invented the pedestrian spectacle in
1989, with its volcano that erupts every fifteen minutes after dark, to
the delight of gathered crowds. When Treasure Island opened in
1993, it upstaged the volcano with a full-fledged pirate battle cli-
maxing with the sinking of a ship—but the battle only takes place a
few times a day.

The authors of *Learning from Las Vegas* long ago groused that "the
Beautification Committee would continue to recommend turning the
Strip into a western Champs- Elysées, obscuring the signs with trees
and raising the humidity with giant fountains." The fountains have
arrived, and the vast sheets of water fronting the Mirage and Treasure
Island are dwarfed by the eight-acre lake at Bellagio, down where the
Dunes used to be, across Flamingo Road from Caesars Palace. Together these four casinos make something altogether new and surpris-
ingly old-fashioned, a wild hybrid of the formal garden and the
pleasure garden spread out along a public thoroughfare. The Mirage's
volcano buried the old Vegas as decisively as Vesuvius did Pompeii,
changing architecture and audiences together. The fountains are everywhere, and it *is* a kind of western Champs-Élysées, in that walking
to look at the architecture and the other walkers has become a
pastime. The Strip is replacing its neon-go-go Americana futurama
vision with Europe, a fun pop-culture version of Europe, a Europe of
architectural greatest hits and boulevardiers in shorts and T-shirts. Is
there anything less peculiar about setting miniature Italian and

Roman temples and bridges in English gardens than in putting up gargantuan ones in the desert, in building volcanoes in eighteenth-century gardens such as Wörlitz in Germany than on boulevards in Nevada? Caesars Palace, with its dark green cypresses, fountains, and classical statuary out front, calls up many of the elements of the formal garden, which was itself an Italian extension of Roman practices adapted by the French, Dutch, and English. Bellagio, with its frontage of fountains, recalls Versailles, whose scale was a demonstration of wealth, power, and triumph over nature. These places are mutant reprises of the landscapes in which walking as scenic pleasure was developed. Vegas has become the successor to Vauxhall, Ranelagh, Tivoli, and all the other pleasure gardens of the past, a place where the unstructured pleasures of walking and looking mingle with highly organized shows (stages for music, theater, and pantomimes were an important part of pleasure gardens, as were areas to dance, eat, drink, and sit). As a Vegas promoter might say, the garden is making a comeback, crossbred with the boulevard, and with them comes pedestrian life.

All the efforts to control who strolls and how suggest that walking may in some way still be subversive. At least it subverts the ideals of entirely privatized space and controlled crowds, and it provides entertainment in which nothing is spent or consumed. Though walking may be an inadvertent side effect of gambling—after all, the casino facades weren't built out of public-spiritedness—the Strip is now a place to walk. And after all, Paris's Champs-Élysées also belongs to tourists and foreigners nowadays, strolling, shopping, eating, drinking, and enjoying the sights. New pedestrian overpasses eliminate the intersection of people and cars where Flamingo Road crosses the Strip, and they are handsome bridges giving some of the best views around. But these bridges are entered and exited from within the casinos, so there may come a time when only the well-dressed can cross the street safely here, and the rabble will have to take their chances with the cars or make a long detour. The Strip is not the Champs-Élysées reborn for other reasons too; it lacks the perfect straightness Le Nôtre gave the older road, the straightness that lets you see far into the distance. It bends and bulges, though there are always the cross-streets—and the bridge over Flamingo Road between Bellagio and

poem. Even if you walk exactly the same route every day—as with a sonnet—the events along the

Caesars provided the best view yet of the desert to the west and Red Rocks. From the other bridge, the one over the Strip from Bellagio to Bally's, I could see—Paris! I had forgotten that a Paris casino was under construction, but there rising out of the dusty soil of the Mojave like an urban mirage, a flâneur's apparition, was the Eiffel Tower, only half finished and half scale but already aggressively straddling what looked like a stumpy Louvre with the Arc de Triomphe jostling it in an antigeographic jumble of architectural greatest hits.

Of course Vegas is reinventing not only the garden, but the city: New York, New York is just down the road from Bellagio, the Tokyo-homage Imperial Palace is up the street, and a much older version of San Francisco—the Barbary Coast—faces Caesars. The 1996 New York, New York is, like the Paris casino, a cluster of famous features; inside is a funny little maze of streets made to look like various Manhattan neighborhoods, complete with street signs, shops (of which only the souvenir and food shops are real, as I found when I foolishly lunged for a bookstore), air conditioners jutting out third-floor windows, and even a graffiti corner—but, of course, without the variety, productive life, dangers, and possibilities of real urban life. Fronted by a Statue of Liberty welcoming gamblers rather than huddled masses yearning to be free, New York, New York is a walk-through souvenir of the city. No longer pocket-size and portable but a destination in itself, it performs a souvenir's function: recalling a few pleasant and reassuringly familiar aspects of a complicated place. I ate a late lunch in New York, New York and drank three pints of water to replenish what had evaporated from me in the desert aridity of my all-day walk.

Back on the boulevard, a young woman from Hong Kong asked me to take a picture of her with the Statue of Liberty behind her and then with the huge golden MGM lion across the street, and she looked ecstatic in both shots. Fat people and thin people, people in baggy shorts and in sleek dresses, a few children and a lot of old people streamed around us, and I handed the camera back and continued south with the crowd, to the last station on these stations of the odds, the Luxor, whose pyramid shape and sphinx say ancient Egypt, but whose shiny glass on which lasers play at night says technology. The newlyweds I had seen before were there in the entryway: she had laid aside her coat and purse to pose for his camera in front of one of the mock-Egyptian

route cannot be imagined to be the same from day to day.... If a poem is each time new, then it is neces-

statues. I wondered about them, about why they had chosen to spend the first hours of their honeymoon strolling the Strip, about what past they brought to this encounter with global fantasy filtered through the Nevada desert's climate and gambling's economy. Who am I to say that because these people who streamed by to my right and my left were Las Vegas tourists they did not have other lives: that this English couple might not take their next vacation in the Lake District, that the old French couple might not live in Paris or near Plum Village where the Vietnamese Buddhist Thich Nhat Hanh teaches walking meditation, that the African-Americans might not have marched in Selma as children, that the beggar in a wheelchair could not have been hit by a car in New Orleans, that the bride and groom might not be Japanese climbers of Mount Fuji, Chinese descendants of mountain hermits, southern Californian executives with treadmills at home, that this Guatemalan handing out helicopter-ride coupons might not walk the stations of the cross in her church or once have promenaded the plaza in her hometown, that bartender going to work might not have gone on the AFL-CIO march across the desert? The history of walking is as expansive as human history, and the most attractive thing about this pedestrian oasis in the middle of suburban blight in the middle of a great desert expanse is that it hints at that history's breadth—not in its fake Rome and Tokyo, but in its Italian and Japanese tourists.

Las Vegas suggests that the thirst for places, for cities and gardens and wilderness, is unslaked, that people will still seek out the experience of wandering about in the open air to examine the architecture, the spectacles, and the stuff for sale, will still hanker after surprises and strangers. That the city as a whole is one of the most pedestrian-unfriendly places in the world suggests something of the problems to be faced, but that its attraction is a pedestrian oasis suggests the possibility of recovering the spaces in which walking is viable. That the space may be privatized to make the liberties of walking, speaking, and demonstrating illegal suggests that the United States is facing as serious a battle over rights-of-way as did English ramblers half a century ago, though this time the struggle is over urban, not rural, space. Equally scary is the widespread willingness to accept simulations of real places, for just as these simulations usually forbid the full exercise of civil liberties, so they banish the full spectrum of sights,

----

sarily an act of discovery, a chance taken, a chance that may lead to fulfillment or disaster.

----

encounters, experiences, that might provoke a poet, a cultural critic, a social reformer, a street photographer.

But the world gets better at the same time it gets worse. Vegas is not an anomaly but an intensification of mainstream culture, and walking will survive outside that mainstream and sometimes reenter it. When the automotive strip and suburb were being developed in the decades after World War II, Martin Luther King was studying Gandhi and reinventing Christian pilgrimage as something politically powerful at one edge of this continent, while at the other Gary Snyder was studying Taoist sages and walking meditation and rethinking the relationship between spirituality and environmentalism. At present, space in which to walk is being defended and sometimes enlarged by pedestrian activist groups springing up in cities across the United States, from Feet First in Seattle and Atlanta's PEDS to Philly Walks and Walk Austin, by the incendiary British-based Reclaim the Streets, by older organizations like the Ramblers' Association and other British insurrections for walking and access, and by pedestrian-favoring urban redesign from Amsterdam to Cambridge, Massachusetts. Walking traditions are maintained by the resurgence of the foot pilgrimage to Santiago de Compostela in Spain and the thriving one in Chimayó, New Mexico, the growing popularity of climbing and mountaineering, the artists working with walking as a medium and the writers with it as a means, the spread of Buddhism with its practices of walking meditation and circumambulating mountains, the newfound secular and religious enthusiasm for labyrinths and mazes. . . .

"This place is a _maze_," grumbled Pat when he found me in Caesars Forum, the arcade attached to the casino. The Forum is the capstone of the arch, the crowning jewel of Vegas's recreation of the past. It is an arcade in exactly the sense Walter Benjamin described Parisian arcades—he quoted an 1852 guidebook that said, "Both sides of these passageways, which are lighted from above, are lined with the most elegant shops, so that such an arcade is a city, even a world, in miniature," and added, "The arcades were a cross between a street and an _interieur_." With its arched roof painted to look like the sky and recessed lighting that changes from day to twilight and back every twenty

---
—A. R. AMMONS, "A POEM IS A WALK"   _Draw an imaginary map. / Put a goal mark where you want to_
---

minutes or so, this one is more so than Benjamin could have imagined.
Its curving "streets" are disorienting and full of distractions: the stores
full of clothes, perfumes, toys, knickknacks, a fountain whose backside
is a huge tropical fish tank, the famous fountain of nubile gods and
goddesses who periodically "come to life" during a simulated thunder-
storm with laser lightning snaking across the skylike dome. I had
visited the arcades of Paris only six months before, and they were
beautiful dead places, like streambeds through which water no longer
runs, with half their shops closed and few wanderers along their
mosaic'd corridors, but Caesars Forum is constantly thronged (as are
the arcades of Bellagio, modeled after Milan's famous Galleria). It is
one of the most financially successful malls in the world, says the *Wall
Street Journal*, adding that a new addition is planned, a re-creation of
a Roman hill town with occasional appearances by horse-drawn
chariots. An arcade was never much more than a mall, and though a
flâneur was supposed to be more contemplative than the average mall
rat, shallow gentlemen are as common as soulful shoppers. "Let's get
out of here," I said to Pat, and we finished our drinks and headed for
Red Rocks.

Red Rocks is as open, as public, as Las Vegas Boulevard, but
nobody is promoting it, just as no one (unless they're selling gear) is
promoting the free activity of walking in preference to the lucrative
industry of cars. While tens of thousands of people wander the Strip,
perhaps a hundred or so at most roam the larger terrain of Red Rocks,
whose spires and buttresses are far taller and more spectacular than
any casino. Many people only drive through or step out long enough
to take a photograph, unwilling to surrender to the slower pace here,
a twilight that comes only once a day, wildlife that does as it pleases,
a place with no human trace to structure one's thoughts but a few
trails, climbing bolts on the rocks, litter, and signs (and an entrenched
tradition of nature-worship). Nothing happens here most of the time,
except seasons, weather, light, and the workings of one's body and
mind.

Musing takes place in a kind of meadowlands of the imagination,
a part of the imagination that has not yet been plowed, developed, or
put to any immediately practical use. Environmentalists are always
arguing that those butterflies, those grasslands, those watershed

----

go. / Go walking on an actual street according to your map. / If there is no street where it should be

----

woodlands, have an utterly necessary function in the grand scheme of things, even if they don't produce a market crop. The same is true of the meadowlands of imagination; time spent there is not work time, yet without that time the mind becomes sterile, dull, domesticated. The fight for free space—for wilderness and for public space—must be accompanied by a fight for free time to spend wandering in that space. Otherwise the individual imagination will be bulldozed over for the chain-store outlets of consumer appetite, true-crime titillations, and celebrity crises. Vegas has not yet decided whether to pave over or encourage that space.

That night we would sleep out near Red Rocks, in an unofficial campground with figures silhouetted against the small fires burning here and there under the starry sky and the glow of Vegas visible over the hill. In the morning we would rendezvous with Paul, a young guide who often drove out from Utah to climb here and who had invited Pat to climb with him. He would lead us along a trail snaking up and down across small arroyos and a dry streambed, past the gorgeous foliage I remembered from earlier trips, junipers with desert mistletoe, tiny-leafed desert oaks, yuccas, manzanitas, and an occasional barrel cactus, all stunted and spread sparsely by the rocky soil, aridity, and scattered boulders in a way that recalls Japanese gardens. Still limping from a fall six months earlier, Pat brought up the rear, while Paul and I talked as we went along about music, climbing, concentration, bicycles, anatomy, apes. When I turned back to look at Las Vegas as I had so often looked toward Red Rocks the day before, he would say, "Don't look back," but I would stare, amazed how thick the city's smog was. The place appeared to be a brown dome with a only few spires murkily visible within it. This state of things whereby the desert could be seen clearly from the city but not the other way round seemed as neat an allegory as I'd ever met. It was as though one could look back from the future to the past but not forward from this ancient place to the future shrouded in trouble, mystery, and fumes.

Paul would lead us off the trail into the brush that led up steep, narrow Juniper Canyon, and I would manage to heave myself up the various shelves where the rock grew more and more gorgeous, sometimes striped red and beige in thin layers, sometimes spotted with pink spots the size of coins, until we were at the foot of the climb. "Olive

Oil: This route ascends obvious crack systems for 700 feet up the south side of the Rose Tower," I would read in Pat's battered *American Alpine Club Climber's Guide* for the region. I lounged and watched them climb with ease up the first few hundred feet and studied the mice, who were less glamorous than the white tigers and dolphins of the Mirage, but livelier. Afterward I would turn around and spend the afternoon wandering in flatter terrain, ambling along the few trails alongside the clear rushing water of Pine Creek, exploring another canyon, turning back to watch the shadows over the hills grow longer and the light thicker and more golden, as though air could turn to honey, honey that would dissolve into the returning night.

Walking has been one of the constellations in the starry sky of human culture, a constellation whose three stars are the body, the imagination, and the wide-open world, and though all three exist independently, it is the lines drawn between them—drawn by the act of walking for cultural purposes—that makes them a constellation. Constellations are not natural phenomena but cultural impositions; the lines drawn between stars are like paths worn by the imagination of those who have gone before. This constellation called walking has a history, the history trod out by all those poets and philosophers and insurrectionaries, by jaywalkers, streetwalkers, pilgrims, tourists, hikers, mountaineers, but whether it has a future depends on whether those connecting paths are traveled still.

---

the city and give flowers to the first person you meet.—YOKO ONO, "MAP PIECE," 1962

Notes

1. TRACING A HEADLAND: An Introduction

5 "An absolutely new prospect is a great happiness": Henry David Thoreau, "Walking," in *The Natural History Essays* (Salt Lake City: Peregrine Smith Books, 1980), 99.

7 It was nuclear weapons that first led: My early writing on walking and on nuclear politics appears in my 1994 book *Savage Dreams: A Journey into the Landscape Wars of the American West* (San Francisco: Sierra Club Books, 1994; Berkeley: University of California Press, 1999).

2. THE MIND AT THREE MILES AN HOUR

15 "I can only meditate when I am walking": Jean-Jacques Rousseau, *The Confessions* (Harmondsworth, England: Penguin Books, 1953), 382.

15 "In one respect, at least": John Thelwall, *The Peripatetic: or, Sketches of the Heart of Nature and Society* (1793; facsimile reprint, New York: Garland Publishing, 1978), 1, 8–9.

15 Felix Grayeff's history: *Aristotle and His School: An Inquiry into the History of the Peripatos* (London: Gerald Duckworth, 1974), 38–39.

17 the Stoics were named: Christopher Thacker, *The History of Gardens* (Berkeley: University of California Press, 1985), 21, is my source for the information on the stoa and the Stoics. Bernard Rudofsky's *Streets for People: A Primer for Americans* (New York: Van Nostrand Reinhold, 1982), also gives a précis of this information.

17 "For recreation I turn": *Selected Letters of Friedrich Nietzsche*, ed. Oscar Levy, trans. Anthony M. Ludovici (New York: Doubleday, 1921), 23.

17 "He used to come": Bertrand Russell, *Portraits from Memory*, quoted in A. J. Ayer, *Wittgenstein* (New York: Random House, 1985), 16.

18 "In order to slacken my pace": Rousseau, *Confessions*, 327.

18 "Behold how luxury": Jean-Jacques Rousseau, "First Discourse" ("Discourse on the Arts and Letters"), in *The First and Second Discourses* (New York: St. Martin's Press, 1964), 46.

19 "wandering in the forests": Rousseau, "Second Discourse," ibid., 137.

20 "I do not remember ever having had": Rousseau, *Confessions*, 64.

20 "Never did I think so much": Ibid., 158.

21 "thinking over subjects": Ibid., 363.

21 Boswell came to visit: "Dialogue with Rousseau," *The Portable Johnson and Boswell*, ed. Louis Kronenberger (New York: Viking, 1947), 417.

22 "Having therefore decided to describe": Jean-Jacques Rousseau, "Second Walk," in *Reveries of the Solitary Walker*, trans. Peter France (Harmondsworth, England: Penguin Books, 1979), 35.

24 "Wherein lay this great contentment?": "Fifth Walk," ibid., 83.

25 "What had at first been": Walter Lowrie, *A Short Life of Kierkegaard* (Princeton, N.J.: Princeton University Press, 1942), 45–46.

26 "Strangely enough, my imagination works best": *Søren Kierkegaard's Journals and Papers*, ed. and trans. Howard V. Hong and Edna H. Hong (Bloomington: Indiana University Press, 1978), 6:113.

26 "In order to bear mental tension": Ibid., 5:271 (1849–1851).

26 "This very moment there is an organ-grinder": Ibid., 5:177 (1841).

26 "Most of *Either/Or* was written only twice": Ibid., 5:341 (1846).

26 "overwhelmed with ideas": Ibid., 6:62–63 (1848).

27 "My atmosphere has been tainted": Ibid., 5:386 (1847).

29 Husserl described walking: "The World of the Living Present and the Constitution of the Surrounding World External to the Organism," in *Edmund Husserl: Shorter Works*, ed. Peter McCormick and Frederick A. Elliston (Notre Dame, Ind.: University of Notre Dame Press, Harvester Press, 1981). I benefited from Edward S. Casey's interpretation of this dense essay in his *The Fate of Place: A Philosophical History* (Berkeley: University of California Press, 1997), 238–50.

30 "If the body is a metaphor": Susan Bordo, "Feminism, Postmodernism, and Gender-Scepticism," in *Feminism/Postmodernism*, ed. Linda J. Nicholson (New York: Routledge, 1990), 145.

3. RISING AND FALLING: The Theorists of Bipedalism

35 "Human walking is a unique activity": John Napier, "The Antiquity of Human Walking," *Scientific American*, April 1967. Napier is one of the earliest to push the history of walking back further into prehuman history and insist on its formative importance there.

36 "point out parallels": Adrienne Zihlman, in "The Paleolithic Glass Ceiling," in *Women in Human Evolution*, ed. Lori D. Hager (London and New York: Routledge, 1997), 99. Zihlman and Dean Falk's readings of Lovejoy and the broader gender politics of human evolution in this book and in Falk's book *Braindance* (New York: Henry Holt, 1992) have been immensely helpful to my own reading.

38 "This is like a modern knee joint" and following: Donald Johanson and Maitland Edey, *Lucy: The Beginnings of Humankind* (New York: Simon and Schuster, 1981), 163. See also C. Owen Lovejoy with Kingsbury G. Heiple and Albert H. Burstein, "The Gait of Australopithicus," *American Journal of Physical Anthropology* 38 (1973): 757–80.

39 "In most primate species": C. Owen Lovejoy, "The Origin of Man," *Science* 211 (1981): 341–50.

39 "Bipedalism . . . figured": "Evolution of Human Walking," *Scientific American,* November 1988.

40 Dean Falk's attack on Lovejoy: "Brain Evolution in Females: An Answer to Mr. Lovejoy," in Hager, *Women in Human Evolution,* 115.

41 Jack Stern and Randall Sussman: Interview with the author, Stonybrook, New York, February 4, 1998. See also their comments in *Origine(s) de la Bipédie chez les Hominidés* (Paris: Editions du CNRS/Cahiers de Paléoanthropologie, 1991) and articles such as "The Locomotor Anatomy of *Australopithicus afarensis,*" *American Journal of Physical Anthropology* 60 (1983). Representations of hominids in deep forest appeared in *National Geographic* in 1997.

43 1991 Conference on the Origins of Bipedalism: The three anthropologists at the Paris conference were Nicole I. Tuttle, Russell H. Tuttle, and David M. Webb; their paper "Laetoli Footprint Trails and the Evolution of Hominid Bipedalism" appears in *Origine(s) de la Bipédie;* the quoted passages appear on 189–90.

44 "One cannot overemphasize": Mary Leakey, *National Geographic,* April 1979, 453.

44 "According to this view": Falk, "Brain Evolution," 115.

44 "these features led to 'whole-body cooling'": Falk, "Brain Evolution," 128, and at length in Falk, *Braindance.* See also E. Wheeler, "The Influence of Bipedalism on the Energy and Water Budgets of Early Hominids," *Journal of Human Evolution* 21 (1991): 117–36.

46 I called up Owen Lovejoy: C. Owen Lovejoy, interview by author, June 23, 1998.

4. THE UPHILL ROAD TO GRACE: Some Pilgrimages

49 Much of the information on Chimayó comes from Elizabeth Kay, *Chimayó Valley Traditions* (Santa Fe: Ancient City Press, 1987), and Don J. Usner, *Sabino's Map: Life in Chimayó's Old Plaza* (Santa Fe: Museum of New Mexico Press, 1995).

50 "These devout and simple people": John Noel, *The Story of Everest* (New York: Blue Ribbon Books, 1927), 108.

54 "All sites of pilgrimage": Victor Turner and Edith Turner, *Image and Pilgrimage in Christian Culture: Anthropological Perspectives* (New York: Columbia University Press, 1978), 41.

55 "Often as she listened to the pilgrims' tales": Leo Tolstoy, *War and Peace*, trans. Ann Dunnigan (New York: Signet Classics, 1965), bk. 2, pt. 3, ch. 26, 589.

55 "When pilgrims begin to walk": Nancy Louise Frey, *Pilgrim Stories: On and Off the Road to Santiago* (Berkeley: University of California Press, 1998), 72.

56 "Liminars are stripped of status and authority": Turner and Turner, *Image and Pilgrimage*, 37.

56 "Wherever you go, there you are": First said by Carl Franz, in his *People's Guide to Mexico*, Greg says.

60 "To remain a wanderer": Introduction, *Peace Pilgrim: Her Life and Work in Her Own Words* (Santa Fe: Ocean Tree Books, 1991), xiii.

60 "a complete willingness": Ibid., 7.

61 "it doesn't show dirt": Ibid., 56.

61 "a comb, a folding toothbrush": Ibid., xiii.

62 "I walk until given shelter": Ibid., 25.

64 "Reverend Charles Billups and other Birmingham ministers": Stephen B. Oates, *Let the Trumpet Sound: A Life of Martin Luther King, Jr.* (New York: Harper and Row, 1982), 236.

65 March of Dimes: Information from telephone conversation with Tony Choppa, April 1998.

67 "At the end of November, 1974": Werner Herzog, *On Walking in Ice* (New York: Tanam Press, 1980), 3.

67 "While I was taking a shit": Ibid., 27.

67 "For one splendid fleeting moment": Ibid., 57.

5. LABYRINTHS AND CADILLACS: **Walking into the Realm of the Symbolic**

76 W. H. Matthews cautions: W. H. Matthews, *Mazes and Labyrinths: Their History and Development* (1922; reprint, New York: Dover, 1970), 66, 69.

76 "Labyrinths . . . are usually in the form of a circle": Lauren Artress, handout at Grace Cathedral, n.d.

81 "each of the speaking characters": Matthews, *Mazes and Labyrinths*, 117.

81 "A garden path": Charles W. Moore, William J. Matchell, and William Turnbull, *The Poetics of Gardens* (Cambridge, Mass.: MIT Press, 1988), 35.

82 "I have a little game": John Finlay, ed., *The Pleasures of Walking* (1934; reprint, New York: Vanguard Press, 1976), 8.

82 "in a kind of out-of-body form": Lucy Lippard, *The Lure of the Local: Senses of Place in a Multicentered Society* (New York: New Press, 1996), 4.

83 "general principles of the mnemonic": Frances Yates, *The Art of Memory* (London: Pimlico, 1992), 18.

6. THE PATH OUT OF THE GARDEN

87 These descriptions of Dorothy Wordsworth occur on pages 132 and 188 of Thomas De Quincey, *Recollections of the Lakes and the Lake Poets* (Harmondsworth, England: Penguin Books, 1970).

87 "Twas a keen frosty morning": William Wordsworth, letter to Samuel Taylor Coleridge, December 24, 1799, in *Letters of William and Dorothy Wordsworth: The Early Years, 1787–1805*, ed. Ernest de Selincourt (Oxford: Clarendon Press, 1967), 273–80. It's worth noting that in this letter Wordsworth refers both to "Taylor's tour," a written account that described the first waterfall they visited, and to the waterfall itself as "a performance as you might expect from some giant gardiner employed by one of Queen Elizabeth's courtiers, if this same giant gardiner had consulted with Spenser," which is to say that his vision was framed in the literary and gardening traditions of England.

88 Wordsworth and his companions are said to have made walking into something new: See, for example, Marion Shoard, *This Land Is Our Land: The Struggle for Britain's Countryside*, 2d. ed. (London: Gaia Books, 1997), 79: "It is to Wordsworth as much as anyone that we also owe the idea that the proper way of communing with nature is by walking through the countryside."

88 "I have always fancied": Christopher Morley, "The Art of Walking" (1917), in Aaron Sussman and Ruth Goode, *The Magic of Walking* (New York: Simon and Schuster, 1967), a cheery evangelistic volume advocating walking for health, providing practical tips, and including an anthology of essays on the subject.

89 use as their demonstration case Carl Moritz: "Yet within less than ten years from the date of Moritz's tour a striking change had taken place, and the fashion of the walking-tour (or pedestrian-tour, as it was then called) had come in. It was the beginning of a movement . . ." (Morris Marples, *Shank's Pony: A Study of Walking* [London: J. M. Dent and Sons, 1959], 31); "to the new phenomenon of the pedestrian tour, and to other less ambitious forms of walking for pleasure . . . established, in the last ten to fifteen years of the eighteenth century" (Robin Jarvis, *Romantic Writing and Pedestrian Travel* [Houndmills, Basingstoke, Hampshire: Macmillan Press, 1997], 4); "removing walking's long-standing implication of necessity and so of poverty and vagrancy" (Anne Wallace, *Walking, Literature and English Culture* [Oxford: Oxford University Press, 1993], 10); "changes in the practice of and attitudes toward travel in general, and walking in particular, which accompany the transport revolution beginning in the mid-eighteenth century" (ibid., 18). They all assert walking is travel; that it is not necessarily so is my argument.

89   "A traveller on foot": Carl Moritz, *Travels of Carl Philipp Moritz in England in 1782: A Reprint of the English Translation of 1795*, with an introduction by E. Matheson (1795; reprint, London: Humphrey Milford, 1924), 37.

90   "I walked with my brother at my side": Dorothy Wordsworth, quoted in Hunter Davies, *William Wordsworth: A Biography* (New York: Antheneum, 1980), 70.

90   "I cannot pass unnoticed": Dorothy Wordsworth to her proud aunt Crackanthorp, April 21, 1794, cited in de Selincourt, *Letters*, 117.

91   "When we walk, we naturally go to the fields and woods": Thoreau, "Walking," 98–99.

91   The eighteenth century created a taste for nature: See Christopher Thacker, *The Wildness Pleases: The Origins of Romanticism* (New York: St. Martin's Press, 1983), 1–2. He writes, "Aristotle claimed that all poetry was 'the imitation of men in action.' By poetry, he implied all forms of art, from sculpture to drama, from epic poetry to history, to painting and even to music. . . . Aristotle's definition of the scope of poetry cuts out many matters which we might consider wholly proper, indeed desirable, as the subject of a work of art. Above all, the depiction of 'nature' is a subject which we, living two centuries after the romantic explosion at the end of the eighteenth century, have come to accept almost without thinking." Naming many landscape paintings, Thacker goes on to say that such subject matter would have seemed incomprehensible, or at least inconsequential, to Aristotle and indeed to any educated onlooker before the transformation of perception that "took place in western Europe in the eighteenth century."

93   "Sixteenth-century doctors stressed": Mark Girouard, *Life in the English Country House: A Social and Architectural History* (New Haven: Yale University Press, 1978), 100.

93   Queen Elizabeth added a raised terrace: Susan Lasdun, *The English Park: Royal, Private & Public* (New York: Vendome Press, 1992), 35.

94   "There is gravel walks and grass and close walks": Celia Fiennes on the gardens at Agnes Burton, *The Journeys of Celia Fiennes*, ed. Christopher Morris (London: Cresset Press, 1949), 90–91.

95   "These avenues provided the shade and shelter for walks": Lasdun, *English Park*, 66.

95   "O glorious Nature!": Shaftesbury in John Dixon Hunt and Peter Willis, *The Genius of the Place: The English Landscape Garden, 1620–1840* (New York: Harper, 1975), 122; also a key text in Thacker, *The Wildness Pleases*, whose title comes from this effusion.

96   "Poetry, Painting, and Gardening": Walpole, quoted in Hunt and Willis, *Genius of the Place*, 11.

97   "asked to be explored": John Dixon Hunt, *The Figure in the Landscape: Poetry, Painting and Gardening during the Eighteenth Century* (Baltimore: Johns Hopkins University Press, 1976), 143.

97 "Whereas the French formal garden was based on a single axial view":
Carolyn Bermingham, *Landscape and Ideology: The English Rustic Tradition,*
1740–1860 (Berkeley: University of California Press, 1986), 12.

98 "into harmony with the age's humanism": Christopher Hussey, *English*
Gardens and Landscapes, 1700–1750 (London: Country Life, 1967), 101.

98 "Within thirty years": *Stowe Landscape Gardens* (Great Britain: National
Trust, 1997), 45.

98 "O lead me to the wide-extended walks": James Thomson, *The Seasons*
(Edinburgh and New York: T. Nelson and Sons, 1860), 139. Kenneth
Johnston, in *The Hidden Wordsworth: Poet, Lover, Rebel, Spy* (New York:
Norton, 1998), calls *The Seasons* the most successful poem of the
century, and Andrew Wilton in *Turner and the Sublime* (Chicago: Uni-
versity of Chicago Press, 1981) documents its impact.

99 "Everyone takes a different way": Pope, letter of 1739, cited in *Stowe*
Landscape Gardens, 66.

99 "or drove about it in cabriolets": Walpole, letter to George Montagu,
July 7, 1770, in *Selected Letters of Horace Walpole* (London: J. M. Dent and
Sons, 1926), 93.

99 "Gardening, as far as Gardening is an Art": Sir Joshua Reynolds, quoted
in Hunt and Willis, *Genius of the Place,* 32.

100 "leapt the fence": Walpole, quoted in Hunt and Willis, *Genius of the Place,* 13.

101 "Within the last sixty years": Wordsworth, *Guide to the Lakes,* ed. Ernest
de Selincourt (Oxford: Oxford University Press, 1977), 69.

101 "greatly compensates for the mediocrity of this park": *Travels of Carl*
Philipp Moritz, 44.

102 "The People of London are as fond of walking": Oliver Goldsmith, *The*
Citizen of the World, vol. 2 of the *Collected Works,* ed. Arthur Friedman
(Oxford: Clarendon Press, 1966), 293.

102 "how to admire an old twisted tree": Jane Austen, *Sense and Sensibility*
(New York: Washington Square Press, 1961), 80.

102 "there is a sense in which". John Barrell, *The Idea of Landscape and the Sense*
of Place (New York: Cambridge University Press, 1972), 4–5.

103 "It is very true": Austen, *Sense and Sensibility,* 83–84.

103 "Were it not for this general deficiency of objects": William Gilpin,
Observations on Several Parts of Great Britain, particularly the Highlands of
Scotland, relative chiefly to picturesque beauty, made in the year 1776 (London:
T. Cadell and W. Davies, 1808), 2:119.

103 "Let us learn, in real scenes, to trace": Richard Payne Knight, "The
Landscape: A Didactic Poem," in *The Genius of the Place,* 344.

103 Thomas Gray's celebrated Lake District tour: Gray wrote about it in
his "Journal in the Lakes," in *The Works of Thomas Gray in Prose and Verse,*
vol. 1, ed. Edmund Gosse (New York: Macmillan, 1902).

105 "They were country ladies": Dorothy Wordsworth, Oct. 16, 1792, in
de Selincourt, *Letters,* 84.

105 "That she should have walked three miles": Jane Austen, *Pride and Prejudice* (Oxford: Oxford University Press/Avenal Books, 1985), 30; "At that moment they were met," 49; "Her figure was elegant," 52; "Miss Bennet, there seemed to be a prettyish kind of a little wilderness," 340; "favourite walk," 164; "More than once did Elizabeth in her ramble," 176; "had never seen a place for which nature had done more," 234; "rapturously cried, 'what delight! what felicity!," 150; "it is not the object of this work," 232; "'My dear Lizzy, where can you have been walking to?,'" 360.

110 "In the morning, I read Mr Knight's *Landscape*": July 27, 1800, reprinted in *Home at Grasmere: Extracts from the Journal of Dorothy Wordsworth and from the Poems of William Wordsworth*, ed. Colette Clark (Harmondsworth, England: Penguin Books, 1978), 53–54.

7. THE LEGS OF WILLIAM WORDSWORTH

112 "His legs were pointedly condemned": Thomas De Quincey, *Recollections of the Lakes and the Lake Poets*, ed. David Wright (Harmondsworth, England: Penguin Books, 1970), 53–54.

113 "Happy in this, that I with nature walked": William Wordsworth, *The Prelude: The Four Texts (1798, 1799, 1805, 1850)*, ed. Jonathan Wordsworth (Harmondsworth, England: Penguin Books, 1995), 322. All quotes are from the 1805 version.

114 "He sees nothing but himself and the universe": Hazlitt, "The Lake School," in *William Hazlitt: Selected Writings* (Harmondsworth, England: Penguin Books, 1970), 218.

114 "more glorious than I had ever beheld": Wordsworth, *Prelude*, 158.

115 "Should the guide I choose": Ibid., 36.

115 "With this act of disobedience": Johnston, *Hidden Wordsworth*, 188.

116 "standing on top of golden hours": Wordsworth, *Prelude*, 226.

116 "each spot of old and recent fame": Ibid., 348.

117 "Oswald had traveled to India": Johnston, *Hidden Wordsworth*, 286.

117 "No region, pervious to human feet": Thomas De Quincey, "Walking Stewart—Edward Irving—William Wordsworth," in *Literary Reminiscences*, vol. 3 of *The Collected Works of Thomas De Quincey* (Boston: Houghton, Osgood and Co., 1880), 597.

117 "I have some thoughts:" in de Selincourt, *Letters*, 153; and "So like a peasant," Wordsworth, *Prelude*, 42.

118 "Throughout that turbulent time": Basil Willey, *The Eighteenth-Century Background* (Boston: Beacon Press, 1961), 205.

118 "that odious class of men called democrats": Wordsworth, letter to a friend, May 23, 1794, in de Selincourt, *Letters*, 119.

118 "The principal object, then, proposed in these poems": Wordsworth, preface to the second edition of *Lyrical Ballads*, in *Anthology of Romanticism*, ed. Ernest Bernbaum (New York: Ronald Press, 1948), 300–301.

119 "I love a public road": Wordsworth, *Prelude*, 496.

120 "Had I been born in a class": Johnston, *Hidden Wordsworth*, 57.

121 "They were surrounded": Hazlitt, "The Lake School," 217.

122 "He won't a man as said a deal to common fwoak": Local quoted in *Wordsworth Among the Peasantry of Westmorland*, cited in Davies, *Wordsworth*, 322.

122 "He would set his head a bit forrad": Andrew J. Bennett, "'Devious Feet': Wordsworth and the Scandal of Narrative Form," *LELH* 59 (1992): 147.

123 "At present he is walking": Dorothy, in a letter to Lady Beaumont, May 1804, in Davies, *Wordsworth*, 166.

123 "almost physiological relation" and following: Seamus Heaney, "The Makings of a Music," in *Preoccupations* (New York: Farrar, Straus and Giroux, 1980), 66, 68.

124 "the lord who owned the ground": Davies, *Wordsworth*, 324. See also Wallace, *Walking, Literature and English Culture*, 117.

124 "The grave old bard": Letter published in the *Manchester Guardian*, October 7, 1887, cited in Howard Hill, *Freedom to Roam: The Struggle for Access to Britain's Moors and Mountains* (Ashbourne, England: Moorland Publishing, 1980), 40. The presence of a deferential "Mr. Justice Coleridge" on the walk and Sir John Wallace at the confrontation make it appear that this is another version of the same event. Late in life Wordsworth, alas, also opposed the building of a railroad that would take tourists to Windermere, crustily remarking that workers could take their holidays closer to home. Though an unkind remark, it is not altogether wrong about the impact of tourism—a century later the Sierra Club would take the phrase "render accessible" out of its mission statement, realizing that people could love the landscape to death with tourism infrastructures and general trampling.

126 "I purpose within a month": Earle Vonard Weller, ed., *Autobiography of John Keats, Compiled from his Letters and Essays* (Stanford: Stanford University Press, 1933), 105.

126 "I should not have consented": Keats, in Marples, *Shank's Pony*, 68.

126 "hunger-bitten girl": Wordsworth, *Prelude*, 374.

8. A Thousand Miles of Conventional Sentiment:
    The Literature of Walking

127 "a processional march": Thomas Hardy, *Tess of the d'Urbervilles* (New York: Bantam Books, 1971), 10.

128 "In the neighborhood of latitude": Aldous Huxley, "Wordsworth in the Tropics," in *Collected Essays* (New York: Bantam Books, 1960), 1.

128 "One of the pleasantest things in the world": William Hazlitt, "On Going a Journey," in *The Lore of the Wanderer*, ed. Geoffrey Goodchild (New York: E. P. Dutton, 1920), 65.

129 "The walks are the unobtrusive connecting thread": Leslie Stephen, "In Praise of Walking," in Finlay, *Pleasures of Walking*, 20.

129 "lameness was too severe": Stephen, "In Praise," 24.

130 "A walking tour should be gone on alone": Robert Louis Stevenson, "Walking Tours," in Goodchild, *Lore of the Wanderer*, 10–11.

130 "I have two doctors": G. M. Trevelyan, "Walking," in Finlay, *Pleasures of Walking*, 57.

131 "Whenever I was with friends": Max Beerbohm, "Going Out on a Walk," in Finlay, *Pleasures of Walking*, 39.

131 "I wish to speak a word for Nature" and following: Thoreau, "Walking," 93–98.

132 "The best thing is to walk": Bruce Chatwin, "It's a Nomad Nomad World," in *Anatomy of Restlessness: Selected Writings, 1969–1989* (New York: Viking, 1996), 103.

132 "How womankind, who are confined": Thoreau, "Walking," 97.

134 "Perhaps walking can be the way to peace": Mort Malkin, "Walk for Peace," *Fellowship* (magazine of the Buddhist Peace Fellowship), July/August 1997, 12. Malkin is the author of *Walk—The Pleasure Exercise* and *Walking—The Weight Loss Exercise*.

134 "Reagan had realized, he told us": Michael Korda, "Prompting the President," *New Yorker*, October 6, 1997, 92.

137 "But why tramp?" Charles F. Lummis, *A Tramp Across the Continent* (Omaha: University of Nebraska Press, 1982), 3.

139 "But strange things do happen": Robyn Davidson, *Tracks* (New York: Pantheon Books, 1980), 191–92.

140 "In properly developed countries, the inhabitants": Alan Booth, *The Roads to Sata: A Two-Thousand-Mile Walk Through Japan* (New York: Viking, 1986), 27.

141 "The Guinness Book of Records defines a walk": Ffyona Campbell, *The Whole Story: A Walk Around the World* (London: Orion Books, 1996), unpaginated preface.

9. MOUNT OBSCURITY AND MOUNT ARRIVAL

144 "the first man to climb a mountain": Kenneth Clark, *Landscape into Art* (Boston: Beacon Press, 1961), 7.

145 "a small mound of rock": Clarence King, *Mountaineering in the Sierra Nevada* (New York: W. W. Norton, 1935), 287.

145   Christian Europe seems to be alone: Both Francis Farquhar's brief bib-
      liography of mountaineering literature and Edwin Bernbaum's *Sacred
      Mountains of the World* (Berkeley: University of California Press, 1997)
      agree on the peculiar European attitude toward mountains before the
      eighteenth century. Edward Whymper also speaks of the legend of the
      Wandering Jew, in Ronald W. Clark, *Six Great Mountaineers* (London:
      Hamish Hamilton, 1956), 14. The terms in which English writers de-
      scribed mountains come from Keith Thomas, *Man and the Natural World:
      Changing Attitudes in England* (Harmondsworth, England: Penguin Books,
      1984), 258–59.

146   drove his chariot up T'ai Shan: The First Emperor's ascent is described
      in Bernbaum, *Sacred Mountains*, 31.

146   "The Chinese phrase for 'going on a pilgrimage'": Gretel Ehrlich, *Ques-
      tions of Heaven: The Chinese Journeys of an American Buddhist* (Boston: Beacon
      Press, 1997), 15.

146   "the vast and very flat valley" and following: *Egeria: Diary of a Pilgrimage*,
      trans. George E. Gingras (New York: Newman Press, 1970), 49–51.

147   "There is nothing to look up to": Cited by Dervla Murphy in her intro-
      duction to Henriette d'Angeville's *My Ascent of Mont Blanc*, trans. Jen-
      nifer Barnes (London: HarperCollins, 1991), xv. The first American
      woman atop Everest (the first woman was Japanese), Stacy Allison,
      similarly declared, "There was nowhere else to climb. I was standing on
      top of the world" (from www.everest.mountainzone.com).

147   "We climbed up through the narrow cleft": Dante, *The Divine Comedy,
      Purgatorio*, canto 4.

148   "To the traveller": Henry David Thoreau, *Walden* (Princeton: Princeton
      University Press, 1973), 290.

148   "To climb up rocks": Charles Edward Montague, "In Hanging Garden
      Gully" (from his book *Fiery Particles*, 1924), excerpted in *Challenge: An An-
      thology of the Literature of Mountaineering*, ed. William Robert Irwin (New
      York: Columbia University Press, 1950), 333.

149   "Because it's there," "We hope to show": Murray Sayle, "The Vulgarity
      of Success," *London Review of Books*, May 7, 1998, 8.

149   "Chomalungma" was taller: See Bernbaum, *Sacred Mountains*, 7.

149   "Whatever Western society regards": Ibid., 236.

150   "And before I started to move": Gwen Moffat, *The Space Below My Feet*
      (Cambridge: Riverside Press, 1961), 66.

151   slowest traverse ever: "We made an epic traverse of the Cuillin Ridge
      at the end of June. The main ridge is seven or eight miles long with
      about sixteen peaks of over three thousand feet strong along the chain.
      The average time taken for the traverse was ten to thirteen hours; some
      parties took twenty-four, others were bringing the record down to fan-
      tastic times, as on the fourteen-peak walk in Snowdonia. We hated

records; we decided to be different. We would take our time, sunbathe, enjoy the views, carry food for two days, sleep out on top of the ridge; we would set up the record for the longest time spent on the Cuillen traverse" (ibid., 101).

151 "turned his keen intelligence": Eric Shipton, Mountain Conquest (New York: American Heritage, 1965), 17.

152 forty-six parties . . . had reached the summit: Ronald W. Clark, A Picture History of Mountaineering (London: Hulton Press, 1956), 31.

153 "The soul has needs," "It was not the puny fame": D'Angeville, My Ascent, xx–xxi.

155 "For half a century": Smoke Blanchard, Walking Up and Down in the World: Memories of a Mountain Rambler (San Francisco: Sierra Club Books, 1985), xv.

156 "On Visiting a Taoist Master": Arthur Cooper, trans., Li Po and Tu Fu (Harmondsworth, England: Penguin Books, 1973), 105.

156 "People ask the way": Cold Mountain: 100 Poems by the T'ang Poet Han-Shan, trans. Burton Watson (New York: Columbia University Press, 1972), poem 82, 100.

156 "Before the sixth century A.D.": Bernbaum, Sacred Mountains, 58.

156 "Every aspect of Shugendō": H. Byron Earhart, A Religious Study of the Mount Haguro Sect of Shugendō: An Example of Japanese Mountain Religion (Tokyo: Sophia University, 1970), 31. Also see Allan G. Grapard, "Flying Mountains and Walkers of Emptiness: Toward a Definition of Sacred Space in Japanese Religions," History of Religion 21, no. 3 (1982).

157 "I . . . set off": Bashō, The Narrow Road to the Deep North and Other Travel Sketches, trans. and ed. Nobuyuki Yuasa (Harmondsworth, England: Penguin Books, 1966), 125. Bashō was climbing Mount Gassan in the north, immediately after climbing better-known, adjacent Mount Haguro, a major focal point of Shugendō.

157 "I had been introduced to the high snow peaks": Gary Snyder, Mountains and Rivers without End (Washington, D.C.: Counterpoint Press, 1996), 153. Snyder also speaks about mountains, spirituality, and his mountaineering in the essay "Blue Mountains Constantly Walking," in The Practice of the Wild (San Francisco: North Point Press, 1990); in Earth House Hold (New York: New Directions, 1969); and most recently in an interview with John O'Grady, in Western American Literature, fall 1998, among many other places.

157 "I was given a chance to see": Snyder, Mountains and Rivers, 156.

157–58 "The closer you get to real matter": David Robertson quoting Kerouac in Dharma Bums, in Real Matter (Salt Lake City: University of Utah Press, 1997), 100.

158 "This sentence states what is perhaps": Ibid., 100, 108.

158 "I translate space from its physical sense": Snyder, interview with O'Grady, 289.

159 "On Climbing the Sierra Matterhorn Again": Gary Snyder, *No Nature: New and Selected Poems* (New York: Pantheon Books, 1992), 362.

10. OF WALKING CLUBS AND LAND WARS

161 "An excursion of this sort": William Colby, *Sierra Club Bulletin*, 1990.

162 outdoor organizations had been proliferating: "The formation of the [British] Alpine Club in 1857 had been followed by the foundation of the Swiss Alpine Club—and of the Societe de Touristes Savoyardes—in 1863. The Italian Alpine Club came later during the same year, while in 1865 there was founded for the further exploration of the Pyrenees the Societe Ramond. The Austrian and the German clubs came in 1869 and the French in 1874, while across the Atlantic the Williamstown Alpine Club had been founded as early as 1863 and was followed in 1873 and 1776 respectively by the White Mountain Club and the Appalachian Mountain Club" (Clark, *Picture History of Mountaineering*, 12). A Ladies' Alpine Club was founded in 1907.

164 "There were solemn hours": Ella M. Sexton, *Sierra Club Bulletin* 4 (1901): 17.

164 "Mr. Colby goes like lightning": Nelson Hackett in oral history transcripts, Sierra Club files, in Bancroft Library, letters of July 5 and July 18, 1908, transcribed at end of Hackett's interview.

167 "We were shocked to discover firsthand": Michael Cohen, *The Pathless Way: John Muir and the American Wilderness* (Madison: University of Wisconsin, 1984), 331.

169 "The Friends of Nature were founded": E-mail from Manfred Pils to the author, October 1998.

171 "On the main thing—rambling": Walter Laqueur, *Young Germany: A History of the German Youth Movement* (New Brunswick and London: Transaction Books, 1984), 33. This and Gerald Masur, *Prophets of Yesterday: Studies in European Culture, 1890–1914* (New York: Macmillan, 1961), Werner Heisenberg, *Physics and Beyond: Encounters and Conversations* (New York: Harper and Row, 1971); and David C. Cassidy, *Uncertainty: The Life and Science of Werner Heisenberg* (New York: W. H. Freeman, 1992), were my principal sources.

172 "After 1919, the militant dictatorships": Masur, *Prophets of Yesterday*, 368

172–73 eighteen million Britons head for the country: Shoard, *This Land Is Our Land*, 264; ten million walkers from *Country Walking* magazine editor in chief Lynn Maxwell, in conversation with the author, May 1998.

173 "Almost a spiritual thing": Roly Smith, conversation with the author, May 1998.

173 accessing the land has been something of a class war: Information on trespassing, poaching, and gamekeepers in various parts of Shoard, *This Land Is Our Land*.

175 Association for the Protection of Ancient Footpaths: Tom Stephenson, *Forbidden Land: The Struggle for Access to Mountain and Moorland*, ed. Ann Holt (Manchester and New York: Manchester University Press, 19), 59.

175 Forest Ramblers' Club: Hill, *Freedom to Roam*, 21.

175 not impressive compared to those of other European countries: Steve Platt, "Land Wars," *New Statesman and Society* 23 (May 10, 1991); Shoard, *This Land Is Our Land*, 451.

177 "Land is not property": James Bryce, quoted in Ann Holt, "Hindsight on Kinder," *Rambling Today*, spring 1995, 17. See also Raphael Samuel, *Theatres of Memory* (London: Verso, 1994), 294: "The Commons, Open Spaces and Footpath Society, founded in 1865—the remote ancestor of the National Trust—was a kind of Liberal front, championing the claims of villagers and commoners against the encroachments of landlords and property developers. . . . "

177 "It is the one thing that is unpleasant": Crichton Porteous, *Derbyshire* (London: Robert Hale Limited, 1950), 33.

177 "a little judicious trespassing": Leslie Stephen, "In Praise," 32.

177 "By the last quarter of the nineteenth century": Hill, *Freedom to Roam*, 24.

178 "Hiking was a major, if unofficial, component": Raphael Samuels, *Theatres of Memory*, 297. The passage continues, "and 'freedom to roam' was a left-wing campaigning issue. It had been given a mass basis, in Edwardian times, by the Clarion League, the 40,000 strong organization of the young who combined Sunday cycle meets with preaching the socialist message on the village green. In the inter-war years it was forwarded by the Woodcraft Folk—a kind of anti-militarist, co-educational version of the Boy Scouts and Girl Guides who combined pacifist advocacy and nature mysticism; by the Youth Hostels Association, formed in 1930; and by that great army of hikers who on high days and holidays went rambling on the mountains and moors. Hiking had a particular appeal to working-class Bohemians, as a mainly intellectual alternative to the dance hall, and one that cost no money."

178 "a genuine hatred of ramblers": Ann Holt, *The Origins and Early Days of the Ramblers' Association*, booklet published by the Ramblers' Association from a speech given April 1995.

179 "Town dwellers lived for weekends": Benny Rothman, *The 1932 Kinder Scout Trespass: A Personal View of the Kinder Scout Mass Trespass* (Altrincham, England: Willow Publishing, 1982), 12.

11. THE SOLITARY STROLLER AND THE CITY

Philip Lopate's essay "The Pen on Foot: The Literature of Walking Around," *Parnassus*, vol. 18, no. 2 and 19, no 1, 1993, pointed me to Edwin Denby's writings and to specific poems of Walt Whitman's.

187 "On Saturday night . . . the city joined in the promenade": Harriet Lane Levy, 920 O'Farrell Street (Berkeley: Heyday Books, 1997), 185–86.

187 Kerouac managed to have two visions on [Market Street]: see *Atlantic Monthly*, reprinting a May 1961 letter, November 1998, 68: "It [*On the Road*] was really a story about two Catholic buddies in search of God. And we found him. I found him in the sky, in Market Street San Francisco (those 2 visions)."

190 how a popular, well-used street is kept safe: Jane Jacobs, *The Death and Life of Great American Cities* (New York: Vintage Books, 1961), throughout the chapter "The Uses of Sidewalks: Safety."

190–91 "What distinguishes the city": Moretti, quoted in Peter Jukes, *A Shout in the Street: An Excursion into the Modern City* (Berkeley: University of California Press, 1991), 184.

192 little more than outdoor salons and ballrooms: *Cities and People* (New Haven and London: Yale University Press, 1985), 166–68, 237–38.

192 "Earlier in the [nineteenth] century," "I hear that pedestrians": Ray Rosenzweig and Elizabeth Blackmar, *The Park and the People: A History of Central Park* (Ithica: Cornell University Press, 1992), 27, 223.

193 "It simply never occurs to us": Bernard Rudofsky, *Streets for People: A Primer for Americans* (New York: Van Nostrand Reinhold, 1982), epigraph quoting his own *Architecture without Architects*.

194 "In ancient Italian towns the narrow main street": Edwin Denby, *Dancers, Buildings and People in the Streets*, introduction by Frank O'Hara (New York: Horizon Press, 1965), 183.

195 "When I am in a serious Humour": Addison in Joseph Addison and Richard Steele, *The Spectator*, Vol. 1 (London: J. M. Dent and Sons, 1907), 96, from *Spectator*, no. 26 (March 30, 1711).

195 "Though you through cleaner allies": John Gay, "Trivia; or, the Art of Walking the Streets of London," book 3, line 126, in *The Abbey Classics: Poems by John Gay* (London: Chapman and Dodd, n.d.), 88.

196 "Here I remark": Ibid., ll. 275–82, 78.

196 "goes forward with the crowd": Wordsworth, *Prelude*, 286.

196 "each charter'd street": The famous opening of William Blake's "London," in *William Blake*, ed. J. Bronowski (Harmondsworth, England: Penguin Books, 1958), 52.

196 one of those desperate London walkers: See Richard Holmes, *Dr. Johnson and Mr. Savage* (New York: Vintage Books, 1993), 44, quoting Sir John Hawkins in the chapter on these walks: "Johnson has told me, that whole nights have been spent by him and Savage in conversations of this kind, not under the hospitable roof of a tavern, where warmth might have invigorated their spirits, and wine dispelled their care; but in a perambulation round the squares of Westminster, St. James's in particular, when all the money they could both raise was less than suf-

ficient to purchase for them the shelter and sordid comforts of a night cellar."

196 "I should have been": James Boswell, *Boswell's London Journal*, ed. Frederick A. Pottle (New York: Signet, 1956), 235.

197 50,000 [prostitutes in London] in 1793: Henry Mayhew, *London Labour and the London Poor*, vol. 4 (1861–62; reprint, New York: Dover Books, 1968), 211, citing Mr. Colquhoun, a police magistrate, and his "tedious investigations."

197 "the circulating harlotry of the Haymarket and Regent Street": Ibid., 213. On 217, "They [the streetwalkers] are to be seen between three and five o'clock in the Burlington Arcade, which is a well known resort of cyprians of the better sort. They are well acquainted with its Paphian intricacies, and will, if their signals are responded to, glide into a friendly bonnet shop, the stairs of which leading to the coenacula or upper chambers are not innocent of their well formed 'bien chaussee' feet. The park is also, as we have said, a favorite promenade, where assignations may be made or acquaintances formed."

197 "Prostitution streetscapes are composed of *strolls*": Richard Symanski, *The Immoral Landscape: Female Prostitution in Western Societies* (Toronto: Butterworths, 1981), 175–76.

197 "think that women who work in whorehouses": Dolores French with Linda Lee, *Working: My Life as a Prostitute* (New York: E. P. Dutton, 1988), 43.

198 "Perception of the new qualities of the modern city": Raymond Williams, *The Country and the City* (New York: Oxford University Press, 1973), 233.

198 "Being myself at that time, of necessity, a peripatetic" and following: De Quincey, *Confessions of an English Opium Eater* (New York: Signet Books, 1966), 42–43.

199 "And this kind of realism," "Few of us understand the street": G. K. Chesterton, *Charles Dickens, a Critical Study* (New York: Dodd, Mead, 1906), 47, 44.

200 "If I couldn't walk fast and far": Dickens to John Forster, cited in Ned Lukacher, *Primal Scenes: Literature, Philosophy, Psychoanalysis* (Ithaca: Cornell University Press, 1986), 288.

200 "I am both a town traveller": Charles Dickens, *The Uncommercial Traveller and Reprinted Pieces Etc.* (Oxford and New York: Oxford University Press, 1958), 1.

200 "So much of my travelling is done on foot," "My walking is of two kinds": Dickens, "Shy Neighborhoods," ibid., 94, 95.

200 "It is one of my fancies": Dickens, "On an Amateur Beat," ibid., 345.

200–1 "Whenever I think I deserve particularly well of myself": Dickens, "The City of the Absent," ibid., 233.

201 "Some years ago, a temporary inability to sleep": Dickens, "Night Walks," ibid., 127.

202 "I would roam the streets": Patti Smith, when asked what she did to prepare to go onstage, *Fresh Air*, National Public Radio, Oct. 3, 1997.

202 "How could I think mountains and climbing romantic?": *The Letters of Virginia Woolf*, vol. 3, *A Change of Perspective*, ed. Nigel Nicholson (London: Hogarth Press, 1975–80), letter to V. Sackville-West, Aug. 19, 1924, 126.

202 "enforce the memories of our own experience": Virginia Woolf, "Street Haunting: A London Adventure," in *The Death of the Moth and Other Essays* (Harmondsworth, England: Penguin Books, 1961), 23.

203 "As we step out of the house," "the shell-like covering": Ibid., 23–24.

204 Two-thirds of all journeys . . . still made on foot: Tony Hiss, editorial, *New York Times*, January 30, 1998.

204 "On the whole North America's Anglo-Saxomania has had a withering effect": Rudofsky, *Streets for People*, 19.

204 "Who often walk'd lonesome walks": Walt Whitman, "Recorders Ages Hence," *Leaves of Grass* (New York: Bantam Books, 1983), 99.

204 "City of orgies, walks and joys": Ibid., 102.

205 "Passing stranger!": Ibid., 103.

205 "the fruited plain": Ken Gonzales-Day, "The Fruited Plain: A History of Queer Space," *Art Issues*, September/October 1997, 17.

206 "dragging themselves through the negro streets," "shoes full of blood": Allen Ginsberg, "Howl," in *The New American Poetry*, ed. Donald M. Allen (New York: Grove Press, 1960), 182, 186.

207 "Strange now to think of you, gone": Allen Ginsberg, *Kaddish and Other Poems, 1958–1960* (San Francisco: City Lights Books, 1961), 7.

207 "where you walked 50 years ago": Ibid., 8.

207 "It was the most extraordinary thing": Brad Gooch, *City Poet: The Life and Times of Frank O'Hara* (New York: Alfred A. Knopf, 1993), 217.

207 "I can't even enjoy a blade of grass": Frank O'Hara, "Meditations in an Emergency," in *The Selected Poems* (New York: Vintage Books, 1974), 87.

207 "I'm becoming": O'Hara, "Walking to Work," ibid., 57.

208 "I'm getting tired of not wearing": O'Hara, "F. (Missive and Walk) I. #53," ibid., 194.

208 "Some nights we'd walk seven or eight hundred blocks": David Wojnarowicz, *Close to the Knives: A Memoir of Disintegration* (New York: Vintage Books, 1991), 5; "long legs and spiky boots," 182; "I had almost died three times," 228; "I'm walking through these hallways," 64; "I walked for hours," 67; "man on second avenue," 70; "I walk this hallway twenty-seven times," 79.

12. Paris, or Botanizing on the Asphalt

213 "Now a landscape, now a room": Walter Benjamin, "On Some Motifs in Baudelaire," in *Reflections: Essays, Aphorisms, Autobiographical Writings* (New York: Schocken Books, 1978), 156.

213 "Not to find one's way in a city," "it had to be in Paris": Walter Benjamin, "A Berlin Chronicle," in *Reflections*, 8, 9.

213 holding an alpenstock before some painted Alps: On mountains, alpenstocks and Benjamin, see his letters of September 13, 1913; July 6–7, 1914; November 8, 1918; and July 20, 1921; and Monme Brodersen, *Walter Benjamin: A Biography* (London: Verso, 1996): "finally a crudely daubed backdrop of the Alps was brought for me. I stand there, bareheaded, with a tortuous smile on my lips, my right hand clasping a walking stick" (12), and "Another taken-for-granted feature of the boy's day-to-day life were the frequent lengthy journeys with the whole family: to the North Sea and the Baltic, to the high peaks of the Risengebirge between Bohemia and Silesia, to Freudenstadt in the Black Forest, and to Switzerland" (13).

214 "I don't think I ever saw him walk": Gershom Sholem, cited in Frederic V. Grunfeld, *Prophets without Honor: A Background to Freud, Kafka, Einstein and Their World* (New York: McGraw Hill, 1979), 233.

214–15 "old Scandinavian": Priscilla Park Ferguson, "The Flâneur: Urbanization and Its Discontents," in *From Exile to Vagrancy: Home and Its Dislocations in 19th Century France*, ed. Suzanne Nash (Albany: State University of New York, 1993), 60, n. 1. See also her *Paris as Revolution* (Berkeley: University of California Press, 1994).

215 "Irish word for 'libertine'": Elizabeth Wilson, "The Invisible Flâneur," *New Left Review* 191 (1992): 93–94.

215 "The crowd is his domain": Charles Baudelaire, "The Painter of Modern Life," *Selected Writings on Art and Artists* (Cambridge: University of Cambridge Press, 1972), 399.

215 "goes botanizing on the asphalt": Walter Benjamin, *Charles Baudelaire: A Lyric Poet in the Era of High Capitalism*, trans. Harry Zohn (London: Verso, 1973), 36.

216 "Arcades," "The flâneurs liked to have the turtles": Ibid., 53–54.

216 he did not exist: On the nonexistence of the flâneur, see Rob Shields, who, in "Fancy Footwork: Walter Benjamin's Notes on the Flâneur," in *The Flâneur*, ed. Keith Tester (London: Routledge, 1994), remarks, "In truth, it must be acknowledged that nineteenth-century visitors and travelogues do not appear to reference flânerie other than as an urban myth. The principal habitat of the flâneur is the novels of Honore de Balzac, Eugene Sue, and Alexandre Dumas."

217 Gérard de Nerval famously took a lobster on walks: Richard Holmes, *Footsteps: Adventures of a Romantic Biographer* (New York: Vintage Books, 1985), 213.

217 "Narrow crevices": Victor Hugo, *Les Misérables*, trans. Charles E. Wilbour (New York: Modern Library, 1992), bk. 12, *Corinth*, chap. 1, 939–40. See also Girouard, *Cities and People*, 200–201: "All visitors commented on these streets, which had no sidewalks, so that pedestrians were con-

stantly in danger of being run down or spattered with mud by fast-moving traffic. 'Walking,' wrote Arthur Young in 1787, 'which in London is so pleasant and so clean, that ladies do it every day, is here a toil and a fatigue to a man, and an impossibility to a well-dress woman.' 'The renowned Tournefort,' according to the Russian traveller Karamzin writing in 1790, 'who had travelled almost the entire world, was crushed to death by a fiacre on his return to Paris because on his travels he had forgotten how to leap in the streets, like a chamois.' In this ambience, browsing in shop windows was not likely to flourish."

217 "which I reached without any other adventure" and following: Frances Trollope, *Paris and the Parisians in 1835* (New York: Harper and Brothers, 1836), 370.

218 "In Paris there are places where people take walks": Muhammed Saffar, *Disorienting Encounters: Travels of a Moroccan Scholar in France, 1845–46*, trans. and ed. Susan Gilson Miller (Berkeley: University of California Press, 1991), 136–37.

218 "long walks and constant affection!": Baudelaire to his mother, May 6, 1861, in Claude Pichois, *Baudelaire* (New York: Viking, 1989), 21.

219 "Whenever I had stopped": Nicholas-Edme Restif de la Bretonne, *Les Nuits de Paris or the Nocturnal Spectator (A Selection)*, trans. Linda Asher and Ellen Fertig, introduction by Jacques Barzun (New York: Random House, 1964), 176.

219 "virgin forest": Susan Buck-Morss, "The Flâneur, the Sandwichman and the Whore: The Politics of Loitering," *New German Critique* 39 (1986): 119: "The popular literature of flanerie may have referred to Paris as a 'virgin forest' (V, 551), but no woman found roaming there alone was expected to be one."

219 "What are the dangers": Benjamin, *Baudelaire*, 39.

220 "Mohicans in spencer jackets": Ibid., 42.

220 "on the Paris pavement": George Sand, *My Life*, trans. Dan Hofstadter (New York: Harper Colophon, 1979), 203–4.

220 "Multitude, solitude": Baudelaire, "Crowds," in *Paris Spleen*, trans. Louis Varese (New York: New Directions, 1947), 20.

221 Haussmann's project: On Haussmann I have been guided by David Pinkney, *Napoleon III and the Rebuilding of Paris* (Princeton: Princeton University Press, 1958), and to a lesser extent by Wolfgang Schivelbusch, *The Railway Journey: The Industrialization of Time and Space in the Nineteenth Century* (Berkeley: University of California Press, 1986). Schivelbusch insists that Haussmann—"the Attila of the straight line"—was entirely utilitarian in his street designs: "It is obvious that the avenues and boulevards were designed to be efficacious army routes, but that function was merely a Bonapartist addendum to the otherwise commercially oriented new system" (181).

221 "Paris is changing!": Charles Baudelaire, "Le Cygne," *The Flowers of Evil*, trans. David M. Dodge for the author.
222 "My Paris, the Paris in which I was born": Jules and Edmond Goncourt, *The Goncourt Journals*, ed. and trans. Lewis Galantiere (New York: Doubleday, Doran, 1937), 93.
223 "For the promenaders, what necessity was there": Schivelbusch, *Railway Journey*, 185 n.
223 "evenings in bed I could not read more": Benjamin, quoted in Susan Buck Morse, *The Dialectics of Seeing: Walter Benjamin and the Arcades Project* (Cambridge, Mass.: MIT Press, 1991), 33.
224 "I still recall the extraordinary role": Andre Breton quoted in the introduction to Louis Aragon, *Paris Peasant*, trans. Simon Watson Taylor (Cambridge, Mass.: Exact Change, 1994), viii.
224 "spoke to this unknown woman": Andre Breton, *Nadja*, trans. Richard Howard (New York: Grove Press, 1960), 64.
224 "Georgette resumed her stroll": Philippe Soupault, *Last Nights of Paris*, trans. William Carlos Williams (Cambridge, Mass.: Exact Change, 1992), 45–46.
225 "Everything is so simple when one knows all the streets": Ibid., 64.
225 "Whenever I happen to be there": Ibid., 80.
225 "famously proposes a detailed 'interpretation'": Michael Sheringham, "City Space, Mental Space, Poetic Space: Paris in Breton, Benjamin and Réda," in *Parisian Fields*, ed. Michael Sheringham (London: Reaktion Books, 1996), 89. Older metaphors of the city as a body existed, but not as a sexual body: in the nineteenth century, parks were often called the "lungs" of the city, and Richard Sennett, in *Flesh and Stone: The Body and the City in Western Civilization* (London and Boston: Faber and Faber, 1994), writes of the bodily metaphors that metaphorized Haussmann's sewers, waterways, and streets as various organs of bodily circulation, necessary for health.
226 "rapt and confused": Djuna Barnes, *Nightwood* (New York: New Directions, 1946), 59–60.
227 "a man who has, with great difficulty": Benjamin, quoted in Grunwald, *Prophets without Honor*, 245.
227 "No one knew the path": Ibid., 248.
228 "whereupon the border officials": Hannah Arendt, introduction, in *Illuminations: Essays and Reflections* (New York: Schocken Books, 1969), 18.
228 "In Paris a stranger feels at home": Ibid., 21.
229 "could set for itself the study": Guy DeBord, "Introduction to a Critique of Urban Geography," in *Situationist International Anthology*, ed. and trans. Ken Knabb (Berkeley: Bureau of Public Secrets, 1981), 5.
230 "The point . . . was to encounter the unknown": Greil Marcus, "Heading for the Hills," *East Bay Express*, February 19, 1999. Marcus writes about

situationism far more extensively in his *Lipstick Traces* (Cambridge, Mass.: Harvard University Press, 1990).

230 "practitioners of the city," "the walking of passers-by": Michel de Certeau, *The Practice of Everyday Life* (Berkeley: University of California Press, 1984), 93, 100.

230–31 "under threat from the tyranny of bad architecture": Jean-Christophe Bailly in Sheringham, "City Space, Mental Space, Poetic Space," *Parisian Fields*, 111.

13. CITIZENS OF THE STREETS: Parties, Processions, and Revolutions

Some of the material here comes from my essays "The Right of the People Peaceably to Assemble in Unusual Clothing: Notes on the Streets of San Francisco" published in *Harvard Design Magazine* in 1998, and "Voices of the Streets," *Camerawork Quarterly*, summer 1995; and my essay on Gulf War activism in *War After War* (San Francisco: City Lights Books, 1991).

236 "the ideal city for riot": Eric Hobsbawm, "Cities and Insurrections," in *Revolutionaries* (New York: Pantheon, 1973), 222. Elizabeth Wilson in *The Sphinx in the City* and Priscilla Parker Ferguson in *Paris as Revolution* also make astute links between the social space and revolutionary potential of a city.

237 "Urban reconstruction, however": Hobsbawm, "Cities and Insurrections," 224.

237 "I take my desires for reality": Angelo Quattrocchi and Tom Nairn, *The Beginning of the End* (London: Verso, 1998), 26.

237 "The difference between rebellion at Columbia and rebellion at the Sorbonne": Mavis Gallant, *Paris Notebooks: Essays and Reviews* (New York: Random House, 1988), 3.

238 the market women . . . had grown accustomed to marching: There are many conflicting versions of the market women's march. I relied on Shirley Elson Roessler's *Out of the Shadows: Women and Politics in the French Revolution, 1789–95* (New York: Peter Land, 1996) for the sequence and details of events, though I also used Michelet's history of the French Revolution (Wynnewood, Pa.: Livingston, 1972), Georges Rude's indispensable *Crowd in the French Revolution* (Oxford: Oxford University Press, 1972), Simon Schama's *Citizens* (New York: Knopf, 1989), and Christopher Hibbert's *The Days of the French Revolution* (New York: Morrow Quill Paperbacks, 1981).

238 "at the discipline, pageantry, and magnitude of the almost daily processions": Rude, *Crowd in the French Revolution*, 66.

239 "she delivered such a blow with her broom": Roessler, *Out of the Shadows*, 18.

240 "their decorated branches amidst the gleaming iron of pikes": Hibbert, *Days*, 104.

241 East Germany was next: On Germany I relied on Timothy Garton Ash, *The Magic Lantern: The Revolutions of 1989 Witnessed in Warsaw, Budapest, Berlin and Prague* (New York: Random House, 1990), and John Borneman, *After the Wall: East Meets West in the New Berlin* (New York: Basic Books, 1991).

241 arrested just for being in the vicinity of disturbances: Borneman, *After the Wall*, 23–24: "In one example . . . a demonstrator was sentenced to six months imprisonment for calling 'No Violence' about fifteen times."

241 "the people acted and the Party reacted": Ash, *Magic Lantern*, 83.

243 "Prague . . . seemed hypnotized, caught in a magical trance": Michael Kukral, *Prague 1989: A Study in Humanistic Geography* (Boulder, Colo.: Eastern European Monographs, 1997), 110.

243 "The government is telling us": Alexander Dubček, quoted in *Time*, December 4, 1989, 21.

243 "The time of massive and daily street demonstrations": Kukral, *Prague 1989*, 95.

244 "Secrecy . . . was a hallmark": Marguerite Guzman Bouvard, *Revolutionizing Motherhood: The Mothers of the Plaza de Mayo* (Wilmington, Del.: Scholarly Resources, 1994), 30.

244 "Much later . . . they described their walks": Bouvard, *Revolutionizing Motherhood*, 70.

245 "They tell me that, while they are marching": Marjorie Agosin, *Circles of Madness: Mothers of the Plaza de Mayo* (Freedonia: White Pine Press, 1992), 43.

248 "People lived in public": Victor Hugo, *1793*, trans. Frank Lee Benedict (New York: Carroll and Graf, 1988), 116.

248 "The revolutionary posters were everywhere": George Orwell, *Homage to Catalonia* (Boston: Beacon Press, 1952), 5.

249 "Revolutionary moments are carnivals in which the individual life celebrates its unification with a regenerated society": Situationist Raoul Vaneigem, quoted in *Do or Die* (Earth First! Britain's newsletter), no. 6 (1997), 4.

14. WALKING AFTER MIDNIGHT: Women, Sex, and Public Space

252 "I used to go out walking": James Joyce, "The Dead," *Dubliners* (New York: Dover, 1991), 149. Anne Wallace, in her *Walking, Literature and English Culture*, pointed me to this use of the term in this text; the Oxford English Dictionary also has a nice section on the phrase.

253 "You have been telling the truth": Glen Petrie, *A Singular Iniquity: The Campaigns of Josephine Butler* (New York: Viking, 1971), 105, where the other details of the Caroline Wyburgh story also appear.

253 "Being born a woman": Sylvia Plath, quoted in Carol Brooks Gardner, *Passing By: Gender and Public Harassment* (Berkeley: University of California Press, 1995), 26. Re gender and travel, see Eric J. Leed, *The Mind of the Traveler: From Gilgamesh to Global Tourism* (New York: Basic Books, 1991), 115–16: "The 'double standard' constructs the spatial domains of interiority (female) and exteriority (male) as domains of, respectively, sexual constraint and sexual freedom. The chastity of women is a technique of inclusion and exclusion, which decrees memberships, rights, and relations between males as well as sanctifying the male line of descent. Women's identification with place in conditions of settlement has been regarded as 'natural,' a result of reproductive necessities that require stability and protection by men; thus the genderization of space. . . . The antithesis between the exteriorizations of men and the interiorizations of women, the superfluity of the sperm and the parsimony of the ovum, has been mapped upon human mobility and come to be considered an element of human nature. But the immobilization of women is a historical achievement. . . . It is this territorialization that makes travel a gendering activity."

255 "who go out onto the street," "Domestic women": Gerda Lerner, *The Creation of Patriarchy* (Oxford: Oxford University Press, 1986), 134, 135–39.

256 "were confined to houses": Sennett, *Flesh and Stone*, 34. Pericles and Xenophon quotes are on pages 68 and 73.

256 "In Greek thought women lack": Mark Wiggins, "Untitled: The Housing of Gender," in *Sexuality and Space*, ed. Beatriz Colomina (Princeton: Princeton University Press, 1992), 335.

257 "The young men strolling on the streets": Joachim Schlor, *Nights in the Big City* (London: Reaktion Books, 1998), 139.

258 "shopping in Les Halles": Petrie, *A Singular Iniquity*, 160.

259 "I asked what the crime was": Ibid., 182.

259 hounded into suicide: The woman was a Mrs. Percy of Aldershot: see the preface and pages 149–54 of Petrie, *A Singular Iniquity*, and Paul McHugh, *Prostitution and Victorian Social Reform* (New York: St. Martin's Press, 1980), 149–51.

259 Lizzie Schauer in Glenna Matthews, *The Rise of Public Woman: Woman's Power and Woman's Place in the United States, 1630–1970* (New York, Oxford: Oxford University Press, 1992), 3.

260 force-feeding the prisoners: My sources were Midge Mackenzie, *Shoulder to Shoulder* (New York: Knopf, 1975), on British suffragists, and Doris Stevens, *Jailed for Freedom: American Women Win the Vote* (1920; new edition, edited by Carol O'Hare, Troutdale, Oregon: New Sage Press, 1995). Djuna Barnes volunteered to be force-fed so she could report on the process.

261 "that women will not feel at ease": B. Houston, "What's Wrong with Sexual Harassment," quoted in June Larkin, "Sexual Terrorism on the Street," in *Sexual Harassment: Contemporary Feminist Perspectives*, ed. Alison M. Thomas and Celia Kitzinger (Buckingham: Open University Press, 1997), 117.

261 Two-thirds of American women are afraid: Jalna Hanmer and Mary Maynard, eds., *Women, Violence and Social Control* (Houndmills, England: MacMillan, 1987), 77.

261 "very worried": Eileen Green, Sandra Hebron, and Diane Woodward, *Women's Leisure, What Leisure* (Houndmills, England: MacMillan, 1990): "One of the most severe restrictions on women's leisure time activities is their fear of being out alone after dark. Many women are afraid to use public transport after dark or late at night whilst for others it's having to walk to bus stops and wait there after dark which deters them. The findings of the second British Crime Survey state that half the women interviewed only went out after dark if accompanied, and 40 percent were 'very worried' about being raped" (89).

262 "If I wrote down every little thing": Larkin, "Sexual Terrorism," 120.

263 "Stanton and Mott . . . began to see similarities": Stevens, *Jailed for Freedom*, 13.

263 Edward Lawson: Abstract of *Kolender, Chief of Police of San Diego, et al., v. Lawson*, 461 U.S. 352, 103 S. Ct. 1855, 75 L. Ed. 2nd 903 (1983).

264 "Throughout the case," and following: Helen Benedict, *Virgin or Vamp: How the Press Covers Sex Crimes* (New York and London: Oxford University Press, 1992), 208.

265 "could leave me speechless": Evelyn C. White, "Black Women and the Wilderness," in *Literature and the Environment: A Reader on Nature and Culture*, ed. Lorraine Anderson, Scott Slovic, John O'Grady (New York: Addison Wesley, 1999), 319.

265 "Had I been older and more mature": Moffat, *Space Below My Feet*, 92.

267 "Could she even get her dinner": Virginia Woolf, *A Room of One's Own* (New York: Harcourt Brace Jovanovich, 1929), 50.

267 "The trick . . . was to identify with Jack Kerouac": Sarah Schulman, *Girls, Visions and Everything* (Seattle: Seal Press, 1986), 17, 97.

267 "That's gay liberation," "Lila walked in the streets": Ibid., 157, 159.

15. Aerobic Sisyphus and the Suburbanized Psyche

272 "the walking city": Kenneth Jackson, *Crabgrass Frontier: The Suburbanization of the United States* (New York: Oxford University Press, 1985), 14–15.

272 Middle-class suburban homes: The source of much of the narrative here is Robert Fishman, *Bourgeois Utopias: The Rise and Fall of Suburbia* (New York: Basic Books, 1987), especially chap. 1, on London's evangelical merchants, and chapter 3, on Manchester's suburbia.

273 "The decision to suburbanize": Ibid., 81–82.

274 "Offices are kept separate": Philip Langdon, A Better Place to Live: Reshaping the American Suburb (Amherst: University of Massachusetts Press, 1994), xi.

275 study that compared the lives of ten-year-olds: Jane Holtz Kay, Asphalt Nation: How the Automobile Took Over America and How We Can Take It Back (New York: Crown Publishers, 1997), 25.

276 more than 1,000 crosswalks have been removed: Gary Richards, "Crossings Disappear in Drive for Safety: Traffic Engineers Say Pedestrians Are in Danger Between the Lines," San Jose Mercury News, November 27, 1998. Traffic engineers in the article blamed pedestrians for their own deaths by automobile in about half the cases and proposed restricting pedestrian access as the solution.

276 "The pedestrian remains the largest single obstacle": Rudofsky, Streets for People, 106.

276 "South Tucson simply has no sidewalks": Lars Eigner, Travels with Lizbeth: Three Years on the Road and on the Streets (New York: Fawcett Columbine, 1993), 18.

277 41 percent of all traffic fatalities: Betsy Thaggard, "Making the Streets a Safer Place," Tube Times, newsletter of the San Francisco Bicycle Coalition, December 1998–January 1999, 5.

277 Giuliani's New York: San Francisco Chronicle; also in Tube Times, 3, citing the Right of Way campaign organized by Time's Up! that is painting memorial stencils on the sites where NYC pedestrians and cyclists have been killed.

277 "that elusive combination": Richard Walker, "Landscape and City Life: Four Ecologies of Residence in the San Francisco Bay Area," Ecumene 2 (1995): 35.

278 "Anyone who has tried to take a stroll at dusk": Mike Davis, "Fortress Los Angeles," in Variations on a Theme Park, ed. Michael Sorkin (New York: Hill and Wang, 1992), 174.

278 "It is extremely regrettable": Søren Kierkegaard's Journals and Papers, ed. and trans. Howard V. Hong and Edna H. Hong (Bloomington: Indiana University Press, 1978), 5:415 (1847).

279 "For most of human history": Life magazine, special millennium issue (1998).

280 "The speed and mathematical directness": Schivelbusch, Railway Journey, 53.

280 "From the elimination of the physical effort of walking": Paul Virilio, The Art of the Motor, trans. Julie Rose (Minneapolis: University of Minnesota Press, 1995), 85.

283 "The Tread-Mill," "It is its monotonous steadiness": James Hardie, The History of the Tread-Mill, containing an account of its origin, construction, operation, effects as it respects the health and morals of the convicts, with their treatment and diet . . . (New York: Samuel Marks, 1824), 16, 18.

284 "always lived by robbery": Robert Graves, *The Greek Myths*, vol. 1 (Harmondsworth, England: Penguin Books, 1957), 168.
284 "Indoors?": Eduardo Galeano, *The Book of Embraces*, trans. Cedric Belfrage (New York, London: W. W. Norton, 1989), 162–63.

16. THE SHAPE OF A WALK

292 "Pollock's near destruction of this tradition": Allan Kaprow, "The Legacy of Jackson Pollock," in *Essays on the Blurring of Art and Life*, ed. Jeff Kelley (Berkeley: University of California Press, 1993), 7.
293 "Emphasizing the body as art": Peter Selz and Kristine Stiles, *Theories and Documents of Contemporary Art: A Sourcebook* (Berkeley: University of California Press, 1996), 679 (introduction to Performance section).
293 "My idea of a piece of sculpture": Lucy R. Lippard, *Overlay: Contemporary Art and the Art of Prehistory* (New York: Pantheon, 1993), 125.
294 "A 1,600-foot wooden trail": Lippard, *Overlay*, 132.
295 "A walk expresses space and freedom": Richard Long, "Five Six Pick Up Sticks, Seven Eight Lay Them Straight," in R. H. Fuchs, *Richard Long* (New York: Solomon R. Guggenheim Museum/Thames and Hudson, 1986), 236. All works described are reproduced in this book.
297 Stanley Brouwn: Brouwn's work is mentioned in Lippard's *Six Years: The Dematerialization of the Art Object* (New York: Praeger Publishers, 1973) and described at length in the essay by Antje Von Graevenitz, "'We Walk on the Planet Earth': The Artist as a Pedestrian: The Work of Stanley Brouwn," *Dutch Art and Architecture*, June 1977. Acconci's *Following Piece* is also described in *Six Years*.
298 Marina Abramović and Ulay: On the performance work of Abramović and Ulay, and on Abramović's sculpture, see Thomas McEvilley, "Abramović/Ulay/ Abramović," *Artforum International*, September 1983; McEvilley's essay in *The Lovers*, catalog/book on the Great Wall Walk (Amsterdam: Stedelijk Museum, 1989); and *Marina Abramović: objects performance video sound* (Oxford: Museum of Modern Art, 1995).
298 "'four dignities'": Gary Snyder, "Blue Mountains Constantly Walking," in *The Practice of the Wild* (San Francisco: North Point Press, 1990), 99.
298 "immobility, silence and watchfulness": McEvilley, "Abramović/Ulay/ Abramović," 54.
299 "When I went to Tibet and the Aborigines": *Marina Abramović*, 63.
299 "mapped out over the millennia": Ibid., 50.
299 "On March 30, 1988": *The Lovers*, 175.
300 "I walk every fucking centimeter," "It took a great number of days": Ibid., 103, 31.
301 "They will discover out of ordinary things": Kaprow, *Essays on the Blurring of Art and Life*, 9.

17. LAS VEGAS, OR THE LONGEST DISTANCE BETWEEN TWO POINTS

304   "The effectiveness of this architecture": J. B. Jackson, "Other-directed Houses," in *Landscapes: Selected Writings of J. B. Jackson,* ed. Ervin H. Jube (University of Massachusetts Press, 1970), 63.

304   "Those streamlined facades": Ibid., 62.

304   more than 30 million a year: A researcher at the Las Vegas Convention Center told me by phone, December 29, 1998.

307   round-the-clock picket: See Sara Mosle, "How the Maids Fought Back," *New Yorker,* February 26 and March 4, 1996, 148–56.

308–9   county passed an ordinance: See the *Las Vegas Review-Journal,* "Petitioners Claim Rights Violated," May 27, 1998; "Clark County Charts Its Strategy to Resurrect Handbill Ordinance," August 18, 1998; "Lawyers to Appeal Handbill Law Ruling," August 26, 1998; "Police Told to Mind Bill of Rights," October 20, 1998.

309   "The theme park": Introduction to Sorkin, *Variations on a Theme Park,* xv.

310   "the Beautification Committee would continue to recommend": Robert Venturi, Denise Scott Brown, and Stephen Izenour, *Learning from Las Vegas: The Forgotten Symbolism of Architectural Form,* rev. ed. (Boston: MIT, 1977), xii.

314   "Both sides of these passageways, which are lighted from above": Benjamin, *Baudelaire,* 37.

316–17   "Olive Oil": Joanne Urioste, *The American Alpine Club Climber's Guide: The Red Rocks of Southern Nevada* (New York: American Alpine Club, 1984), 131.

I₊ dex

Sources for Foot Quotations

I. THE PACE OF THOUGHTS

Honoré de Balzac, quoted in note 15 in Andrew J. Bennett, "Devious Feet: Wordsworth and the Scandal of Narrative Form," *ELH*, spring 1992.

Lucy R. Lippard, *Overlay: Contemporary Art and the Art of Prehistory* (New York: Pantheon, 1993).

Gary Snyder, "Blue Mountains Constantly Walking," *The Practice of the Wild*, (San Francisco: North Point Press, 1990), 98–99.

Virginia Woolf, *Moments of Being* (New York: Harcourt, Brace, Jovanovich, 1976), 82.

Wallace Stevens, "Of the Surface of Things," *The Collected Poems* (New York: Vintage Books, 1982), 57.

John Buchan, in William Robert Irwin, ed., *Challenge: An Anthology of the Literature of Mountaineering* (New York: Columbia University Press, 1950), 354.

Leon Rosenfeld, from his papers, in Niels Bladel, *Harmony and Unity: The Life of Niels Bohr* (Berlin, New York: Science Tech, Springer-Verlag, 1998), 195.

John Keats, letter, in Aaron Sussman and Ruth Goode, *The Magic of Walking* (New York: Simon and Schuster, 1967), 355.

Voltaire, letter to Rousseau, August 30, 1755, in Gavin de Beer, *Jean-Jacques Rousseau and His World* (New York: Putnam, 1972), 42.

Sigmund Freud, *Civilization and Its Discontents*, cited in Ivan Illich, *H2O and the Waters of Forgetfulness* (Dallas: Dallas Institute of Humanities and Culture, 1985), 34.

Samuel Beckett, cited in *The Nation*, July 28–August 4, 1997, 30.

Effie Gray Ruskin, in John Dixon Hunt, *The Wider Sea: A Life of John Ruskin* (London: Dent, 1982), 201.

John 11:10 and Psalms 26:1–12, King James Bible.

Allan G. Grapard, "Flying Mountains and Walkers of Emptiness: Toward a Definition of Sacred Space in Japanese Religions," *History of Religion* 21, no. 3 (1982), 206.

The Three Pillars of Zen, ed. Philip Kapleau (Garden City, N.Y.: Anchor Press, 1980), 33–34.

Paul Shepard, *Nature and Madness* (San Francisco: Sierra Club Books, 1982), 161.

Thomas Merton, in Nancy Louise Frey, *Pilgrim Stories: On and Off the Road to Santiago* (Berkeley: University of California Press, 1998), 79.

Gloria Anzaldua, *Borderlands/La Frontera* (San Francisco: Aunt Lute Books, 1987), 3, 16.

Paul Klee, *Pedagogical Sketchbook*, 1925, cited in *The Oxford Dictionary of Quotations*, (Oxford: Oxford University Press, 1979), 305.

Charles Baudelaire, "Le Soleil," *Baudelaire*, selected and translated by Francis Scarfe (Harmondsworth, England: Penguin Books, 1964), 13.

Kirk Savage, "The Past in the Present," *Harvard Design Magazine*, fall 1999, 19.

Pablo Neruda, "Walking Around," *The Vintage Anthology of Contemporary World Poetry* (New York: Vintage Books, 1966), 527.

II. FROM THE GARDEN TO THE WILD

Alexander Pope, "Epistle to Miss Blount," *The Norton Anthology of English Literature*, vol. 1, 3rd ed. (New York: W. W. Norton, 1974), 2174.

Maria Edgeworth, *Belinda* (Oxford: Oxford University Editions, 1994), 90.

Thomas Gray, writing in 1769, "Journal in the Lakes," *Collected Works of Gray in Prose and Verse*, vol. 1, ed. Edmund Gosse (London: MacMillan and Co., 1902), 252.

E. P. Thompson, *Making History: Writings on History and Culture* (New York: New Press, 1995), 3.

Mary Shelley, *Frankenstein*, in *Three Gothic Novels* (Harmondsworth, England: Penguin Books, 1968), 360.

Johann Wolfgang von Goethe, *The Sorrow of Young Werther*, ed. and trans. Victor Lange (New York: Holt, Rinehart and Winston, 1949). "He who does not know," 4; "If in such moments I find," 52; "Ossian has superseded Homer," 83; "At noon I went to walk," 91.

Hunter Davies, *William Wordsworth: A Biography* (New York: Atheneum, 1980), 213.

Benjamin Haydon, cited in James Fenton, "A Lesson from Michelangelo," *New York Review of Books*, 1996.

Amos Bronson Alcott, cited in Carlos Baker, *Emerson Among the Eccentrics* (New York: Viking, 1996), 305–6.

Henry David Thoreau, "A Walk to Wachusett," *The Natural History Essays* (Salt Lake City: Peregrine Smith Books, 1980), 48.

Bertrand Russell, *The Autobiography of Bertrand Russell* (New York: Bantam Books, 1978), 78–79.

E. M. Forster, *Howards End* (Harmondsworth, England: Penguin Books, 1992), 181.

Elias Canetti, *Crowds and Power* (New York: Viking, 1963), 31.

Morris Marples, *Shank's Pony: A Study of Walking* (London: J. M. Dent and Sons, 1959), 190.

David Roberts, *Moments of Doubt: and Other Mountaineering Writings* (Seattle: Mountaineers, 1986), 186.

Hamish Brown, *Hamish's Mountain Walk: The First Traverse of All the Scottish Monroes in One Journey* (London: Victor Gollancz, 1978), 356–57. (A monroe is a Scottish peak over 3,000 feet.)

Gary Snyder, "Blue Mountains Constantly Walking," *The Practice of the Wild*, 113.

Gilles Deleuze and Felix Guattari, *Nomadalogy*, transl. Brian Massumi (New York: Semiotexte, 1986), 36–37.

Friedrich Engels, *The Condition of the Working Class in England* (Harmondsworth, England: Penguin Books, 1987), 173.

Charles Dickens, *Bleak House* (New York: New American Library, 1964), 517.

III. LIVES OF THE STREETS

J. B. Jackson, "The Stranger's Path," in *Landscapes: Selected Essays of J. B. Jackson* (Boston: University of Massachusetts Press, 1970), 102.

Horace Walpole, letter to George Montagu on a "ridotto" at Vauxhall Gardens, May 11, 1769, in *Letters of Horace Walpole* (London: J. M. Dent, 1926), 92.

Patrick Delany, in Carole Fabricant, *Swift's Landscape* (Baltimore: Johns Hopkins University Press, 1992).

Elena Poniatowska, "In the Street" (on the homeless children of Mexico City), *Doubletake*, winter 1998, 118–19.

F. Bloch, *Types du Boulevard*, cited in Margaret Cohen, *Profane Illumination: Walter Benjamin and the Paris of Surrealist Revolution* (Berkeley: University of California Press, 1995), 84.

Victor Hugo, *Les Misérables*, trans. Charles E. Wilbour (New York: Modern Library, 1992), 506.

Jules and Edmond Goncourt, October 26, 1856, *The Goncourt Journals*, ed. and transl. Lewis Galantiere (Doubleday, Doran, 1937), 38.

Andre Castelot, *The Turbulent City: Paris 1783–1871* (New York: Harper & Row, 1962), 186.

Richard Ellmann, *James Joyce* (New York: Oxford University Press, 1982), 518.

Victor Hugo, *Les Misérables*, 106–7.

Walter Benjamin, *Charles Baudelaire: A Lyric Poet in the Era of High Capitalism*, trans. Harry Zohn (London: Verso, 1973), 60.

Edgar Allan Poe, "The Murders in the Rue Morgue," in *The Fall of the House of Usher and Other Tales* (New York: New American Library, 1966), 53.

Alexis de Tocqueville, *Democracy in America* (New York: HarperPerennial, 1969), 108.

Subcommandante Marcos, cited in *Utne Reader*, May–June 1998, 55.

Paul Monette, "The Politics of Silence," in *The Writing Life* (New York: Random House, 1995), 210.

Martha Shelley, *Haggadah: A Celebration of Freedom* (San Francisco: Aunt Lute Books, 1997), 19.

Jan Goodwin: *Price of Honor: Muslim Women Lift the Veil of Silence on the Islamic World* (Boston: Little, Brown, 1994), 161.

Mary Wortley Montagu, August 1712, letter to her future husband, *Letters of Mary Wortley Montagu* (London: J. M. Dent, n.d.), 32.

Song of Solomon 7:1, King James Bible.

London passerby to a streetwalker, in Richard Symanski, *The Immoral Landscape: Female Prostitution in Western Societies* (Toronto: Butterworths, 1981), 164.

Emily Post, *Etiquette* (1922 ed.), quoted by Edmund Wilson in "Books of Etiquette & Emily Post," *Classics and Commercials* (New York: Vintage, 1962), 378.

Harper's Bazaar, July 1997, 18.

Elizabeth Wilson, *The Sphinx in the City* (Berkeley: University of California Press, 1992), 16.

IV. PAST THE END OF THE ROAD

Nancy Louise Frey, *Pilgrim Stories*, 132.

Luz Benedict, in Edna Ferber, *Giant* (Garden City, N.Y.: Doubleday, 1952), 153.

Norman Klein, *History of Forgetting* (London: Verso, 1997), 118.

Brett Pulley, *New York Times*, Sunday, November 8, 1998, 3.

Mick LaSalle, *San Francisco Chronicle*, January 5, 1999, E1.

Hannah Nyala, *Point Last Seen* (Boston: Beacon Press, 1997), 1–3.

Paul Virilio, *Speed and Politics*, trans. Mark Polizzotti (New York: Semiotexte, 1986), 144.

Ivan Illich, *Whole Earth Review*, summer 1997.

Lee, "a Catholic American" pilgrim, in Nancy Louise Frey, *Pilgrim Stories*, 74–75.

A. R. Ammons, "A Poem Is a Walk," *Epoch* 18, no. 1 (1968), 116.

Yoko Ono, "Map Piece," summer 1962. © 1962 Yoko Ono. From the exhibition *Searchlight: Consciousness at the Millennium*, CCAC Institute, San Francisco, 1999.